Colloids in Cosmetics and Personal Care

Edited by
Tharwat F. Tadros

Colloids and Interface Science Series

Colloid Stability

The Role of Surface Forces, Part I
Volume 1

2007
ISBN 978-3-527-31462-1

Colloid Stability

The Role of Surface Forces, Part II
Volume 2

2007
ISBN 978-3-527-31503-1

Colloid Stability and Applications in Pharmacy

Volume 3

2007
ISBN 978-3-527-31463-8

Colloids in Cosmetics and Personal Care

Volume 4

2007
ISBN 978-3-527-31464-5

Colloids in Agrochemicals

Volume 5

2007
ISBN 978-3-527-31465-2

Colloids in Paints

Volume 6

2007
ISBN 978-3-527-31466-9

Colloids and Interface Science Series
Volume 4

Colloids in Cosmetics and Personal Care

Edited by
Tharwat F. Tadros

WILEY-VCH

WILEY-VCH Verlag GmbH & Co. KGaA

The Editor

Prof. Dr. Tharwat F. Tadros
89 Nash Grove Lane
Wokingham, Berkshire RG40 4HE
Great Britain

Library of Congress Card No.: applied for

British Library Cataloguing-in-Publication Data
A catalogue record for this book is available from the British Library

Bibliographic information published by the Deutsche Nationalbibliothek
Die Deutsche Nationalbibliothek lists this publication in the Deutsche Nationalbibliografie; detailed bibliographic data are available in the Internet at http://dnb.d-nb.de.

Printed in the Federal Republic of Germany
Printed on acid-free paper

Cover design Adam Design, Weinheim
Composition Asco Typesetters, Hong Kong
Printing Strauss GmbH, Mörlenbach
Bookbinding Litges & Dopf GmbH, Heppenheim

ISBN 978-3-527-31464-5

Contents

Colloids and Interface Science Series, Vol. 4
Colloids in Cosmetics and Personal Care. Edited by Tharwat F. Tadros

ISBN: 978-3-527-31464-5

Preface

Cosmetic and personal care formulations consist of complex systems of emulsions, suspensions and their mixtures (suspoemulsions). Several cosmetic systems are also formulated as nano-emulsions and nano-suspensions (covering the size range from 20 to 200 nm). These formulations also contain self-assembly structures such as micelles, liposomes, liquid crystalline phases, etc. Understanding the basic colloid and interface science principles will enable one to analyze the complex interactions in these complex formulations. This will also lead to a more rational approach to their formulations, control of their long-term physical stability and achieve the required sensory attributes. This volume addresses some of these basic principles and their application.

The first chapter gives an overview of the colloid aspects of cosmetic formulations with particular reference to polymeric surfactants that have been applied to obtain systems with a long shelf-life and the required sensory attributes. Four main topics are covered: (1) interaction forces between particles or droplets in a dispersion; (2) a description of the system stability in terms of these interaction forces; (3) self-assembly structures and their role in stabilization, skin feel, moisturization and delivery of actives; and (4) use of polymeric surfactants for stabilization of emulsions, nano-emulsions and multiple emulsions.

Chapter 2 deals specifically with the use of hydrophobically modified inulin polymeric surfactants for stabilization of nano-emulsions, while the next chapter deals with the integration of polymeric surfactants in cosmetic formulations for the enhancement of performance and stability. Chapter 4 discusses the application of colloid and interface science principles to the optimization of sunscreen dispersions, and Chapter 5 describes the use of hydrophobically modified polymers (associative thickeners) for the control of rheology of surfactant systems with particular reference to shampoo formulations. Cosmetic formulations based on the liquid crystalline phases of the lamellar type ("Oleosomes" and "Hydrosomes"), their rheological characteristics and sensory evaluation are described in Chapter 6. Chapter 7 deals with the application of a surfactant/biopolymer system for stabilization of emulsions. The rheological characteristics of the resulting systems are correlated with some of their sensory attributes. Chapter 8 attempts to correlate the "body butter" texture and structure of cosmetic emulsions to their rheological characteristics. The latter have been analyzed at a fundamental level

Colloids and Interface Science Series, Vol. 4
Colloids in Cosmetics and Personal Care. Edited by Tharwat F. Tadros

ISBN: 978-3-527-31464-5

and could be correlated with some of the sensory attributes. An overview is given in Chapter 9 of interparticle interactions in color cosmetics. The fundamental principles of preparation of pigment dispersions are described. Chapter 10 deals with the specific topic of starch dispersions.

Chapter 11 gives a comparison of the skin hydration potential of an emulsion based on a cationic surfactant with one based on an anionic surfactant. Higher skin hydration was detected from the cationic emulsion, especially in the initial stages. The impact of urea on the colloidal structure of alkylpolyglucoside emulsions is described in Chapter 12. Both physico-chemical and *in vitro/in vivo* characterization are described. Chapter 13 presents models for calculation of sun protection factors and parameters characterizing the UVA protection of cosmetic sunscreens.

The text gives a comprehensive overview of several applications of colloid and interface science principles in personal care and cosmetic formulations and should be particularly valuable for fundamental studies of the complex interactions in the various cosmetic disperse systems. It will also provide the reader with knowledge on how to relate the rheological characteristics of these complex systems with some of the sensory attributes. Using such fundamental knowledge will enable the formulation scientist to arrive at the right recipe in a shorter period of time. A great deal of time could then be saved in sensory evaluation, which can be related to rheological measurements.

January 2008 *Tharwat Tadros*

List of Contributors

Karl Booten
ORAFTI Bio Based Chemicals
Aandorenstraat 1
3300 Tienen
Belgium

Ignác Capek
Slovak Academy of Science
Polymer Institute
Dúbravská cesta 9
84236 Bratislava
Slovakia
and
Trenčín University of A. Dubček
Faculty of Industrial Technologies
Ul. I. Krasku 30
02001 Púchov
Slovakia

Bernd Herzog
Ciba Specialty Chemicals Inc.
79630 Grenzach-Whylen
Germany

Steven Housley
Croda Research and Development
Wilton Centre
Redcar TS10 4RF
United Kingdom

Biljana Jancic
University of Belgrade
Faculty of Pharmacy
Institute of Pharmaceutical
Chemistry and Drug Analysis
Vojvode Stepe 450
11000 Belgrade
Serbia

Lorna M. Kessell
Croda Research and Development
Wilton Centre
Redcar TS10 4RF
United Kingdom

Martine Lemmens
ORAFTI Bio Based Chemicals
Aandorenstraat 1
3300 Tienen
Belgium

Sandra Léonard
Croda Research and Development
Wilton Centre
Redcar TS10 4RF
United Kingdom

Bart Levecke
ORAFTI Bio Based Chemicals
Aandorenstraat 1
3300 Tienen
Belgium

Colloids and Interface Science Series, Vol. 4
Colloids in Cosmetics and Personal Care. Edited by Tharwat F. Tadros

ISBN: 978-3-527-31464-5

Jela Milic
University of Belgrade
Faculty of Pharmacy
Institute of Pharmaceutical
Technology and Cosmetology
Vojvode Stepe 450
11000 Belgrade
Serbia

Benjamin J. Naden
Croda Research and Development
Wilton Centre
Redcar TS10 4RF
United Kingdom

Frederico Roschzttardtz
Uniqema Research and Development
P.O. Box 2
2800 AA Gouda
The Netherlands

Snezana Savic
University of Belgrade
Faculty of Pharmacy
Institute of Pharmaceutical
Technology and Cosmetology
Vojvode Stepe 450
11000 Belgrade
Serbia

Tharwat Tadros
89 Nash Grove Lane
Wokingham, Berkshire RG40 4HE
United Kingdom

Marie-Claire Taelman
2 Lindestraat
9790 Wortegem-Petegem
Belgium

Slobodanka Tamburic
University of the Arts London
London College of Fashion
Cosmetic Science
20 John Prince's Street
London W1G OBJ
United Kingdom

Ian R. Tooley
Croda Research and Development
Wilton Centre
Redcar TS10 4RF
United Kingdom

Cornelis Verboom
Uniqema Research and Development
P.O. Box 2
2800 AA Gouda
The Netherlands

Gordana Vuleta
University of Belgrade
Faculty of Pharmacy
Institute of Pharmaceutical
Technology and Cosmetology
Vojvode Stepe 450
11000 Belgrade
Serbia

Vincent Wortel
Uniqema Research and Development
P.O. Box 2
2800 AA Gouda
The Netherlands

1
Colloid Aspects of Cosmetic Formulations with Particular Reference to Polymeric Surfactants

Tharwat F. Tadros

Abstract

The use of polymeric surfactants for the stabilization of cosmetic and personal care formulations is described in terms of their adsorption and conformation at the solid/liquid and liquid/liquid interface. The most effective polymeric surfactants are the A–B, A–B–A block and BA_n or AB_n graft types (where B is the anchor chain and A is the stabilizing chain). The mechanism by which these polymeric surfactants stabiles suspensions and emulsions is briefly discussed in terms of their interaction when particles or droplets approach. This provides very strong repulsion, which is referred to as steric stabilization. Particular attention is given to a recently developed graft copolymer AB_n based on inulin (which is extracted from chicory roots) that is hydrophobized by grafting several alkyl groups (B) onto the linear polyfructose chain (A). This polymeric surfactant is referred to as hydrophobically modified inulin (HMI) and is commercially available as INUTEC® SP1 (ORAFTI, Belgium). It is used for the stabilization of oil-in-water (O/W) emulsions both in aqueous media and in the presence of high electrolyte concentrations. The emulsions remained stable for more than one year at room temperature and at 50 °C. INUTEC® SP1 is also effective in reducing Ostwald ripening in nano-emulsions. It could also be applied for the preparation of W/O/W and O/W/O multiple emulsions and for stabilization of liposomes and vesicles. Based on these fundamental studies, INUTEC® SP1 could be applied for the preparation of stable personal care formulations. The amount of polymeric surfactant required for maintenance of stability (for more than one year at ambient temperature) was relatively low (of the order of 1 w/w% based on the oil phase). In addition, the polymeric surfactant showed no skin irritation, no stickiness or greasiness and it gave an excellent skin-feel.

For the optimum formulation of cosmetic preparations, colloid and interface principles have to be applied. The most effective stabilizers against flocculation and coalescence are polymeric surfactants of the A–B, A–B–A block and BA_n or AB_n graft types (where B is the anchor chain and A is the stabilizing chain).

Colloids and Interface Science Series, Vol. 4
Colloids in Cosmetics and Personal Care. Edited by Tharwat F. Tadros

ISBN: 978-3-527-31464-5

Polymeric surfactants also reduce Ostwald ripening in nano-emulsions. They are also applied for the stabilization of multiple emulsions of both the W/O/W and the O/W/O types. Polymeric surfactants are also used for stabilization of liposomes and vesicles. These benefits of polymeric surfactants justify their application in cosmetic and personal care preparations. Apart from their excellent stabilization effect, they can also eliminate any skin irritation.

1.1
Introduction

Cosmetic and toiletry products are generally designed to deliver a function benefit and to enhance the psychological well-being of consumers by increasing their esthetic appeal. Thus, many cosmetic formulations are used to clean hair, skin, etc., and impart a pleasant odor, make the skin feel smooth and provide moisturizing agents, provide protection against sunburn, etc. In many cases, cosmetic formulations are designed to provide a protective, occlusive surface layer, which either prevents the penetration of unwanted foreign matter or moderates the loss of water from the skin [1, 2]. In order to have consumer appeal, cosmetic formulations must meet stringent esthetic standards such as texture, consistency, pleasing color and fragrance and convenience of application. This results in most cases in complex systems consisting of several components of oil, water, surfactants, coloring agents, fragrants, preservatives, vitamins, etc. In recent years, there has been considerable effort in introducing novel cosmetic formulations that provide great beneficial effects to the customer, such as sunscreens, liposomes and other ingredients that may maintain healthy skin and provide protection against drying, irritation, etc.

Since cosmetic products come into close contact with various organs and tissues of the human body, a most important consideration for choosing ingredients to be used in these formulations is their medical safety. Many of the cosmetic preparations are left on the skin after application for indefinite periods and, therefore, the ingredients used must not cause any allergy, sensitization or irritation. The ingredients used must be free of any impurities that have toxic effects.

One of the main areas of interest of cosmetic formulations is their interaction with the skin [3]. The top layer of the skin, which is the man barrier to water loss, is the stratum corneum, which protects the body from chemical and biological attack [4]. This layer is very thin, approximately 30 μm, consists of ~10% by weight of lipids that are organized in bilayer structures (liquid crystalline), and at high water content is soft and transparent. A schematic representation of the layered structure of the stratum corneum, suggested by Elias et al. [5], is given in Figure 1.1. In this picture, ceramides were considered as the structure-forming elements, but later work by Friberg and Osborne [6] showed the fatty acids to be the essential compounds for the layered structure and that a considerable part of the lipids are located in the space between the methyl groups. When a cosmetic formulation is applied to the skin, it will interact with the stratum corneum and

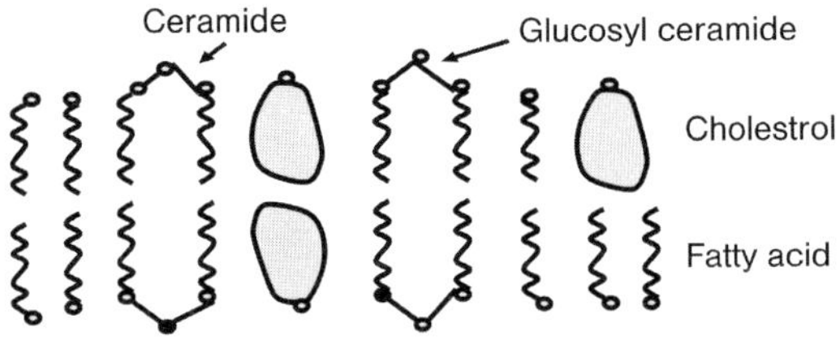

Figure 1.1 Schematic representation of the "bilayer" structure of the stratum corneum.

it is essential to maintain the "liquid-like" nature of the bilayers and prevent any crystallization of the lipids. This happens when the water content is reduced below a certain level. This crystallization has a drastic effect on the appearance and smoothness of the skin ("dry" skin feeling).

To achieve the above criteria, "complex" multiphase systems are formulated: (1) oil-in-water (O/W) emulsions; (2) water-in-oil (W/O) emulsions; (3) solid/liquid dispersions (suspensions); (4) emulsion–suspension mixtures (suspoemulsions); (5) nanoemulsions; (6) nanosuspensions; (7) multiple emulsions. All these disperse systems contain "self-assembly" structures: (1) micelles (spherical, rod-shaped, lamellar); (2) liquid crystalline phases (hexagonal, cubic or lamellar); (3) liposomes (multilamellar bilayers) or vesicles (single bilayers). They also contain "thickeners" (polymers or particulate dispersions) to control their rheology.

The above complex multiphase systems require a fundamental understanding of the colloidal interactions between the various components. Understanding these interactions enables the formulation scientist to arrive at the optimum composition for a particular application. The fundamental principles involved also help in predicting the long-term physical stability of the formulations. Below a summary of some of the most commonly used formulations in cosmetics is given [7].

1. *Lotions:* These are usually (O/W emulsions that are formulated in such a way (see below the section on cosmetic emulsions) as to give a shear thinning system. The emulsion will have a high viscosity at low shear rates ($0.1\ s^{-1}$) in the region of few hundred Pa s, but the viscosity decreases very rapidly with increase in shear rate, reaching values of a few Pa s at shear rates greater than $1\ s^{-1}$.
2. *Hand creams:* These are formulated as O/W or W/O emulsions with special surfactant systems and/or thickeners to give a viscosity profile similar to that of lotions, but with orders of magnitude greater viscosities. The viscosity at low shear rates ($<0.1\ s^{-1}$) can reach thousands of Pa s and they retain a relatively high viscosity at high shear rates (of the order of few hundred Pa s at shear rates $>1\ s^{-1}$). These systems are sometimes described to have a "body" mostly in the form of a gel-network structure that may be achieved by the use of surfactant mixtures to form liquid crystalline structures. In some case, thickeners (hydrocolloids) are added to enhance the gel network structure.

3. *Lipsticks:* These are suspensions of pigments in a molten vehicle. Surfactants are also used in their formulation. The product should show good thermal stability during storage and rheologically it should behave as a viscoelastic solid. In other words, the lipstick should show small deformation at low stresses and this deformation should recover on removal of the stress. Such information could be obtained using creep measurements.
4. *Nail polishes:* These are pigment suspensions in a volatile non-aqueous solvent. The system should be thixotropic. On application by the brush it should show proper flow for an even coating but should have sufficient viscosity to avoid "dripping". After application, "gelling" should occur on a controlled time scale. If "gelling" is too fast, the coating may leave "brush marks" (uneven coating). If "gelling" is too slow, the nail polish may drip. The relaxation time of the thixotropic system should be accurately controlled to ensure good leveling, and this requires the use of surfactants.
5. *Shampoos:* These are normally a "gelled" surfactant solution of well-defined associated structures, e.g. rod-shaped micelles. A thickener such as a polysaccharide may be added to increase the relaxation time of the system. The interaction between the surfactants and polymers is of great importance.
6. *Antiperspirants:* These are suspensions of solid actives in a surfactant vehicle. Other ingredients such as polymers that provide good skin feel are added. The rheology of the system should be controlled to avoid particle sedimentation. This is achieved by addition of thickeners. Shear thinning of the final product is essential to ensure good spreadability. In stick application, a "semi-solid" system is produced.
7. *Foundations:* These are complex systems consisting of a suspension–emulsion system (sometimes referred to as suspoemulsions). Pigment particles are usually dispersed in the continuous phase of an O/W or W/O emulsion. Volatile oils such as cyclomethicone are usually used. The system should be thixotropic to ensure uniformity of the film and good leveling.

The overview in this chapter, which is by no means exhaustive, will deal with the following topics: (1) interaction forces between particles or droplets in a dispersion and their combination; (2) description of stability in terms of the interaction forces; (3) self-assembly structures and their role in stabilization, skin feel, moisturization and delivery of actives; and (4) use of polymeric surfactants for stabilization of nanoemulsions, multiple emulsions, liposomes and vesicles.

1.2 Interaction Forces and Their Combination

Three main interaction forces can be distinguished: (1) van der Waals attraction; (2) double layer repulsion; and (3) steric interaction. These interaction forces and their combination are briefly described below [8].

The van der Waals attraction is mainly due to the London dispersion forces, which arise from charge fluctuations in the atoms or molecules. For an assembly of atoms or molecules (particles or droplets), the attractive forces can be summed, resulting in long-range attraction. The attractive force or energy for two particles or droplets increases with decrease in separation distance between them and at short distances it reaches very high values. In the absence of any repulsive force, the particles or droplets in a dispersion will aggregate, forming strong flocs that cannot be redispersed by shaking.

The van der Waals attraction between two spherical particles or droplets each of radius R separated by a surface-to-surface distance of separation h, is given by the following expression (when $h \ll R$):

$$V_A = -\frac{AR}{12h} \tag{1}$$

where A is the effective Hamaker constant, given by

$$A = ({A_{11}}^{1/2} - {A_{22}}^{1/2})^2 \tag{2}$$

where A_{11} and A_{22} are the Hamaker constants of particles or droplets and medium, respectively.

The Hamaker constant A of any material is given by

$$A = \pi q^2 \beta \tag{3}$$

where q is the number of atoms or molecules per unit volume and β is the London dispersion constant (that is related to the polarizability of the atoms or molecules).

To counteract this attraction, one needs a repulsive force that operates at intermediate distances of separation between the particles. With particles or droplets containing a charge repulsion occurs as a result of formation of electrical double layers [9]. Repulsion results from charge separation and formation of electrical double layers, e.g. when using ionic surfactants. At low electrolyte concentrations ($<10^{-2}$ mol dm^{-3} NaCl) the double layers extend to several nanometers in solution. When two particles or droplets approach a distance of separation that becomes smaller than twice the double-layer extension, double-layer overlap occurs, resulting in strong repulsion. The repulsive force V_{el} is given by the following expression [10]:

$$V_{el} = \frac{4\pi\varepsilon_r\varepsilon_0 R^2 {\psi_0}^2 \exp(-\kappa h)}{2R + h} \tag{4}$$

where ε_r is the relative permittivity (78.6 for water at 25 °C), ε_0 is the permittivity of free space, R is the particle or droplet radius, ψ_0 is the surface potential (that is approximately equal to the measurable zeta potential) and κ is the Debye–Hückel parameter that is related to the number of ions n_0 per unit volume (of each type present in solution) and the valency of the ions Z_i (note that $1/\kappa$ is a measure of the double-layer extension and is referred to as the "thickness of the double layer"):

$$\frac{1}{\kappa} = \left(\frac{\varepsilon_r \varepsilon_0 kT}{2 n_0 Z_i^2 e^2}\right)^{1/2} \tag{5}$$

where k is Boltzmann's constant and T is the absolute temperature.

The magnitude of repulsion increases with increase in zeta potential and decrease in electrolyte concentration and decrease in valency of the counter and co-ions.

A more effective repulsion is due to the presence of adsorbed nonionic surfactants or polymers [11, 12]. These molecules consist of hydrophobic chains which adsorb strongly on hydrophobic particles or oil droplets and hydrophilic chains which are strongly solvated by the molecules of the medium. One can establish a thickness for the solvated (hydrated) chain. When two particles or droplets approach a distance of separation that is smaller than twice the adsorbed layer thickness, repulsion occurs as a result of two main effects: (1) unfavorable mixing of the solvated chains, which results in an increase in the osmotic pressure in the overlap region (solvent molecules diffuse, separating the particles or droplets), and is referred to as the mixing interaction, G_{mix}; and (2) a reduction in configurational entropy of the chains on significant overlap, which is referred to as the elastic interaction, G_{el}.

G_{mix} is given by the following expression [13, 14]:

$$\frac{G_{mix}}{kT} = \left(\frac{2V_2^2}{V_1}\right) \nu_2^2 \left(\frac{1}{2} - \chi\right) \left(\delta - \frac{h}{2}\right)^2 \left(3R + 2\delta + \frac{h}{2}\right) \tag{6}$$

where k is Boltzmann's constant, T is the absolute temperature, V_2 is the molar volume of polymer, V_1 is the molar volume of solvent, ν_2 is the number of polymer chains per unit area, χ is the Flory–Huggins interaction parameter and δ is the hydrodynamic thickness of the adsorbed layer.

The sign of G_{mix} depends on the value of the Flory–Huggins interaction parameter χ: if $\chi < 0.5$, G_{mix} is positive and one obtains repulsion; if $\chi > 0.5$, G_{mix} is negative and one obtains attraction; if $\chi = 0.5$, $G_{mix} = 0$ and this is referred to as the θ-condition.

The elastic interaction is given by the following expression [15]:

$$\frac{G_{el}}{kT} = 2\nu_2 \ln\left[\frac{\Omega(h)}{\Omega(\infty)}\right] = 2\nu_2 R_{el}(h) \tag{7}$$

where $\Omega(h)$ is the number of configurations of the chains at separation distance h and $\Omega(\infty)$ is the value at $h = \infty$. $R_{el}(h)$ is a geometric function whose form

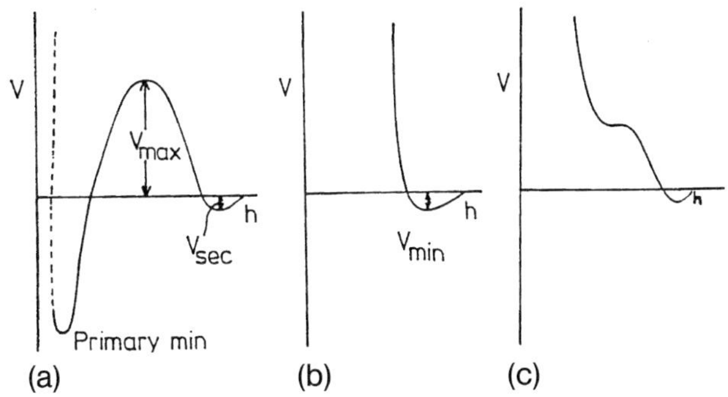

Figure 1.2 Energy–distance curves for electrostatic (a), steric (b) and electrosteric (c) systems.

depends on the chain segment distribution at the surface of the particle or droplet.

Combination of van der Waals attraction with double-layer repulsion forms the basis of the theory of colloid stability due to Deyaguin, Landau, Verwey and Overbeek (DLVO theory) [16, 17]. The force–distance curve according to the DLVO theory is represented schematically in Figure 1.2a. This shows two minima and one maximum. The minimum at long separation distances (secondary minimum, a few kT units) results in weak and reversible flocculation. This could be useful is some applications, e.g. reduction of formation of hard sediments or cream layers. The minimum at short distances (primary minimum, several hundred kT units) results in very strong (irreversible) flocculation. The maximum at intermediate distances (energy barrier) prevents aggregation into the primary minimum. To maintain kinetic stability of the dispersion (with long-term stability against strong flocculation) the energy barrier should be $>25kT$. The height of the energy barrier increases with decrease in electrolyte concentration, decrease in valency of the ions and increase of the surface or zeta potential.

Combination of van der Waals attraction with steric repulsion (combination of mixing and elastic interaction) forms the basis of the theory of steric stabilization [18]. Figure 1.2b gives a schematic representation of the force–distance curve of sterically stabilized systems. This force–distance curve shows a shallow minimum at a separation distance h comparable to twice the adsorbed layer thickness (2δ) and when $h < 2\delta$, very strong repulsion occurs. Unlike the V–h curve predicted by the DLVO theory (which shows two minima), the V–h curve of sterically stabilized systems shows only one minimum whose depth depends on the particle or droplet radius R, the Hamaker constant A and the adsorbed layer thickness δ. At given R and A, the depth of the minimum decreases with increase in the adsorbed layer thickness δ. When the latter exceeds a certain value (particularly with small particles or droplets) the minimum depth can become $<kT$ and the dispersion approaches thermodynamic stability. This forms the basis of the stability of nanodispersions.

Combination of the van der Waals attraction with double-layer and steric repulsion is illustrated schematically in Figure 1.2c and this is sometimes referred to as electrosteric stabilization, as produced for example by the use of polyelectrolytes. This V–h curve has a minimum at long distances of separation, a shallow maximum at intermediate distances (due to double-layer repulsion) and a steep rise in repulsion at smaller h values (due to steric repulsion).

These energy–distance curves can be applied to describe some of the structures (states) produced in suspensions and emulsions. Figure 1.3 shows a schematic representation of the various states that may be produced in a suspension. One also has to consider the effect of gravity, which is very important when the particle size is relatively large (say >1 μm) and the density difference between the particles and the medium is significant (>0.1).

States (a) to (c) in Figure 1.3 represent the case for colloidally stable suspensions. In other words, the net interaction in the suspension is repulsive. Only state (a) with very small particles is physically stable. In this case the Brownian diffusion can overcome the gravity force and no sedimentation occurs; this is the

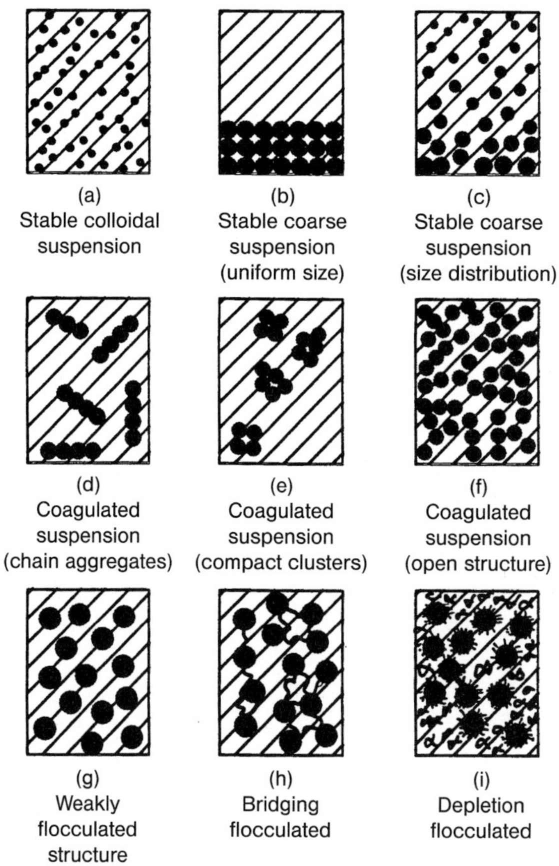

Figure 1.3 Different states of suspensions.

case with nanosuspensions (with size range 20–200 nm):

$$kT > \frac{4}{3}\pi R^3 \Delta\rho g L \tag{8}$$

where R is the particle radius, $\Delta\rho$ is the buoyancy (difference between particle density and that of the medium), g is the acceleration due to gravity and L is the height of the container.

States (b) and (c) are physically unstable (showing settling and formation of hard sediments), even though the system is colloidally stable. In this case the gravity force exceeds the Brownian diffusion:

$$kT \ll \frac{4}{3}\pi R^3 \Delta\rho g L \tag{9}$$

States (d) to (f) are strongly flocculated systems. In other words, the net interaction between the particles is attractive with a deep primary minimum. In state (d), chain aggregates are produced particularly under conditions of no stirring. These aggregates sediment under gravity, forming an "open" structure with the particles strongly held together. State (e) represents the case of formation of compact clusters which will also sediment forming a more "compact" structure again with the particles strongly held together. State (f) is the case of a highly concentrated suspension with the particles forming a strong three-dimensional "gel" structure that extends through the whole volume of the suspension. Such strongly flocculated structure (which is sometimes described as "one-floc") may undergo some contraction and some of the continuous phase may appear at the top, a phenomenon described as syneresis. Clearly, all these strongly flocculated structures must be avoided since the suspension cannot be redispersed on shaking.

The most important cases are those of (g) and (h), which represent reversible weakly flocculated systems. State (g) is the case of secondary minimum flocculation that prevents the formation of hard sediments. These weakly flocculated structures can be redispersed on shaking or on application and they sometimes show thixotropy (reduction of viscosity on application of shear and recovery of the viscosity when the shear is stopped). State (h) is produced by the addition of a weakly adsorbed high molecular weight polymer that causes bridging between the particles. Under conditions of incomplete coverage of the particles by the polymer chains, the latter become simultaneously adsorbed on two or more particles. If the adsorption of the polymeric chain is not strong, these polymer bridges can be broken under shear and the suspension may also show thixotropy. State (i) is a weakly flocculated suspension produced by the addition of "free" nonadsorbing polymer. Addition of a nonadsorbing polymer to a sterically stabilized suspension results in the formation of depletion zones (that are free of the polymer chains) around the particles. The free polymer chains cannot approach the surface of the particles since this will reduce entropy that is not compensated

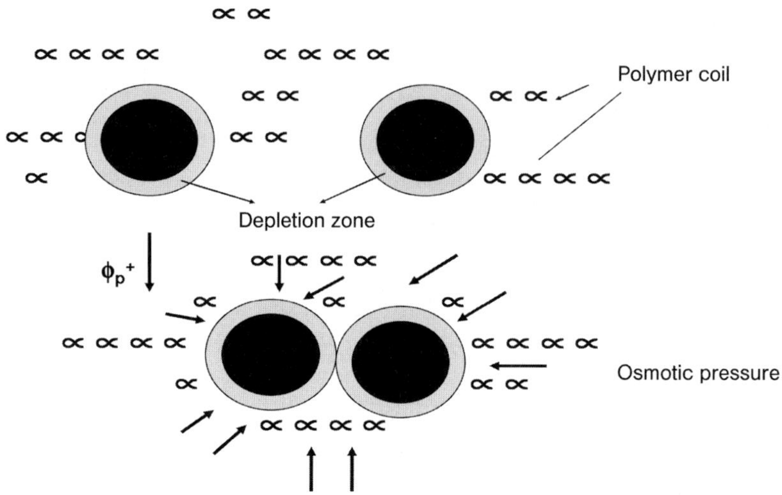

Figure 1.4 Schematic representation of depletion flocculation.

by an adsorption energy. On increasing the free polymer concentration or volume fraction ϕ_p above a critical value ϕ_p^+, the depletion zones overlap and the polymer chains become "squeezed out" from between the particles. This results in an increase in the osmotic pressure outside the particles, resulting in a weak attraction that is referred to as depletion flocculation. A schematic representation of depletion flocculation is shown in Figure 1.4.

The magnitude of the depletion attraction energy G_{dep} is proportional to the polymer volume fraction ϕ_p and the molecular weight of the free polymer M. The range of depletion attraction is determined by the thickness Δ of the depletion zone, which is roughly equal to the radius of gyration of the free polymer, R_g.

G_{dep} is given by the following expression:

$$G_{dep} = \frac{2\pi R \Delta^2}{V_1} (\mu_1 - \mu_1{}^0) \left(1 + \frac{2\Delta}{R}\right) \tag{10}$$

where V_1 is the molar volume of the solvent, μ_1 the chemical potential of the solvent in the presence of free polymer with volume fraction ϕ_p and $\mu_1{}^0$ the chemical potential of the solvent in the absence of free polymer.

The different states of emulsions are illustrated schematically in Figure 1.5.

The states of emulsions represented in Figure 1.5 have some common features with suspensions. Creaming or sedimentation results from gravity, in which case the emulsion separates. If the emulsion droplet size is reduced to say 20–200 nm, the Brownian diffusion can overcome the gravity force and no separation occurs. This is the case with nanoemulsions. Emulsion flocculation can occur when there is not sufficient repulsion. Flocculation can be weak or strong depending on the magnitude of the attractive energy. Ostwald ripening of emulsions can

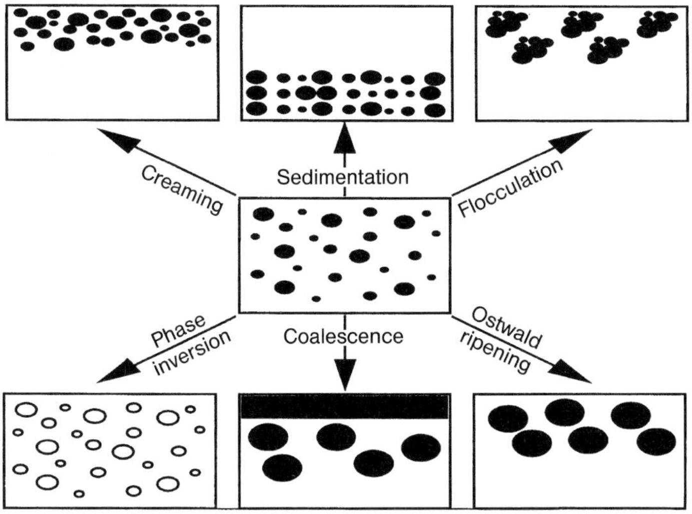

Figure 1.5 Different states of emulsions.

occur if the oil solubility is significant. The smaller droplets (with high radius of curvature) have higher solubility than larger droplets. This results in diffusion of the oil molecules from the small to the large droplets, resulting in an increase in the droplet size. Emulsion coalescence is the result of thinning and disruption of the liquid film between the droplets with the ultimate oil separation. Phase inversion can occur above a critical volume fraction of the disperse phase.

A number of the above instability problems with suspensions, emulsions and suspoemulsions can be overcome by using polymeric surfactants, which will be discussed later. For example, strong flocculation, coalescence and Ostwald ripening can be reduced or eliminated by the use of specially designed polymeric surfactants. Creaming or sedimentation can be eliminated by the use of "thickeners" that are sometimes referred to as "rheology modifiers".

1.3 Self-Assembly Structures in Cosmetic Formulations

Surfactant micelles and bilayers are the building blocks of most self-assembly structures. One can divide the phase structures into two main groups [19]: (1) those that are built of limited or discrete self-assemblies, which may be characterized roughly as spherical, prolate or cylindrical, and (2) infinite or unlimited self-assemblies whereby the aggregates are connected over macroscopic distances in one, two or three dimensions. The hexagonal phase (see below) is an example of one-dimensional continuity, the lamellar phase of two-dimensional continuity, whereas the bicontinuous cubic phase and the sponge phase (see later) are exam-

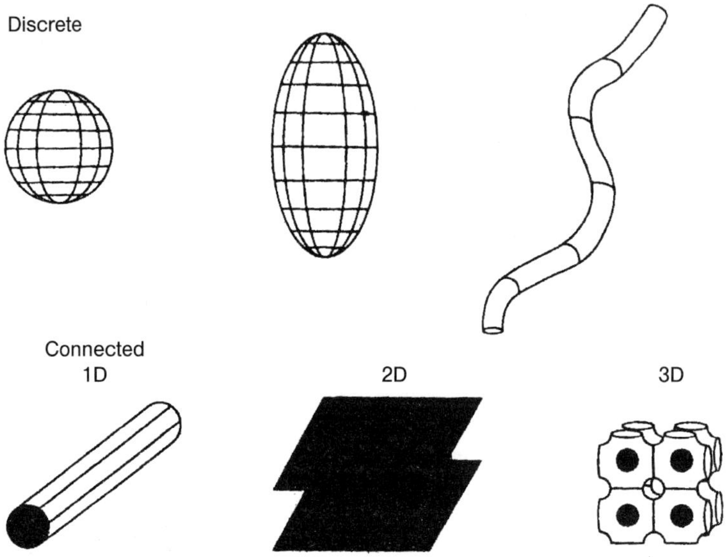

Figure 1.6 Schematic representation of self-assembly structures.

ples of three-dimensional continuity. These two types are illustrated schematically in Figure 1.6.

1.4 Structure of Liquid Crystalline Phases

The above-mentioned unlimited self-assembly structures in 1D, 2D or 3D are referred to as liquid crystalline structures. The last type behave as fluids and are usually highly viscous. At the same time, X-ray studies of these phases yield a small number of relatively sharp lines which resemble those produced by crystals [20]. Since they are fluids they are less ordered than crystals, but because of the X-ray lines and their high viscosity it is also apparent that they are more ordered than ordinary liquids. Thus, the term liquid crystalline phase is very appropriate for describing these self-assembled structures. Below, a brief description of the various liquid crystalline structures that can be produced with surfactants is given and Table 1.1 shows the most commonly used notation to describe these systems.

1.4.1 Hexagonal Phase

This phase is built up of (infinitely) long cylindrical micelles arranged in a hexagonal pattern, with each micelle being surrounded by six other micelles, as

Table 1.1 Notation of the most common liquid crystalline structures.

Phase structure	Abbreviation	Notation
Micellar	mic	L_1, S
Reversed micellar	rev mic	L_2, S
Hexagonal	hex	H_1, E, M_1, middle
Reversed hexagonal	rev hex	H_2, F, M_2
Cubic (normal micellar)	cub_m	I_1, S_{1c}
Cubic (reversed micelle)	cub_m	I_2
Cubic (normal bicontinuous)	cub_b	I_1, V_1
Cubic (reversed bicontinuous)	cub_b	I_2, V_2
Lamellar	lam	L_α, D, G, neat
Gel	gel	L_β
Sponge phase (reversed)	spo	L_3 (normal), L_4

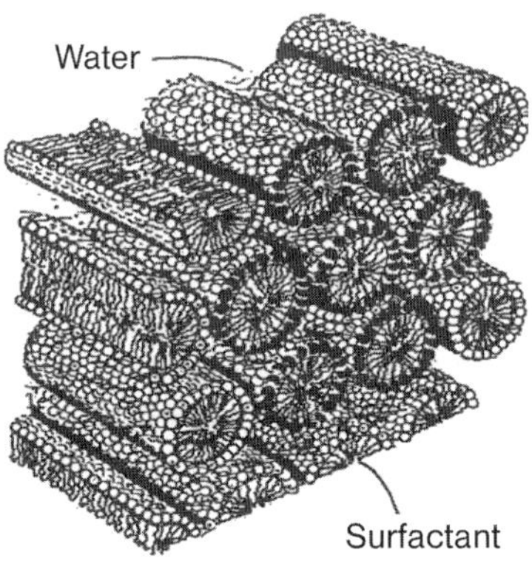

Figure 1.7 Schematic representation of the hexagonal phase.

shown schematically in Figure 1.7. The radius of the circular cross-section (which may be somewhat deformed) is again close to the surfactant molecule length [21].

1.4.2 Micellar Cubic Phase

This phase is built up of a regular packing of small micelles, which have similar properties to small micelles in the solution phase. However, the micelles are short prolates (axial ratio 1–2) rather than spheres, since this allows better packing. The micellar cubic phase is highly viscous. A schematic representation of the micellar cubic phase [22] is shown in Figure 1.8.

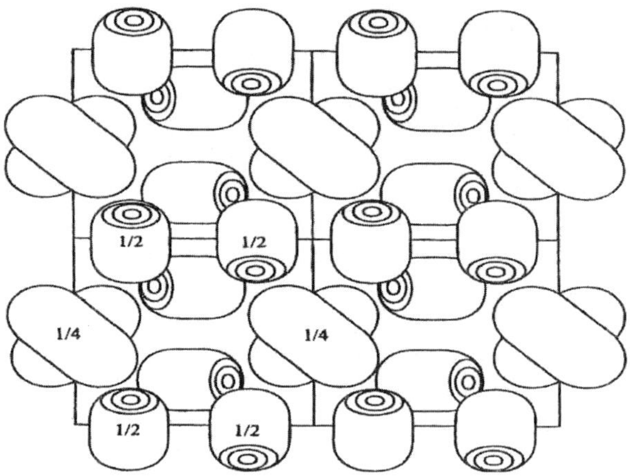

Figure 1.8 Representation of the micellar cubic phase.

1.4.3 Lamellar Phase

This phase is built of bilayers of surfactant molecules alternating with water layers. The thickness of the bilayers is somewhat smaller than twice the surfactant molecule length. The thickness of the water layer can vary over wide ranges, depending on the nature of the surfactant. The surfactant bilayer can range from being stiff and planar to being very flexible and undulating. A schematic representation of the lamellar phase [21] is shown in Figure 1.9.

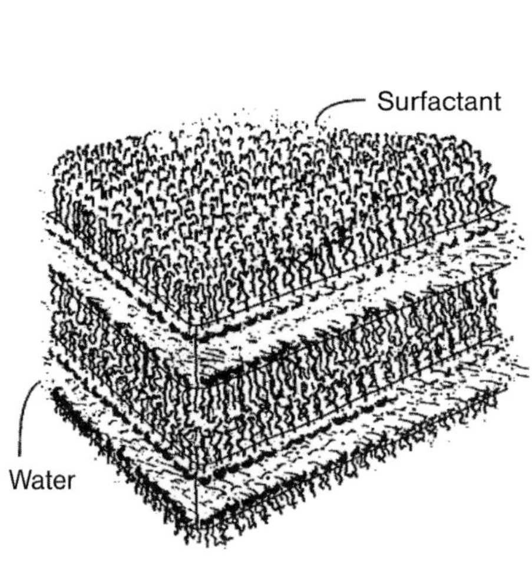

Figure 1.9 Schematic representation of the lamellar phase [7].

1.4.4
Discontinuous Cubic Phases

These phases can be a number of different structures, where the surfactant molecules form aggregates that penetrate space, forming a porous connected structure in three dimensions. They can be considered as structures formed by connecting rod-like micelles (branched micelles) or bilayer structures [23].

1.4.5
Reversed Structures

Except for the lamellar phase, which is symmetrical around the middle of the bilayer, the different structures have a reversed counterpart in which the polar and non-polar parts have changed roles. For example, a hexagonal phase is built up of hexagonally packed water cylinders surrounded by the polar head groups of the surfactant molecules and a continuum of the hydrophobic parts. Similarly, reversed (micellar-type) cubic phases and reversed micelles consist of globular water cores surrounded by surfactant molecules. The radii of the water cores are typically in the range 2–10 nm.

1.5
Driving Force for Formation of Liquid Crystalline Phases

One of the simplest methods for predicting the shape of an aggregated structure is based on the critical packing parameter P [8].

For a spherical micelle with radius r and containing n molecules each with volume v and cross-sectional area a_0:

$$n = \frac{4\pi r^3}{3v} = \frac{4\pi r^2}{a_0} \tag{11}$$

$$a_0 = \frac{3v}{r} \tag{12}$$

The cross-sectional area of the hydrocarbon tail, a, is given by

$$a = \frac{v}{l_c} \tag{13}$$

where l_c is the extended length of the hydrocarbon tail.

$$P = \frac{a}{a_0} = \frac{1}{3}\frac{r}{l_c} \tag{14}$$

Since $r < l_c$, then $P < 1/3$.

For a cylindrical micelle with radius r and length d:

$$n = \frac{\pi r^r d}{v} = \frac{2\pi r d}{a_0} \tag{15}$$

$$a_0 = \frac{2v}{r} \tag{16}$$

$$P = \frac{a}{a_0} = \frac{1}{2}\frac{r}{l_c} \tag{17}$$

Since $r < l_c$, $1/3 < P < 1/2$. For liposomes and vesicles $1 > P > 2/3$; for lamellar micelles $P \approx 1$; and for reverse micelles $P > 1$.

The packing parameter can be controlled by using mixtures of surfactants to arrive at the most desirable structure.

The most useful liquid crystalline structures in personal care applications are those of the lamellar phase. These lamellar phases can be produced in emulsion systems by using a combination of surfactants with various HLB numbers and choosing the right oil (emollient). In many cases, liposomes and vesicles are also produced by using lipids of various compositions. Two main types of lamellar liquid crystalline structures can be produced: "oleosomes" and "hydrosomes" (Figure 1.10).

Several advantages of lamellar liquid crystalline phases in cosmetics can be quoted: (1) they produce an effective barrier against coalescence; (2) they can produce "gel networks" that provide the right consistency for application in addition to preventing creaming or sedimentation; (3) they can influence the delivery of active ingredients of both the lipophilic and hydrophilic types; (4) since they mimic the skin structure (in particular the stratum corneum), they can offer prolonged hydration potential.

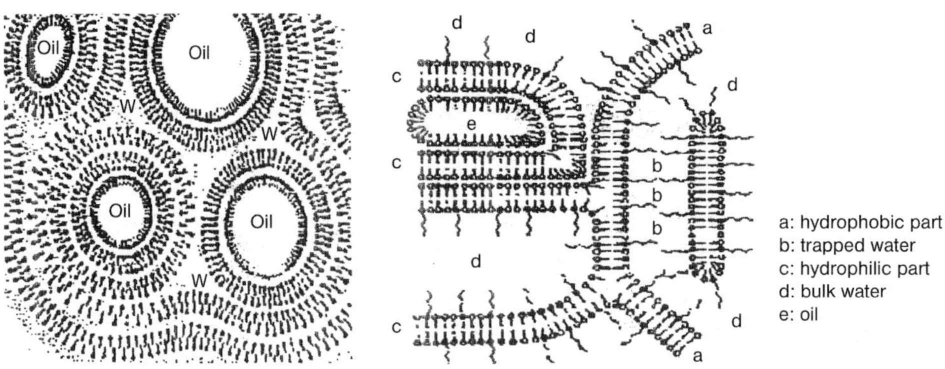

Figure 1.10 Schematic representation of "oleosomes" and "hydrosomes".

1.6 Polymeric Surfactants in Cosmetic Formulations

Polymeric surfactants of the A–B, A–B–A block or BA_n (or AB_n) graft types (where B is the "anchor" chain and A is the "stabilizing" chain) offer more robust stabilizing systems for dispersions (suspensions and emulsions) in cosmetics: (1) the high molecular weight of the surfactant (>1000) ensures strong adsorption of the molecule (no desorption); (2) the strong hydration of the A chain(s) ensures effective steric stabilization; (3) a lower emulsifier or dispersant concentration is sufficient (usually one order of magnitude lower than low molecular weight surfactants); (4) this lower concentration and high molecular weight of the material ensure the absence of any skin irritation.

One of the earliest polymeric surfactants used is the A–B–A block copolymer of poly(ethylene oxide) (PEO, A) and propylene oxide (PPO, B): Pluronics, Synperonic PE or Poloxamers. These are not ideal since adsorption by the PPO chain is not strong.

Recently, ORAFTI (Belgium) developed a polymeric surfactant based on inulin (a natural, linear polyfructose molecule produced from chicory roots) [24]. By grafting several alkyl chains on the polyfructose chain, a graft copolymer was produced (Figure 1.11).

The alkyl chains are strongly adsorbed at the oil or particle surface, leaving loops of polyfructose in the aqueous continuous phase (Figure 1.12). The polyfructose loops extend in solution (giving a layer thickness in the region of 10 nm) and they are highly solvated by the water molecules (solvation forces). The loops remain hydrated at high temperatures (>50 °C) and also in the presence of high electrolyte concentrations (up to 4 mol dm^{-3} NaCl and 1.5 mol dm^{-3} $MgSO_4$. Several O/W emulsions were prepared using INUTEC SP1 at a concentration of 1% for a 50:50 v/v emulsion. Hydrocarbon and silicone oils were used and the emulsions were prepared in water, 2 mol dm^{-3} NaCl and 1 mol dm^{-3} $MgSO_4$. All emulsions were stable against coalescence at room temperature and 50 °C for more than 1 year. The high stability of the emulsions is due to the unfavorable mixing of the strongly hydrated polyfructose loops (osmotic repulsion).

The multipoint anchoring of the polymer chains also ensures strong elastic (entropic) repulsion. This provides enhanced steric stabilization.

Evidence for the high stability of emulsions when using INUTEC SP1 has recently been obtained [25] from disjoining pressure measurements between two

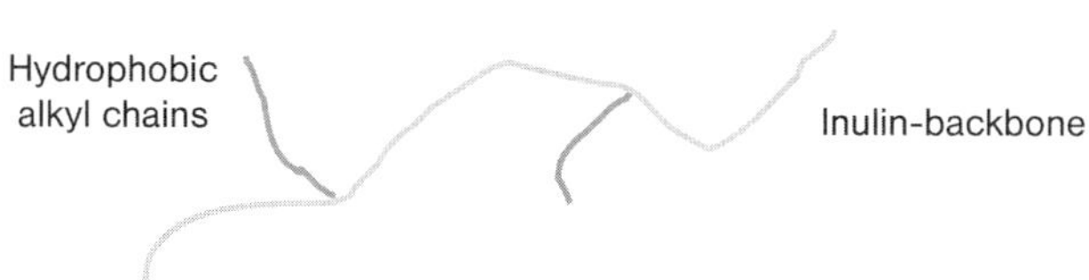

Figure 1.11 Hydrophobically modified inulin (HMI): INUTEC SP1.

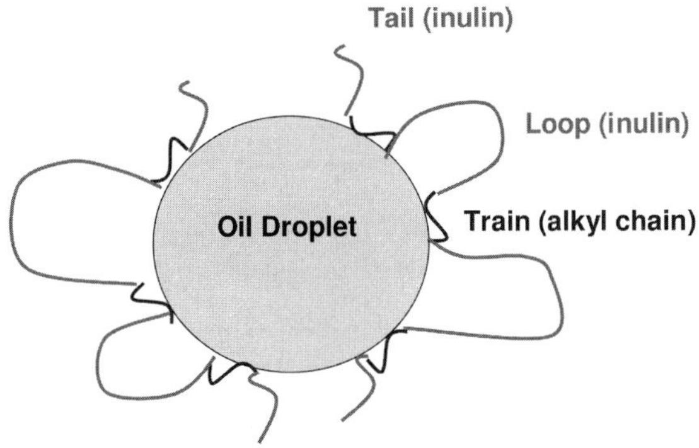

Figure 1.12 Schematic representation of the adsorption and conformation of INUTEC SP1 on oil droplets in aqueous medium.

oil droplets containing adsorbed polymer surfactant both in water and in high electrolyte solutions. A schematic representation of the measuring cell developed by Exerowa and Kruglyakov [26] is shown in Figure 1.13. A porous plate is used to produce a thin film with radius r between two oil droplets and the capillary pressure can be gradually increased to values reaching 45 kPa.

Figure 1.14 shows the variation of disjoining pressure with film thickness at various NaCl concentrations. It can be seen that by increasing the capillary pressure a stable Newton black film (NBF) is obtained at a film thickness of ~7 nm. The lack of rupture of the NBF up to the highest pressure applied, namely 4.5×10^4 Pa, clearly indicates the high stability of the liquid film in the presence of high NaCl concentrations (2 mol dm^{-3}). This result is consistent with the high emulsion stability obtained at high electrolyte concentrations and high temperature. Emulsions of Isopar M in water are very stable under such conditions and this could be accounted for by the high stability of the NBF. The droplet size of 50:50 O/W emulsions prepared using 2% INUTEC SP1 is in the range 1–10 μm. This corresponds to a capillary pressure of $\sim 3 \times 10^4$ Pa for 1-μm drops and $\sim 3 \times 10^3$ Pa for 10-μm drops. These capillary pressures are lower than those to which the NBF has been subjected and this clearly indicates the high stability obtained against coalescence in these emulsions.

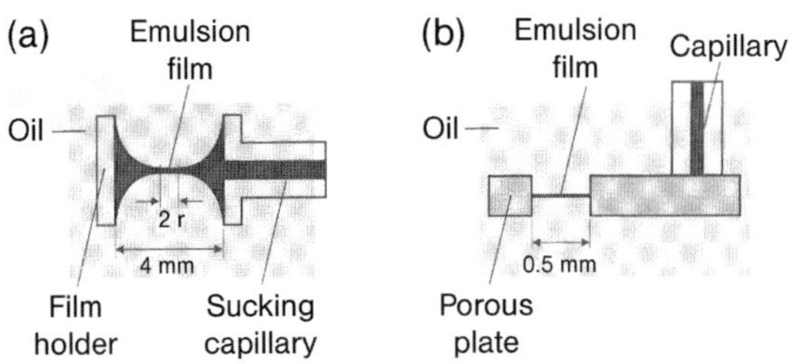

Figure 1.13 Schematic representation of Emulsion film stability measurement.

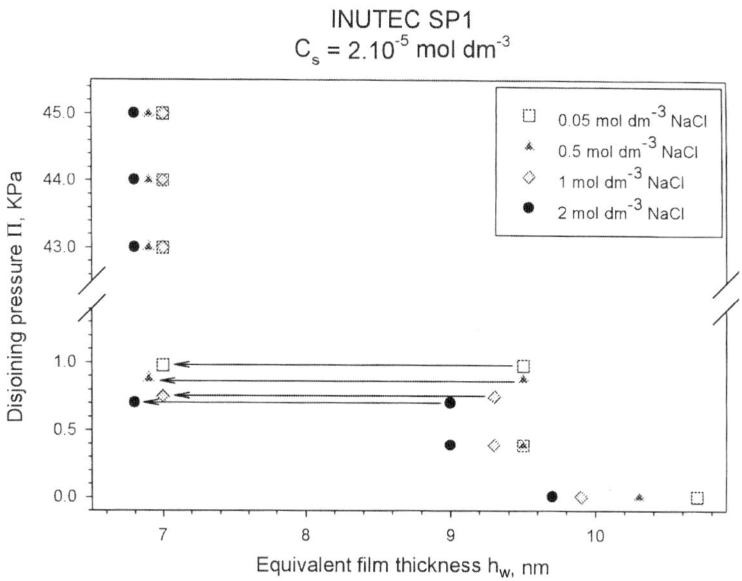

Figure 1.14 Variation of disjoining pressure with film thickness at various NaCl concentrations.

The graft copolymer INUTEC SP1 can also be used for the stabilization of hydrophobic particles in aqueous media. The alkyl chains are strongly adsorbed on the particle surface with multi-point attachment leaving the strongly hydrated polyfructose loops and tails dangling in solution, thus providing an effective steric barrier. Evidence for this high stability obtained using INUTEC SP1 has been obtained using atomic force microscopy (AFM) measurements [27] between a hydrophobically modified glass sphere and a plate both containing an adsorbed layer of INUTEC SP1. Results were obtained both in water and in various Na_2SO_4 solutions. Figure 1.15 shows the variation of force with separation distance

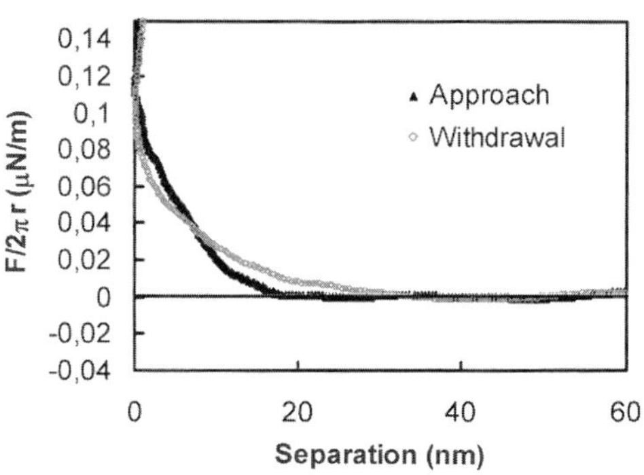

Figure 1.15 Force–distance curves between hydrophobized glass surfaces containing adsorbed INUTEC SP1 in water.

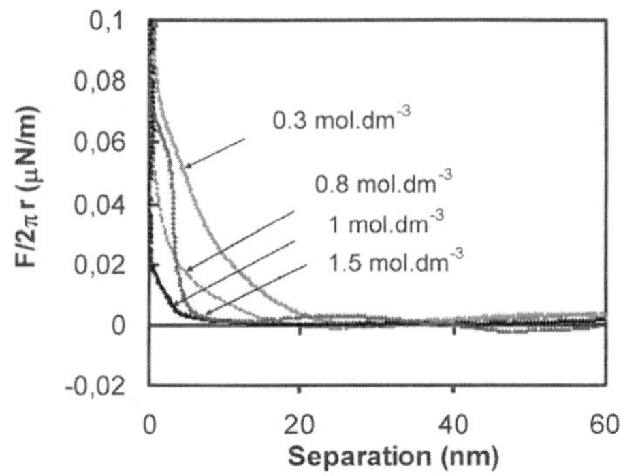

Figure 1.16 Force–distance curves for hydrophobized glass surfaces containing adsorbed INUTEC SP1 at various Na_2SO_4 concentrations.

between the glass sphere and plate in aqueous solutions containing INUTEC SP1 at the saturation adsorption concentration. The results at various Na_2SO_4 concentrations are shown in Figure 1.16.

It can be seen from Figure 1.15 that the force between the hydrophobized glass surface containing adsorbed INUTEC SP1 starts to increase at a separation distance of ~20 nm, which corresponds to an adsorbed layer thickness of ~10 nm. The above thickness is maintained in 0.3 mol dm^{-3} Na_2SO_4 (Figure 1.16). With increasing Na_2SO_4 concentration the adsorbed layer thickness decreases, reaching ~3 nm in the presence of 1.5 mol dm^{-3} Na_2SO_4. Even at such a high electrolyte concentration, the interaction is still repulsive.

1.7
Polymeric Surfactants for Stabilization of Nanoemulsions

Nanoemulsions are transparent or translucent systems in the size range 20–200 nm [28]. Whether the system is transparent or translucent depends on the droplet size, the volume fraction of the oil and the refractive index difference between the droplets and the medium. Nanoemulsions having diameters < 50 nm appear transparent when the oil volume fraction is <0.2 and the refractive index difference between the droplets and the medium is not large. With increase in droplet diameter and oil volume fraction the system may appear translucent and at higher oil volume fractions the system may become turbid.

Nanoemulsions are only kinetically stable. They have to be distinguished from microemulsions (that cover the size range 5–50 nm), which are mostly transparent and thermodynamically stable. The long-term physical stability of nanoemulsions (with no apparent flocculation or coalescence) makes them unique and they are sometimes referred to as "approaching thermodynamic stability". The inherently high colloid stability of nanoemulsions can be well understood from consideration of their steric stabilization (when using nonionic surfactants and/

or polymers) and how this is affected by the ratio of the adsorbed layer thickness to droplet radius, as will be discussed below.

Unless adequately prepared (to control the droplet size distribution) and stabilized against Ostwald ripening (that occurs when the oil has some finite solubility in the continuous medium), nanoemulsions may show an increase in the droplet size and an initially transparent system may become turbid on storage.

The attraction of nanoemulsions for application in personal care and cosmetics is due to the following advantages: (1) the very small droplet size causes a large reduction in the gravity force and the Brownian motion may be sufficient for overcoming gravity; this means that no creaming or sedimentation occurs on storage; (2) the small droplet size also prevents any flocculation of the droplets; weak flocculation is prevented and this enables the system to remain dispersed with no separation; (3) the small droplets also prevent their coalescence, since these droplets are non-deformable and hence surface fluctuations are prevented; in addition, the significant surfactant film thickness (relative to droplet radius) prevents any thinning or disruption of the liquid film between the droplets; (4) nanoemulsions are suitable for efficient delivery of active ingredients through the skin – the large surface area of the emulsion system allows rapid penetration of actives; (5) due to their small size, nanoemulsions can penetrate through the "rough" skin surface and this enhances penetration of actives; (6) the transparent nature of the system, their fluidity (at reasonable oil concentrations) and the absence of any thickeners may give them a pleasant esthetic character and skin feel; (7) unlike microemulsions (which require a high surfactant concentration, usually in the region of 20% and higher), nanoemulsions can be prepared using reasonable surfactant concentrations; for a 20% O/W nanoemulsion, a surfactant concentration in the range 5–10% may be sufficient; (8) the small size of the droplets allows them to deposit uniformly on substrates; wetting, spreading and penetration may be also enhanced as a result of the low surface tension of the whole system and the low interfacial tension of the O/W droplets; (9) nanoemulsions can be applied for delivery of fragrants which may be incorporated in many personal care products; this could also be applied in perfumes which are desirable to be formulated alcohol free; (10) nanoemulsions may be applied as a substitute for liposomes and vesicles (which are much less stable) and it is possible in some cases to build lamellar liquid crystalline phases around the nanoemulsion droplets.

The inherently high colloid stability of nanoemulsions when using polymeric surfactants is due to their steric stabilization. The mechanism of steric stabilization was discussed before. As shown in Figure 1.2a, the energy distance curve shows a shallow attractive minimum at a separation distance comparable to twice the adsorbed layer thickness 2δ. This minimum decreases in magnitude as the ratio of adsorbed layer thickness to droplet size increases. With nanoemulsions the ratio of adsorbed layer thickness to droplet radius (δ/R) is relatively large (0.1–0.2) compared with macroemulsions. This is illustrated schematically in Figure 1.17, which shows the reduction in $G_{\min}$ with increase in δ/R.

These systems approach thermodynamic stability against flocculation and/or coalescence. The very small size of the droplets and the dense adsorbed layers

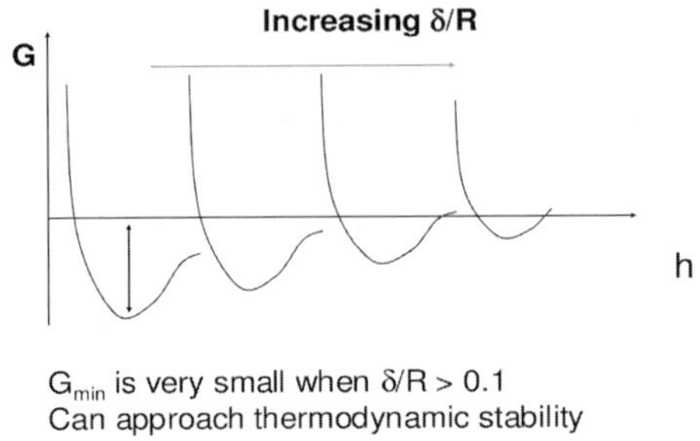

Figure 1.17 Importance of the ratio of adsorbed layer thickness to particle size.

ensure lack of deformation of the interface and lack of thinning, and disruption of the liquid film between the droplets and hence coalescence is also prevented.

One of the main problems with nanoemulsions is Ostwald ripening, which results from the difference in solubility between small and large droplets [28]. The difference in chemical potential of dispersed phase droplets between different sized droplets was given by Lord Kelvin:

$$c(r) = c(\infty)\,\exp\left(\frac{2\gamma V_m}{rRT}\right) \tag{18}$$

where $c(r)$ is the solubility surrounding a particle of radius r, $c(\infty)$ is the bulk phase solubility and V_m is the molar volume of the dispersed phase. The quantity $2\gamma V_m/RT$ is termed the characteristic length. It has an order of ~1 nm or less, indicating that the difference in solubility of a 1-μm droplet is of the order of 0.1% or less.

Theoretically, Ostwald ripening should lead to condensation of all droplets into a single drop (i.e. phase separation). This does not occur in practice since the rate of growth decreases with increase in droplet size.

For two droplets of radii r_1 and r_2 (where $r_1 < r_2$):

$$\frac{RT}{V_m}\ln\left[\frac{c(r_1)}{c(r_2)}\right] = 2\gamma\left(\frac{1}{r_1} - \frac{1}{r_2}\right) \tag{19}$$

Equation (19) shows that the larger the difference between r_1 and r_2, the higher is the rate of Ostwald ripening.

Ostwald ripening can be quantitatively assessed from plots of the cube of the radius versus time t [28]:

$$r^3 = \frac{8}{9}\left[\frac{c(\infty)\gamma V_m}{\rho RT}\right]t \tag{20}$$

where D is the diffusion coefficient of the disperse phase in the continuous phase.

Ostwald ripening can be reduced by incorporation of a second component which is insoluble in the continuous phase (e.g. squalane). In this case, significant partitioning between different droplets occurs, with the component having low solubility in the continuous phase expected to be concentrated in the smaller droplets. During Ostwald ripening in a two-component disperse phase system, equilibrium is established when the difference in chemical potential between different sized droplets (which results from curvature effects) is balanced by the difference in chemical potential resulting from partitioning of the two components. If the secondary component has zero solubility in the continuous phase, the size distribution will not deviate from the initial one (the growth rate is equal to zero). In the case of limited solubility of the secondary component, the distribution is the same as governed by Eq. (19), i.e. a mixture growth rate is obtained which is still lower than that of the more soluble component.

The above method is of limited application since one requires a highly insoluble oil as the second phase which is miscible with the primary phase.

Another method for reducing Ostwald ripening depends on modification of the interfacial film at the O/W interface. According to Eq. (19), a decrease in γ results in a reduction of Ostwald ripening. However, this alone is not sufficient since one has to reduce γ by several orders of magnitude. It has been suggested that by using surfactants which are strongly adsorbed at the O/W interface (i.e. polymeric surfactants) and which do not desorb during ripening, the rate could besignificantly reduced. An increase in the surface dilatational modulus and decrease in γ would be observed for the shrinking drops. The difference in γ between the droplets would balance the difference in capillary pressure (i.e. curvature effects).

To achieve the above effect, it is useful to use A–B–A block copolymers that are soluble in the oil phase and insoluble in the continuous phase. A strongly adsorbed polymeric surfactant that has limited solubility in the aqueous phase can also be used [e.g. hydrophobically modified inulin, INUTEC SP1 (ORAFTI, Belgium], as will be discussed below.

Two methods may be applied for the preparation of nanoemulsions (covering the droplet radius size range 20–200 nm): use of high-pressure homogenizers (aided by appropriate choice of surfactants and cosurfactants) or application of the phase inversion temperature (PIT) concept. The production of small droplets (submicron) requires the application of high energy. The process of emulsification is generally inefficient, as illustrated below. Simple calculations show that the mechanical energy required for emulsification exceeds the interfacial energy by several orders of magnitude. For example, to produce a nanoemulsion at $\varphi = 0.1$ with an average radius R of 200 nm, using a surfactant that gives an interfacial tension $\gamma = 10$ mN m^{-1}, the net increase in surface free energy is $A\gamma = 3\varphi\gamma/R = 1.5 \times 10^4$ J m^{-3}. The mechanical energy required in a homogenizer is 1.5×10^7 J m^{-3}, i.e. an efficiency of 0.1%. The rest of the energy (99.9%) is dissipated as heat.

The intensity of the process or the effectiveness in making small droplets is often governed by the net power density [$\varepsilon(t)$]:

$$p = \varepsilon(t)\,\mathrm{d}t \tag{21}$$

where t is the time during which emulsification occurs.

Break-up of droplets will only occur at high ε values, which means that the energy dissipated at low ε levels is wasted. Batch processes are generally less efficient than continuous processes. This shows why with a stirrer in a large vessel, most of the energy applies at low intensity is dissipated as heat. In a homogenizer, p is simply equal to the homogenizer pressure.

Several procedures may be applied to enhance the efficiency of emulsification when producing nanoemulsions: (1) one should optimize the efficiency of agitation by increasing ε and decreasing the dissipation time; (2) the nanoemulsion is preferably prepared at high volume faction of the disperse phase and subsequently diluted; however, very high φ values may result in coalescence during emulsification; (3) add more surfactant, whereby creating a smaller γ_{eff} and possibly diminishing recoalescence; (4) use a surfactant mixture that shows a greater reduction in γ than the individual components; (5) if possible dissolve the surfactant in the oil phase, which produces smaller droplets; (6) it may be useful to emulsify in steps of increasing intensity, particularly with nanoemulsions having highly viscous disperse phase.

Low-energy techniques may be applied for the preparation of nanoemulsions. Two methods can be applied: (1) the emulsifier is dissolved in the oil phase and the aqueous phase is gradually added; initially a W/O emulsion is produced but at a critical volume fraction of the aqueous phase inversion occurs and the resulting O/W system may form sufficiently small droplets in the nano-size range; (2) the phase inversion temperature (PIT) technique, which is by far the most suitable method for producing a nanoemulsion; it is limited to systems that contain an ethoxylated surfactant.

When an O/W emulsion prepared using a nonionic surfactant of the ethoxylate type is heated, then at a critical temperature (the PIT), the emulsion inverts to a W/O emulsion. At the PIT the hydrophilic and lipophilic components of the surfactant are exactly balanced and the PIT is sometimes referred to as the HLB temperature. At the PIT the droplet size reaches a minimum and the interfacial tension also reaches a minimum. However, the small droplets are unstable and they coalesce very rapidly.

By rapid cooling of the emulsion that is prepared at a temperature near the PIT, very stable nanoemulsion droplets could be produced. Near the HLB temperature, the interfacial tension reaches a minimum.

Several experiments were carried to investigate the methods of preparation of nanoemulsions and their stability. The first method applied the PIT principle for the preparation of nanoemulsions. Experiments were carried out using hexadecane as the oil phase and Brij 30 ($C_{12}EO_4$) as the nonionic emulsifier. The HLB temperature was determined using conductivity measurements, whereby 10^{-2} mol dm^{-3} NaCl was added to the aqueous phase (to increase the sensitivity of the conductibility measurements).

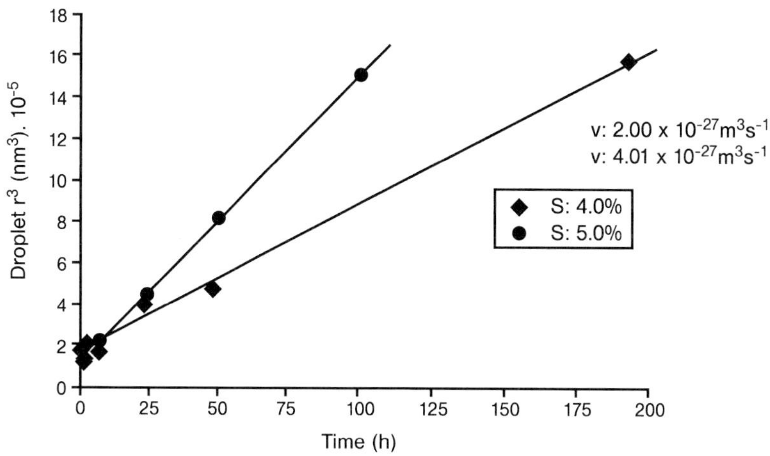

Figure 1.18 Variation of r^3 with time for hexadecane–water emulsions prepared using the PIT method.

Nanoemulsions were prepared by rapid cooling of the system to 25 °C. The droplet diameter was determined using photon correlation spectroscopy (PCS). At 4 and 5% surfactant, the average droplet diameter was 116 and 95 nm, respectively. However, the nanoemulsions showed significant polydispersity (polydispersity index of 0.29 and 0.09 at 4 and 5% surfactant, respectively). Nanoemulsions could not be produced when the surfactant concentration was reduced to below 4%. Nanoemulsions were then prepared using a high-pressure homogenizer (Emulsiflex) and these were smaller in size and much less polydisperse. For example, using 4% surfactant and 20% O/W emulsion, the average droplet diameter was 69 nm with a very low polydispersity index.

Figure 1.18 shows the variation of r^3 with time t for 20:80 O/W nanoemulsions at two $C_{12}EO_4$ concentrations prepared by the PIT method. It can be seen that the emulsion containing the higher surfactant concentration gives a higher rate of Ostwald ripening. This may be due to solubilization of the oil by the surfactant micelles.

Since the driving force for Ostwald ripening is the difference in solubility between smaller and larger droplets, one would expect that the narrower the droplet size distribution, the slower the rate. This is illustrated in Figure 1.19, which shows the variation of r^3 with time for nanoemulsions prepared using the PIT method and the homogenizer. It can be seen that the rate of Ostwald ripening is smaller for nanoemulsions prepared using the homogenizer when compared with the rate obtained using the PIT method.

Further evidence for Ostwald ripening was obtained by using a more soluble oil, namely a branched hexadecane (Arlamol HD). The results are shown in Figure 1.20 for nanoemulsions prepared using 4% surfactant. It can be seen that

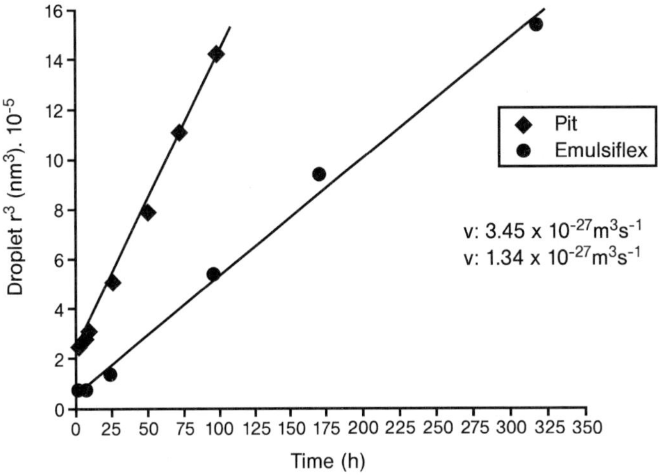

Figure 1.19 Comparison of Ostwald ripening using the PIT method and the Emulsiflex.

the more soluble oil (Arlamol HD) give a higher rate of Ostwald ripening when compared with a less soluble oil such as hexadecane.

As mentioned above, polymeric surfactants can reduce Ostwald ripening by enhancing the Gibbs elasticity at the O/W interface. Hydrophobically modified inulin (INUTEC SP1) is ideal for reduction of Ostwald ripening due to its strong adsorption and its limited solubility in the aqueous phase (no desorption occurs). This is illustrated in Figure 1.21, which shows plots of R^3 versus time for 20% v/v silicone O/W emulsions at two concentrations of INUTEC SP1. The concen-

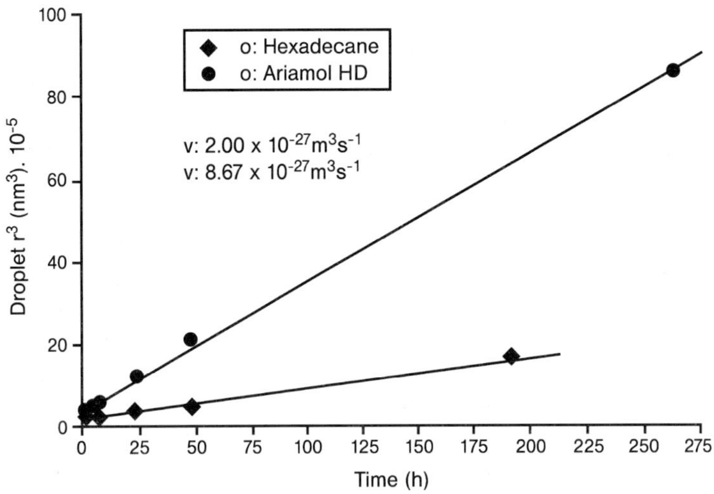

Figure 1.20 Ostwald ripening for hexadecane and Arlamol HD nanoemulsions.

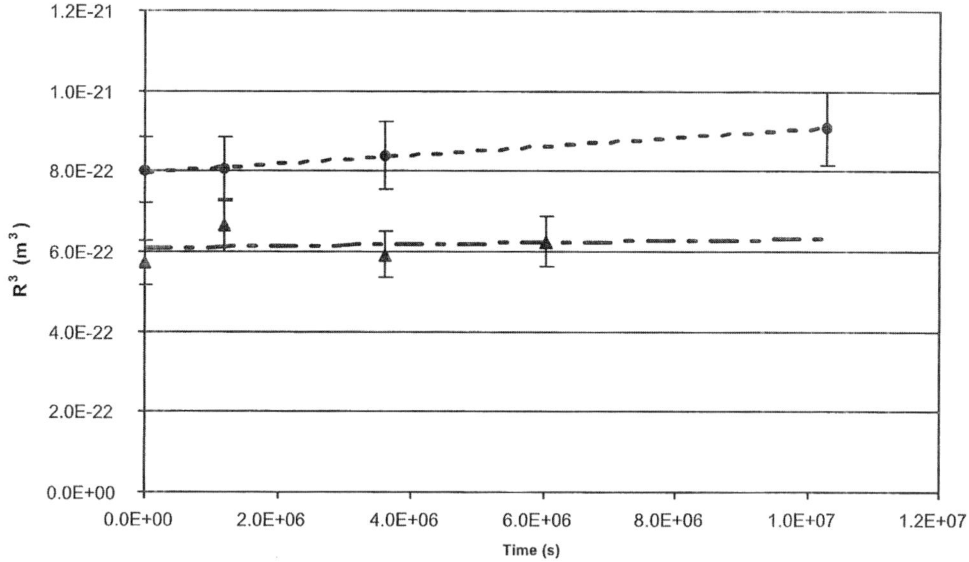

Figure 1.21 R^3 versus time for nanoemulsions at 1.6 and 2.4% HMI.

tration of INUTEC SP1 is much lower than that required when using nonionic surfactants.

The rate of Ostwald ripening is 1.1×10^{-29} and 2.4×10^{-30} m^3 s^{-1} at 1.6 and 2.4% INUTEC SP1, respectively. These rates are about three orders of magnitude lower than those obtained using a nonionic surfactant. Addition of 5% glycerol was found to decrease the rate of Ostwald ripening in some nanoemulsions.

The above systems of nanoemulsions are attractive for cosmetic applications: (1) low viscosity for application in sprayables; (2) efficient delivery of active ingredients through the skin; (3) ability to penetrate through the "rough" skin surface.

Various nanoemulsions with hydrocarbon oils of different solubility were prepared using INUTEC SP1. Figure 1.22 shows plots of r^3 versus t for nanoemulsions of the hydrocarbon oils that were stored at 50 °C. It can be seen that both paraffinum liquidum with low and high viscosity give almost a zero slope, indicating the absence of Ostwald ripening in this case. This is not surprising since both oils have very low solubility and the hydrophobically modified inulin, INUTEC SP1, strongly adsorbs at the interface, giving high elasticity that reduces both Ostwald ripening and coalescence.

With the more soluble hydrocarbon oils, namely isohexadecane, there is an increase in r^3 with time, giving a rate of Ostwald ripening of 4.1×10^{-27} m^3 s^{-1}. The rate for this oil is almost three orders of a magnitude lower than that obtained with a nonionic surfactant, namely laureth-4 (C_{12}-alkyl chain with 4 mol of ethylene oxide) when the nanoemulsion was stored at 50 °C. This clearly shows the effectiveness of INUTEC SP1 in reducing Ostwald ripening. This reduction can be attributed to the enhancement of the Gibbs dilatational

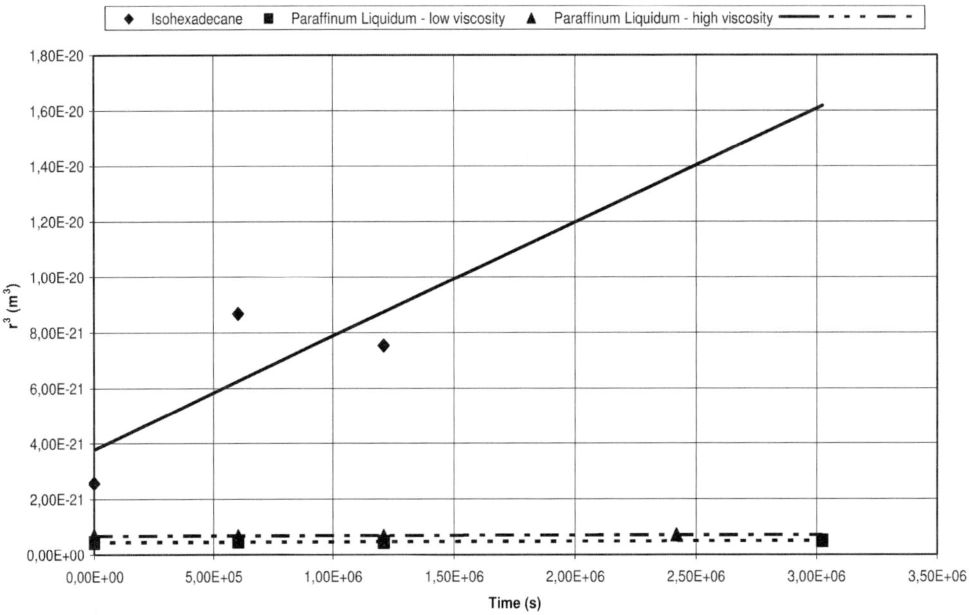

Figure 1.22 r^3 versus t for nanoemulsions based on hydrocarbon oils.

elasticity which results from the multi-point attachment of the polymeric surfactant with several alkyl groups to the oil droplets. This results in a decrease in the molecular diffusion of the oil from the smaller to the larger droplets.

1.8 Polymeric Surfactants in Multiple Emulsions

Multiple emulsions are complex systems of emulsions of emulsions [29, 30]: water-in-oil-in-water (W/O/W) and oil-in-water-in-oil (O/W/O). The W/O/W multiple emulsions are the most commonly used systems in personal care products. Multiple emulsions are ideal systems for application in cosmetics: (1) one can dissolve actives in three different compartments; (2) they can be used for controlled and sustained release; (3) they can be applied as creams by using thickeners in the outer continuous phase.

Multiple emulsions are conveniently prepared by a two-step process. For W/O/W, a W/O emulsion is first prepared using a low-HLB polymeric surfactant using a high-speed stirrer to produce droplets of ~1 μm. The W/O emulsion is then emulsified in an aqueous solution containing a high-HLB polymeric surfactant using a low-speed stirrer to produce droplets of 10–100 μm.

To prepare a stable multiple emulsion, the following criteria must be satisfied: (1) two emulsifiers with low and high HLB numbers to produce the primary W/O emulsion and the final W/O/W multiple emulsion; (2) polymeric emulsifiers that provide steric stabilization are necessary to maintain the long-term physical stability; (3) optimum osmotic balance for W/O/W between the internal water droplets and outer continuous phase, which can be achieved by using electrolytes or non-electrolytes.

Multiple emulsions are conveniently prepared using a two step process: a W/O system is first prepared by emulsification of the aqueous phase (which may contain an electrolyte to control the osmotic pressure) into an oil solution of the polymeric surfactant with the low HLB number – a high-speed stirrer is used to produce droplets of ~1 μm. The droplet size of the primary emulsion can be determined using dynamic light scattering. The primary W/O emulsion is then emulsified into an aqueous solution (of an electrolyte to control the osmotic pressure) containing the polymeric surfactant with high HLB number – in this case a low-speed stirrer is used to produce multiple emulsion droplets in the range 10–100 μm. The droplet size of the multiple emulsion can be determined using optical microscopy (with image analysis) or using light diffraction techniques (Malvern Mastersizer). A schematic representation of the preparation of W/O/W multiple emulsions is shown in Figure 1.23.

A W/O/W multiple emulsion was prepared using two polymeric surfactants. A W/O emulsion was prepared using an A–B–A block copolymer of poly(hydroxystearic acid) (PHS, A) and poly(ethylene oxide) (PEO, B), i.e. PHS–PEO–PHS (Arlacel P135, UNIQEMA). This emulsion was prepared using a high-speed mixer giving droplet sizes in the region of 1 μm. The W/O emulsion was then emulsified in an aqueous solution of INUTEC SP1 using low-speed stirring to produce multiple emulsion droplets in the range 10–100 μm. The osmotic balance was achieved using 0.1 mol dm^{-3} $MgCl_2$ in the internal water droplets and

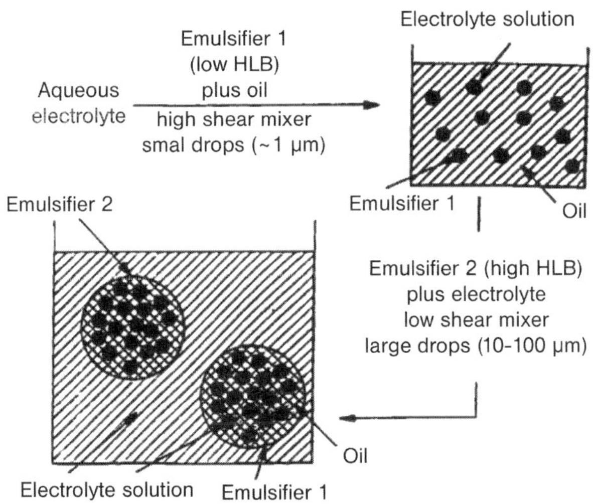

Figure 1.23 Scheme for preparation of W/O/W multiple emulsion.

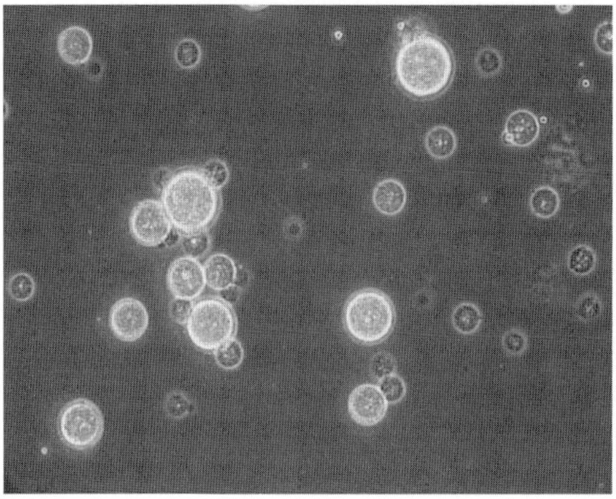

Figure 1.24 Photomicrograph of the W/O/W multiple emulsion.

outside continuous phase. The multiple emulsion was stored at room temperature and 50 °C and photomicrographs were taken at various intervals of time. The multiple emulsion was very stable for several months. A photomicrograph of the W/O/W multiple emulsion is shown in Figure 1.24.

An O/W/O multiple emulsion was made by first preparing a nanoemulsion using INUTEC SP1. The nanoemulsion was then emulsified into an oil solution of Arlacel P135 using a low-speed stirrer. The O/W/O multiple emulsion was stored at room temperature and 50 °C and photomicrographs were taken at various intervals of time. The O/W/O multiple emulsion was stable for several months both at room temperature and 50 °C. A photomicrograph of the O/W/O multiple emulsion is shown in Figure 1.25.

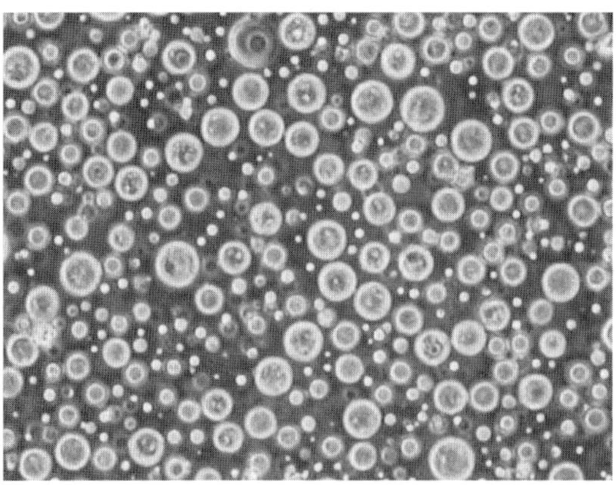

Figure 1.25 Photomicrograph of the O/W/O multiple emulsion.

1.9
Polymeric Surfactants for Stabilization of Liposomes and Vesicles

Liposomes are multilamellar structures consisting of several bilayers of lipids (several μm). They are produced by simply shaking an aqueous solutions of phospholipids, e.g. egg lecithin. When sonicated, these multilayer structures produce unilamellar structures (with a size range of 25–50 nm) that are referred to as liposomes. A schematic diagram of liposomes and vesicles is given in Figure 1.26. Glycerol-containing phospholipids are used for the preparation of liposomes and vesicles: phosphatidylcholine, phosphatidylserine, phosphatidylethanolamine, phosphatidylanisitol, phosphatidylglycerol, phosphatidic acid and cholesterol. In most preparations, a mixture of lipids is used to obtain the optimum structure.

The free energy for an amphiphile in a spherical vesicle of outer and inner radii R_1 and R_2 depends on (1) γ, the interfacial tension between hydrocarbon and water; (2) n_1, n_2, the number of molecules in the outer an inner layers; (3) e, the charge on the polar head group; (4) D, the thickness of the head group; and (5) v, the hydrocarbon volume per amphiphile (taken to be constant) [31].

The minimum free energy, $\mu_N{}^0$, configuration per amphiphile for a particular aggregation number N is given by [31]

$$\mu_N{}^0(\min) = 2a_0\gamma\left(1 - \frac{2\pi Dt}{Na_0}\right) \tag{22}$$

where a_0 is the surface area per amphiphile in a planer bilayer ($N = \infty$).

Several conclusions can be drawn from the thermodynamic analysis of vesicle formation: (1) $\mu_N{}^0$ is slightly lower than $\mu_N{}^0(\min)$ ($= 2a_0\gamma$); (2) since a spherical vesicles has much lower aggregation number N than a planer bilayer, then spherical vesicles are more favored over planer bilayers; (3) $a_1 < a_0 < a_2$; (4) for vesicles with a radius $> R_1{}^c$, there are no packing constraints – these vesicles are not favored over smaller vesicles which have lower N; (5) the vesicle size distribution is nearly Gaussian, with a narrow range; for example, vesicles produced from phospatidylcholine (egg lecithin) have $R_1 \approx 10.5 \pm 0.4$ nm – the maximum hydrocarbon chain length is ~1.75 nm; (6) once formed, vesicles are homogeneous

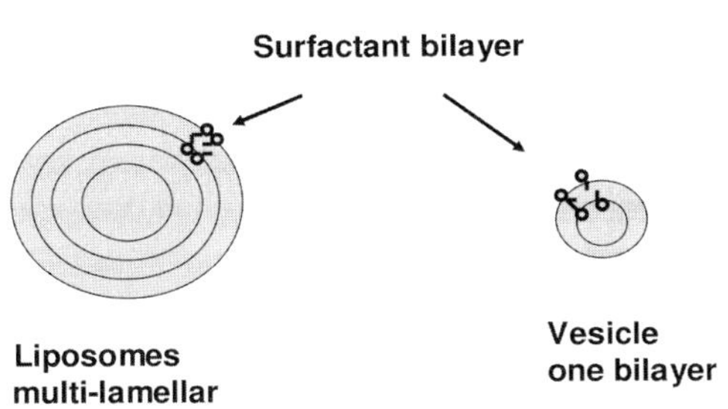

Figure 1.26 Schematic representation of liposomes and vesicles.

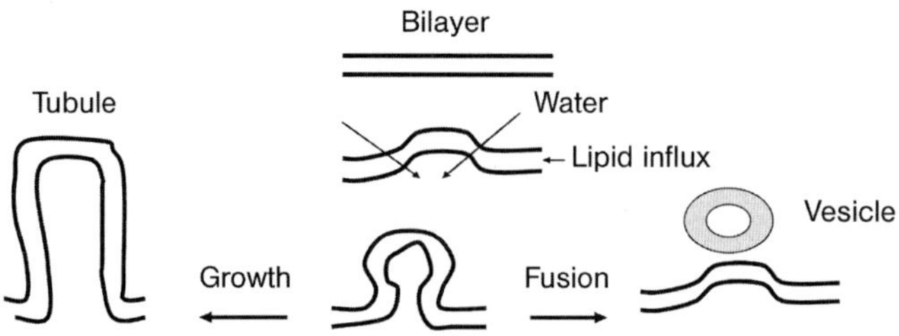

Figure 1.27 Mechanism of spontaneous formation of a vesicle from a bilayer.

and stable and they are not affected by the length of time and strength of sonication; and (7) sonication is necessary in most cases to break up the lipid bilayers which are first produced when the phospholipid is dispersed into water.

A schematic representation of the formation of bilayers and their break-up into vesicles is shown in Figure 1.27.

Liposomes and vesicles are ideal systems for cosmetic applications. They offer a convenient method for solubilizing nonpolar active substances in the hydrocarbon core of the bilayer. Polar substances can also be intercalated in the aqueous layer between the bilayer. They will also form lamellar liquid crystalline phases and they do not disrupt the stratum corneum. No facilitated transdermal transport is possible, thus eliminating skin irritation. Phospholipid liposomes can be used as *in vitro* indicators for studying skin irritation by surfactants.

The main problem with liposomes and vesicles is their physical instability on storage. Polymeric surfactants of the A–B–A block type (such as Pluronics PEO–PPO–PEO) can be used to stabilize the liposomes and vesicles [32]. The PPO chain becomes incorporated in the hydrocarbon bilayer, leaving the PEO chain in the aqueous internal and external phases, thus providing steric stabilization. The graft copolymer of INUTEC SP1 can also be used to stabilize the liposomes and vesicles. The alkyl chains are incorporated in the hydrocarbon bilayers leaving polyfructose loops in the aqueous internal and external phases. This provides an effective steric barrier and hence the long-term stability of the liposomes and

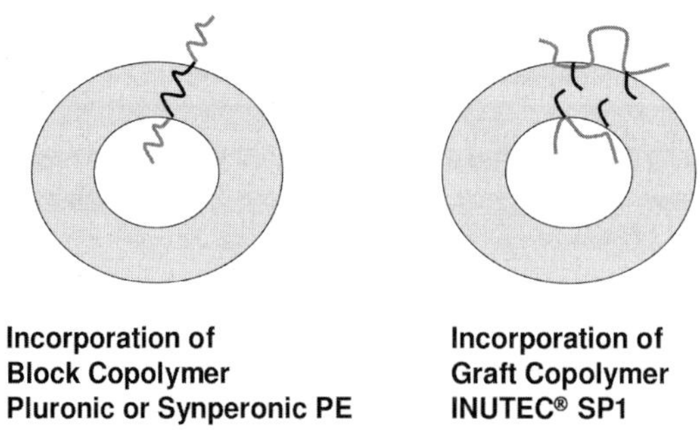

Figure 1.28 Incorporation of block and graft copolymers into the vesicle bilayer.

vesicles can be maintained. In addition, the rigidity of the lipid–polymer bilayer is greatly increased and this prevents the breakdown of the liposomes and vesicles into lamellar structures. A schematic representation of the incorporation of the A–B–A block copolymer (Pluronic or Synperonic PE) into the vesicle structure is given in Figure 1.28. The same figure also shows the incorporation of the graft copolymer INUTEC SP1 into the vesicle bilayer.

1.10 Conclusions

For optimum formulation of cosmetic preparations, one needs to apply the colloid and interface principles. Three main stabilization mechanisms can be identified: electrostatic, steric and electrosteric. The physical states of suspensions and emulsions can be described in terms of the interaction forces between the particles or droplets. Most cosmetic formulation contain self-assembled structures or liquid crystalline phases. The most useful type of liquid crystals is the lamellar phase, which provides an effective barrier against coalescence of the emulsions. These lamellar liquid crystalline structures can enhance penetration of lipophilic and hydrophilic active ingredients. They also provide effective and prolonged moisturization. Polymeric surfactants provide effective stabilization against flocculation and coalescence. They also reduce Ostwald ripening in nanoemulsions. Polymeric surfactants are also applied for stabilization of multiple emulsions of both the W/O/W and O/W/O types. Polymeric surfactants are also used for the stabilization of liposomes and vesicles. The above benefits of polymeric surfactants justify their application in cosmetic preparations. Apart from their excellent stabilization effect, they also eliminate any skin irritation.

References

1 M.M. Breuer, in *Encyclopedia of Emulsion Technology*, Vol. 2, Chap. 7, P. Becher (ed.), Marcel Dekker, New York, 1985.
2 S. Harry, in *Cosmeticology*, J.B. Wilkinson, R.J. Moore (eds.), Chemical Publishing, New York, 1981.
3 S.E. Friberg, *J. Soc. Cosmet. Chem.*, **41**, 155 (1990).
4 A.M. Kligman, in *Biology of the Stratum Corneum in Epidermis*, W. Montagna (ed.), Academic Press, New York, 1964, pp. 421–446.
5 P.M. Elias, B.E. Brown, P.T. Fritsch, R.J. Gorke, G.M. Goay, R.J. White, *J. Invest. Dermatol.*, **73**, 339 (1979).
6 S.E. Friberg, D.W. Osborne, *J. Dispers. Sci. Technol.*, **6**, 485 (1985).
7 S.C. Vick, *Soaps Cosmet. Chem. Spec.*, 36 (1984).
8 Th. Tadros, *Applied Surfactants*, Wiley-VCH, Weinheim, 2005.
9 J. Lyklema, Structure of the solid/liquid interface and the electrical double layer, in *Solid/Liquid Dispersions*, Th.F. Tadros (ed.), Academic Press, London, 1987.
10 B.H. Bijesterbosch, Stability of solid/liquid dispersions, in *Solid/Liquid Dispersions*, Th.F. Tadros (ed.), Academic Press, London, 1987.
11 Th.F. Tadros, Polymer adsorption and dispersion stability, in *The Effect of Polymers on Dispersion Properties*, Th.F. Tadros (ed.), Academic Press, London, 1981.

12 D.H. Napper, *Polymeric Stabilization of Colloidal Dispersions*, Academic Press, London, 1983.

13 P.J. Flory, W.R. Krigbaum, *J. Chem. Phys.*, **18**, 1086 (1950).

14 E.W. Fischer, *Kolloid Z.*, **160**, 120 (1958).

15 E.L. Mackor, J.H. van der Waals, *J. Colloid Sci.*, **7**, 535 (1951).

16 H.R. Kruyt (ed.), *Colloid Science*, Vol. I, Elsevier, Amsterdam, 1952.

17 E.J.W. Verwey, J.Th.G. Overbeek, *Theory of Stability of Lyophobic Colloids*, Elsevier, Amsterdam, 1948.

18 F.Th. Hesselink, A. Vrij, J.Th.G. Overbeek, *J. Phys. Chem.*, **75**, 2094 (1971).

19 K. Holmberg, B. Jonsson, B. Kronberg, B. Lindman, *Surfactants and Polymers in Aqueous Solution*, Wiley, New York, 2003.

20 R.G. Laughlin, *The Aqueous Phase Behavior of Surfactants*, Academic Press, London (1994).

21 K. Fontell, *Mol. Cryst. Liquid Cryst.*, **63**, 59 (1981).

22 K. Fontell, C. Fox, E. Hanson, *Mol. Cryst. Liquid Cryst.*, **1**, 9 (1985).

23 D.F. Evans, H. Wennerstrom, *The Colloid Domain, Where Physics, Chemistry and Biology Meet*, Wiley, New York, 1994.

24 Th.F. Tadros, A. Vandamme, B. Levecke, K. Booten, C.V. Stevens, *Adv. Colloid Interface Sci.*, **108–109**, 207 (2004).

25 D. Exerowa, G. Gotchev, Kolarov, Khr. Khristov, B. Levecke, Th. Tadros, *Langmuir*, **23**, 1711 (2007).

26 D. Exerowa, P.M. Kruglyakov, *Foam and Foam Films*, Elsevier, Amsterdam, 1998.

27 J. Nestor, J. Esquena, C. Solans, P.F. Luckham, M. Musoke, B. Levecke, K. Booten, Th.F. Tadros, *J. Colloid Interface Sci.*, **311**, 430 (2007).

28 Tharwat Tadros, P. Izquierdo, J. Esquena, C. Solans, *Adv. Colloid Interface Sci.*, **108–109**, 303 (2004).

29 D. Attwood, A.T. Florence, *Surfactant Systems, Their Chemistry, Pharmacy and Biology*, Chapman and Hall, New York, 1983.

30 J.L. Grossierd, M. Seiller (eds.), *Multiple Emulsions: Structure, Properties and Application*, Editions de Santé, Paris, 1997.

31 J.N. Isrealachvili, D.J. Mitchell, B.W. Ninham, *J. Chem. Soc., Faraday Trans. 2*, **72**, 1525 (1976).

32 K. Kostarelos, Th.F. Tadros, P.F. Luckham, *Langmuir*, **15**, 369 (1999).

2
Formulation and Stabilization of Nanoemulsions Using Hydrophobically Modified Inulin (Polyfructose) Polymeric Surfactant

Tharwat F. Tadros, Martine Lemmens, Bart Levecke, and Karl Booten

Abstract

Nano-emulsions, which cover a size range of 50–200 nm [1, 2], have recently been applied in personal care and cosmetic formulations. Their inherent stability is due to the steric stabilization produced by the use of non-ionic surfactants and/or polymers. However, one of the most serious instability problems is Ostwald ripening (disproportionation), which arises from the difference in solubility between small and larger droplets [3, 4]. On storage, the smaller droplets, which have higher solubility, dissolve and become deposited on the larger ones. This results in a shift in the droplet size distribution to larger values and the system may undergo some creaming or sedimentation. In this chapter, we will show that by using hydrophobically modified inulin (HMI), INUTEC® SP1, Ostwald ripening can be significantly reduced. Once stabilized, nano-emulsions offer many advantages in personal care application: efficient delivery of active ingredients through the skin, transparency (provided the refractive index of the droplets is not much larger than that of the continuous phase), low viscosity which may give a pleasant skin-feel, uniform deposition on rough skin, etc. The HMI reduces the Ostwald ripening rate of nano-emulsions when compared with non-ionic surfactants such as laureth-4. This is due to the strong adsorption of INUTEC® SP1 at the oil/water interface (by multi-point attachment) and enhancement of the Gibbs dilational elasticity, both of which reduce the diffusion of oil molecules from the smaller to the larger droplets. The present study also showed that the nature of the oil phase has a big influence on the rate, with the more soluble and more polar oils giving the highest Ostwald ripening rate. However, in all cases, when using INUTEC® SP1, the rates are reasonably low allowing this polymeric surfactant to be used in the formulation of nano-emulsions for personal care applications.

Colloids and Interface Science Series, Vol. 4
Colloids in Cosmetics and Personal Care. Edited by Tharwat F. Tadros

ISBN: 978-3-527-31464-5

2.1 Introduction

Nanoemulsions are transparent, translucent or turbid systems mostly covering the size range 20–200 nm [1, 2]. Whether the system is transparent, translucent or turbid depends on the droplet radius, the refractive index difference between the oil and continuous phase and the volume fraction of the oil. This can be easily understood from the dependence of turbidity τ on the above three parameters,

$$\tau = KNV^2 \tag{1}$$

where K is an optical constant that is given by the difference in refractive index of the oil and the medium or continuous phase $(n_{\text{oil}} - n_{\text{med}})$ and N is the number of droplets per unit volume each with a volume V. It is clear from Eq. (1) that the smaller the value of $(n_{\text{oil}} - n_{\text{med}})$, the smaller is the value of N (i.e. the lower the oil volume fraction), and the smaller the value of V (or droplet radius), the smaller is the value of τ, and these are the conditions to produce transparent nanoemulsions.

Nanoemulsions have also been referred to as miniemulsions [3, 4]. Unlike microemulsions (which are also transparent or translucent and thermodynamically stable), nanoemulsions are only kinetically stable. However, the long-term physical stability of nanoemulsions (with no apparent flocculation or coalescence) makes them unique and they are sometimes referred to as "approaching thermodynamic stability".

The inherently high colloid stability of nanoemulsions can be well understood from consideration of their steric stabilization (when using nonionic surfactants and/or polymers) and how this is affected by the ratio of the adsorbed layer thickness to the droplet radius as will be discussed below.

Unless adequately prepared (to control the droplet size distribution) and stabilized against Ostwald ripening (which occurs when the oil has some finite solubility in the continuous medium), nanoemulsions may lose their transparency with time as a result of the increase in droplet size.

The attraction of nanoemulsions for application in personal care and cosmetics and also in health care is due to the following advantages. (1) The very small droplet size causes a large reduction in the gravity force and the Brownian motion may be sufficient for overcoming gravity. This means that no creaming or sedimentation occurs on storage. (2) The small droplet size also prevents any flocculation of the droplets. Weak flocculation is prevented and this enables the system to remain dispersed with no separation. (3) The small droplets also prevent their coalescence, since these droplets are non-deformable and hence surface fluctuations are prevented. In addition, the significant surfactant film thickness (relative to droplet radius) prevents any thinning or disruption of the liquid film between the droplets. (4) Nanoemulsions are suitable for efficient delivery of active ingredients through the skin. The large surface area of the emulsion system allows rapid penetration of active ingredients. (5) Due to their small size, nanoemulsions can penetrate through the "rough" skin surface and this enhances penetration of active ingredients. (6) The transparent nature of the system, their fluidity (at reasonable oil concentrations) and the absence of any thickeners may give them a pleasant esthetic character and skin feel. (7) Unlike microemulsions (which require a high surfactant concentration, usually in the region of 20% and higher), nanoemulsions can be prepared using reasonable surfactant concentrations. For a 20% oil-in-water (O/W) nanoemulsion, a surfactant concentration in the region of 5–10% may be sufficient. With polymeric surfactants the concentration required to prepare a 20% O/W nanoemulsion can be significantly lower (in the region of ~2%). (8) The small size of the droplets allows them to deposit uniformly on substrates. Wetting, spreading and penetration may be also enhanced as a result of the low surface tension of the whole system and the low interfacial tension of the O/W droplets. (9) Nanoemulsions can be applied for delivery of fragrances which may be incorporated in many personal care products. This could also be applied in perfumes, which are desirable to be formulated alcohol free. (10) Nanoemulsions may be applied as a substitute for liposomes and vesicles (which are much less stable) and it is possible in some cases to build lamellar liquid crystalline phases around the nanoemulsion droplets.

In spite of the above advantages, nanoemulsions have only attracted interest in recent years for the following reasons. (1) Preparation of nanoemulsions in many cases requires special application techniques, such as the use of high-pressure homogenizers and ultrasonics. Such equipments (such as the microfluidizer) became available only in recent years. (2) There is a perception in the personal care and cosmetics industry that nanoemulsions are expensive to produce. Expensive equipment is required in addition to the use of high concentrations of emulsifiers. (3) Lack of understanding of the mechanism of production of submicron droplets and the role of surfactants and cosurfactants. (4) Lack of demonstration of the benefits that can be obtained from using nanoemulsions when compared with the classical macroemulsion systems. (5) Lack of understanding of the interfacial chemistry that is involved in production of nanoemulsions. (6) Lack of knowledge on the mechanism of Ostwald ripening, which is perhaps the most serious instability problem with nanoemulsions. (7) Lack of knowledge of the ingredients that may be incorporated to overcome Ostwald ripening. For example, addition of a second oil phase with very low solubility and/or incorporation of polymeric surfactants that strongly adsorb at the O/W interface (which are also insoluble in the aqueous medium). (8) Fear of introduction of new systems without full evaluation of the cost and benefits.

In spite of the above difficulties, several companies have introduced nanoemulsions in the market and, within the next few years, the benefits will be evaluated. Nanoemulsions have been used in the pharmaceutical field as drug delivery systems [7].

Acceptance of nanoemulsions as a new type of formulation depends on customer perception and acceptability. With the advent of new instruments for high-pressure homogenizers and the competition between various manufacturers, the cost of production of nanoemulsions will decrease and may approach that of classical macroemulsions.

Fundamental research into investigating the role of surfactants in the process [5, 6] will lead to optimized emulsifier systems and more economical use of surfactants will emerge. In this chapter we will demonstrate that by using a polymeric surfactant one can stabilize nanoemulsions against flocculation, coalescence and in particular Ostwald ripening. The polymeric surfactant used in this study is a graft copolymer of inulin (a natural polysaccharide of linear polyfructose) on which several alkyl groups are randomly grafted. This graft copolymer adsorbs very strongly at the O/W interface by multi-point attachment with several alkyl groups (which may also dissolve in the oil phase), leaving strongly hydrated loops and tails of linear polyfructose that showed very effective steric stabilization. In addition, the strong adsorption of the polymeric surfactant and its low solubility in the continuous phase produce a high dilatational elasticity and this reduces Ostwald ripening significantly.

2.2 Materials

Several oils were used which varied in their polarity and solubility: paraffinic oils, namely isohexadecane, paraffinum liquidum low viscosity and paraffinum liquidum high viscosity, isopropyl alkylates, namely isopropyl myristate, isopropyl palmitate and isopropyl stearate, natural oils, namely squalene, squalane, *Ricinus communis*, *Macadamia ternifolia* and *Buxis chinensis*, silicone oils, namely dimethicone (50 cSt), phenyltrimethicone and cyclopentasiloxane, esters, namely butyl stearate, caprylic capric triglyceride, ethylhexyl palmitate, cetearyl ethylhexanoate, cetearyl isononanoate and C_{12-15}-alkyl benzoate, PPG-15 stearyl ether and polydecene.

The hydrophobically modified graft copolymer based on inulin [8–10] referred to as INUTEC SP1 was supplied by ORAFTI Bio Based Chemicals (Belgium). Inulin is a natural product obtained from chicory roots. It is a linear polyfructose chain with a glucose end. When extracted from chicory roots, inulin has a wide range of chain lengths ranging from 2 to 65 fructose units. It is fractionated to obtain a molecule with a narrow molecular weight distribution with a degree of polymerization >23, and this is commercially available as INUTEC N25. The latter molecule is used to prepare the copolymers by random grafting of C_{12}-alkyl chains (using alkyl isocyanate) on the inulin backbone. It has an average molecular weight of ~5000 Da.

2.3
Preparation of Nanoemulsions

An O/W emulsion was first prepared using a high-speed stirrer (Ultra-Turrax) following the procedure described previously [11]. The emulsion was subjected to high-pressure homogenization using a microfluidizer (Microfluidics, USA). In all cases the pressure used was 700 bar and homogenization was carried out for 1 min.

2.4
Determination of Droplet Diameter

The *Z*-average droplet diameter of the nanoemulsion was measured using dynamic light scattering, usually referred to as photon correlation spectroscopy (PCS). For this purpose, an HPPS instrument (supplied by Malvern, UK) was used. The method is based on measurement of the intensity fluctation of scattered light (using laser light) as the droplets undergo Brownian motion [12].

When a light beam passes through a nanoemulsion, an oscillating dipole moment is induced in the droplets, thereby reradiating the light. Due to the random position of the droplets, the intensity of scattered light will, at any instant, appear as a random diffraction or "speckle" pattern. As the droplets undergo Brownian motion, the random configuration of the pattern will therefore fluctuate such that the time taken for an intensity maximum to become a minimum, i.e. the coherence time, corresponds exactly to the time required for the droplet to move one wavelength. Using a photomultiplier of active area about the diffraction maximum, i.e. one coherence area, this intensity fluctuation can be measured. The analog output is digitized using a digital correlator that measures the photocount (or intensity) correlation function of the scattered light. The photocount correlation function $G^{(2)}(\tau)$ is given by the equation

$$G^{(2)}(\tau) = B\{1 + \gamma^2[g^{(1)}(\tau)]^2\} \tag{2}$$

where τ is the correlation delay time. The correlator compares $G^{(2)}(\tau)$ for many values of τ. B is the background value to which $G^{(2)}(\tau)$ decays at long delay times, $g^{(1)}(\tau)$ is the normalized correlation function of the scattered electric field and γ is a constant ($\sim$1).

For monodisperse non-interacting droplets,

$$g^{(1)} = \exp(-\Gamma\tau) \tag{3}$$

where Γ is the decay rate or inverse coherence time, which is related to the translational diffusion coefficient D by the equation

$$\Gamma = DK^2 \tag{4}$$

where K is the scattering vector:

$$K = \frac{4\pi n}{\lambda} \sin\left(\frac{\theta}{2}\right) \tag{5}$$

λ is the wavelength of light in vacuum, n is the refractive index of the solution and θ is the scattering angle.

The droplet radius R can be calculated from D using the Stokes–Einstein equation:

$$D = \frac{kT}{6\pi\eta_0 R} \tag{6}$$

where η_0 is the viscosity of the medium.

The above analysis is valid for dilute monodisperse droplets. With many nanoemulsions, the droplets are not perfectly monodisperse (usually with a narrow size distribution) and the light scattering results are analyzed for polydispersity (the data are expressed as an average size and a polydispersity index that gives information on the deviation from the average size).

2.5 Steric Stabilization of Nanoemulsions and the Role of the Adsorbed Layer Thickness

Since the nanoemulsions studied are prepared using the graft copolymer INUTEC SP1, it is necessary to consider the interaction forces between droplets containing the polymeric surfactant adsorbed layers (steric stabilization). This has been described in detail in several reviews and textbooks, so only a summary is given here [13, 14].

When two droplets, each containing an adsorbed layer of thickness δ, approach to a distance of separation h, whereby h becomes less than 2δ, repulsion occurs as result of two main effects. The first is unfavorable mixing of the stabilizing chains A of the adsorbed layers, when these are in good solvent conditions. This is referred to as the mixing (osmotic interaction, G_{mix}, and is given by the following expression:

$$\frac{G_{mix}}{kT} = \frac{4\pi}{3V_1}\phi_2{}^2\left(\frac{1}{2}-\chi\right)\left(\delta-\frac{h}{2}\right)^2\left(3a+2\delta+\frac{h}{2}\right) \tag{7}$$

where k is Boltzmann's constant, T is the absolute temperature, V_1 is the molar volume of the solvent, ϕ_2 is the volume fraction of the polymer (the A chains) in the adsorbed layer and χ is the Flory–Huggins (polymer–solvent interaction) parameter.

It can be seen that G_{mix} depends on three main parameters: the volume fraction of the A chains in the adsorbed layer (the denser the layer, the higher is the value of G_{mix}), the Flory–Huggins interaction parameter χ (for G_{mix} to remain positive, i.e. repulsive, χ should be lower than $\frac{1}{2}$) and the adsorbed layer thickness δ.

The second effect is a reduction in the configurational entropy of the chains on significant overlap. This referred to as elastic (entropic) interaction and is given by the expression

$$G_{el} = 2\nu_2 \ln\left[\frac{\Omega(h)}{\Omega(\infty)}\right] \tag{8}$$

where ν_2 is the number of chains per unit area, $\Omega(h)$ is the configurational entropy of the chains at a separation distance h and $\Omega(\infty)$ is the configurational entropy at infinite distance of separation.

Combination of G_{mix} and G_{el} with the van der Waals attraction G_A gives the total energy of interaction, G_T:

$$G_T = G_{mix} + G_{el} + G_A \tag{9}$$

Figure 2.1 gives a schematic representation of the variation of G_{mix}, G_{el}, G_A and G_T with h. As can be seen, G_{mix} increases very rapidly with decrease in h as soon as $h < 2\delta$, G_{el} increases very rapidly with decrease in h when $h < \delta$ and G_T shows one minimum, G_{min}, and increases very rapidly with decrease in h when $h < 2\delta$.

The magnitude of G_{min} depends on the particle radius R, the Hamaker constant A and the adsorbed layer thickness δ. As an illustration, Figure 2.2 shows the variation of G_T with h at various ratios δ/R. It can be seen that the depth of the minimum decreases with increasing δ/R. This is the basis of the high kinetic stability of nanoemulsions. With nanoemulsions having a radius in the region of 50 nm and an adsorbed layer thickness of say 10 nm, the value of δ/R is 0.2. This

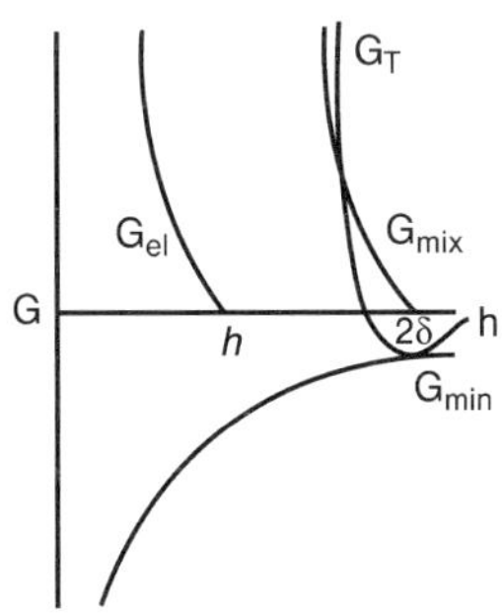

Figure 2.1 Variation of G_{mix}, G_{el}, G_A and G_T with h.

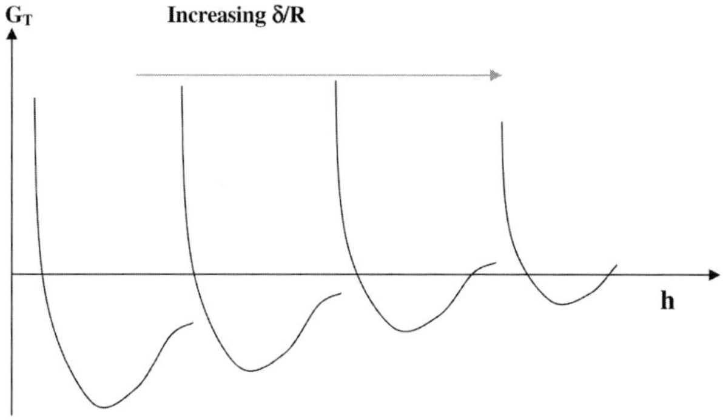

Figure 2.2 Variation of G_T with h with increasing δ/R.

high value (when compared with the situation with macroemulsions where δ/R is at least an order of magnitude lower) results in a very shallow minimum (which could be less than kT).

The above situation results in very high stability with no flocculation (weak or strong). In addition, the very small size of the droplets and the dense adsorbed layers ensures lack of deformation of the interface, lack of thinning and disruption of the liquid film between the droplets and hence coalescence is also prevented. The only instability problem with nanoemulsions is Ostwald ripening, which is discussed below.

2.6 Ostwald Ripening

One of the main problems with nanoemulsions is Ostwald ripening, which results from the difference in solubility between small and large droplets. The difference in the chemical potential of dispersed phase droplets between different sized droplets as given by Lord Kelvin [15] is

$$c(r) = c(\infty)\exp\left(\frac{2\gamma V_{\mathrm{m}}}{rRT}\right) \tag{10}$$

where $c(r)$ is the solubility surrounding a particle of radius r, $c(\infty)$ is the bulk phase solubility and V_{m} is the molar volume of the dispersed phase.

The quantity $(2\gamma V_{\mathrm{m}}/RT)$ is termed the characteristic length. It has an order of ~1 nm or less, indicating that the difference in solubility of a 1-μm droplet is of the order of 0.1% or less.

Theoretically, Ostwald ripening should lead to condensation of all droplets into a single drop (i.e. phase separation). This does not occur in practice since the rate of growth decreases with increase in droplet size.

For two droplets of radii r_1 and r_2 (where $r_1 < r_2$),

$$\left(\frac{RT}{V_m}\right) \ln\left[\frac{c(r_1)}{c(r_2)}\right] = 2\gamma\left(\frac{1}{r_1} - \frac{1}{r_2}\right) \tag{11}$$

Equation (11) shows that the larger the difference between r_1 and r_2, the higher is the rate of Ostwald ripening.

Ostwald ripening can be quantitatively assessed from plots of the cube of the radius versus time t [the Lifshitz–Slesov–Wagner (LSW) theory] [16, 17]:

$$r^3 = \frac{8}{9}\left[\frac{c(\infty)\gamma V_m D}{\rho RT}\right] t \tag{12}$$

where D is the diffusion coefficient of the disperse phase in the continuous phase and ρ is the density of the disperse phase.

Several methods may be applied to reduce Ostwald ripening [18–20]:

1. Addition of a second disperse phase component which is insoluble in the continuous phase (e.g. squalene). In this case, significant partitioning between different droplets occurs, with the component having low solubility in the continuous phase expected to be concentrated in the smaller droplets. During Ostwald ripening in a two-component disperse phase system, equilibrium is established when the difference in chemical potential between different sized droplets (which results from curvature effects) is balanced by the difference in chemical potential resulting from partitioning of the two components. If the secondary component has zero solubility in the continuous phase, the size distribution will not deviate from the initial one (the growth rate is equal to zero). In the case of limited solubility of the secondary component, the distribution is the same as governed by Eq. (12), i.e. a mixture growth rate is obtained which is still lower than that of the more soluble component. This method is of limited application since one requires a highly insoluble oil as the second phase which is miscible with the primary phase.
2. Modification of the interfacial film at the O/W interface: according to Eq. (12), a decrease in γ results in a reduction in Ostwald ripening. However, this alone is not sufficient since one has to reduce γ by several orders of magnitude. Walstra [21] suggested that by using surfactants which are strongly adsorbed at the O/W interface (i.e. polymeric surfactants) and which do not desorb during ripening, the rate could be significantly reduced. An increase in the surface dilatational modulus ε and decrease in γ would be observed for the shrinking drops. The difference in ε between the droplets would balance the difference in capillary pressure (i.e. curvature effects).

To achieve the above effect, it is useful to use block or graft copolymers such as INUTEC SP1 that are strongly "anchored" to the oil phase (with multi-point attachment) and with low solubility in the continuous phase. The polymeric surfactant should enhance the lowering of γ by any additional emulsifier. In other words, the emulsifier and the polymeric surfactant should show synergy in lowering γ.

2.7 Results and Discussion

As mentioned above, the influence of droplet size on oil solubility is given by the Kelvin equation [Eq. (10)], which shows that the solubility of the oil increases with decrease in droplet radius. With nanoemulsions that cover the size range 50–200 nm, the solubility of the oil will be high for all droplets. However, the smaller droplets will have higher solubility than the larger droplets. Therefore, on storage, oil molecules will diffuse from the smaller to the larger droplets and this will lead to a shift in the droplet size distribution to larger values.

The rate of Ostwald ripening is given by the LSV equation [Eq. (12)], which shows that a plot of r^3 versus t gives a straight line and the slope is the rate of Ostwald ripening. Figure 2.3 shows plots of r^3 versus t for nanoemulsions of the hydrocarbon oils that were stored at 50 °C. It can be seen that both parraffinum

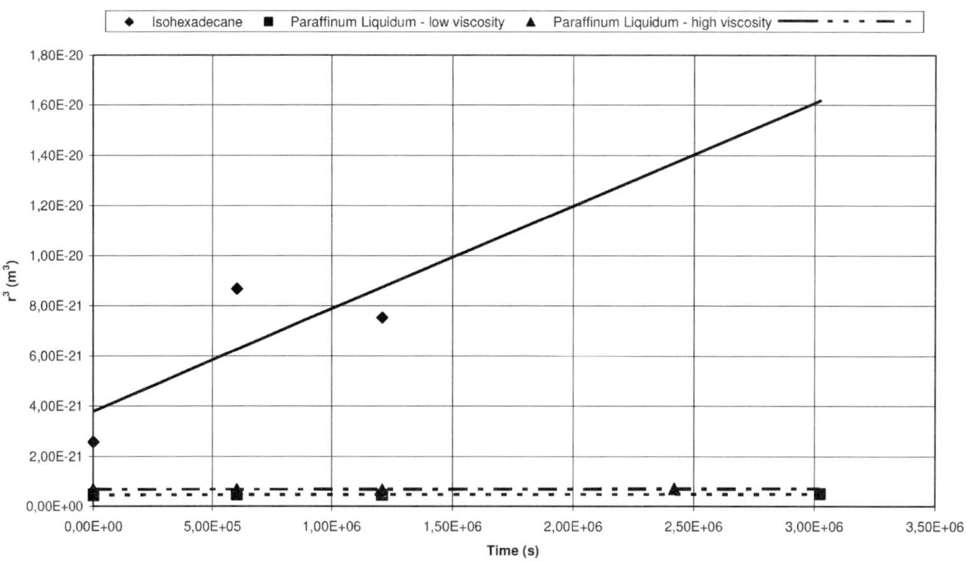

Figure 2.3 r^3 versus t for nanoemulsions based on hydrocarbon oils.

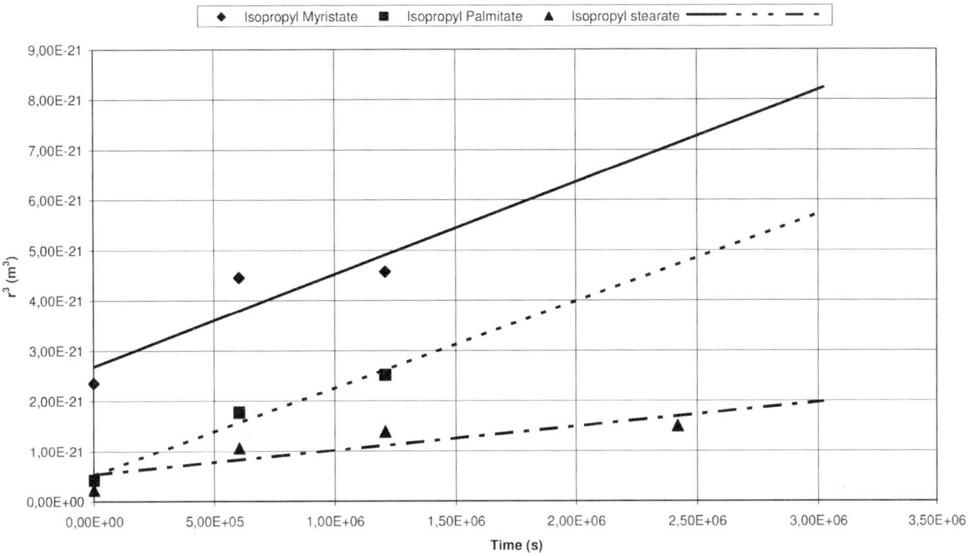

Figure 2.4 r^3 versus t for nanoemulsions based on isopropylalkylate.

liquidum with low and high viscosity give almost a zero slope, indicating the absence of Ostwald ripening in this case. This is not surprising since both oils have very low solubility and the hydrophobically modified inulin, INUTEC SP1, strongly adsorbs at the interface, giving high elasticity that reduces both Ostwald ripening and coalescence. However, with the more soluble hydrocarbon oils, namely isohexadecane, there is an increase in r^3 with time, giving a rate of Ostwald ripening of 4.1×10^{-27} m^3 s^{-1}. The rate for this oil is almost three orders of a magnitude lower than that obtained with a nonionic surfactant, namely laureth-4 (C_{12}-alkyl chain with 4 mol ethylene oxide) [9]. This clearly shows the effectiveness of INUTEC SP1 in reduding Ostwald ripening. This reduction can be attributed to the enhancement of the Gibbs -dilatational elasticity [10], which results from the multi-point attachment of the polymeric surfactant with several alkyl groups to the oil droplets. This results in a reduction in the molecular diffusion of the oil from the smaller to the larger droplets.

Figure 2.4 shows the results for the isopropylalkylate O/W nanoemulsions. As with the hydrocarbon oils, there is a significant reduction in the Ostwald ripening rate with increase in the alkyl chain length of the oil. The rate constants are 1.8×10^{-27}, 1.7×10^{-27} and 4.8×10^{-28} m^3 s^{-1}, respectively.

Figure 2.5 shows the $r^3 - t$ plots for nanoemulsions based on natural oils. In all cases, the Ostwald ripening rate is very low. However, a comparison between squalene and squalane shows that rate is relatively higher for squalene (unsaturated oil) compared with squalane (with lower solubility). The Ostwald ripening rate for these natural oils is given in Table 2.1.

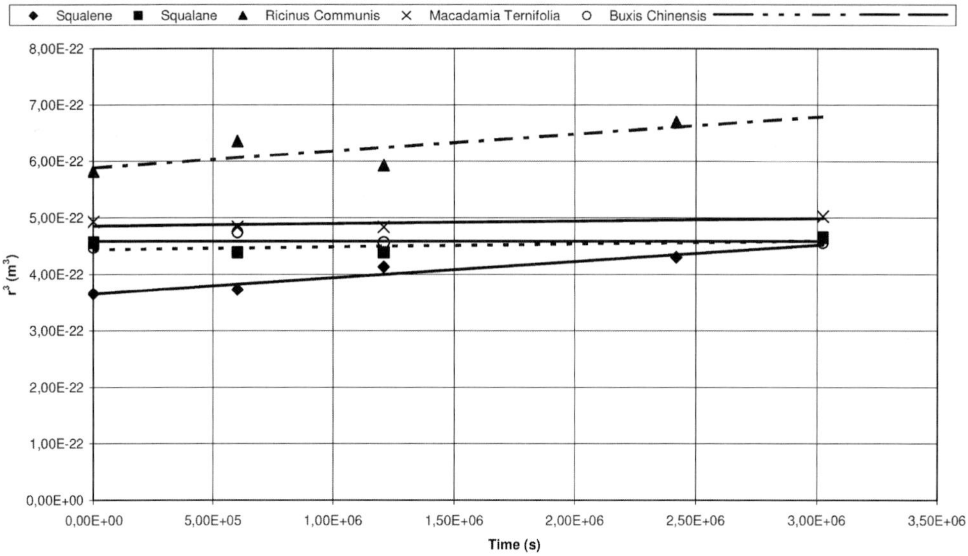

Figure 2.5 r^3 versus t for nanoemulsions based on natural oils.

Figure 2.6 shows the results based on silicone oils. Both dimethicone and phenyltrimethicone give an Ostwald ripening rate close to zero, whereas cyclopentasiloxane gives a rate of 5.6×10^{-28} m^3 s^{-1}.

Figure 2.7 shows the results for nanoemulsions based on esters and the Ostwald ripening rates are given in Table 2.2. C_{12-15}-alkyl benzoate seems to give the highest rate.

Figure 2.8 gives a comparison for two nanoemulsions based on polydecene, a highly insoluble nonpolar oil, and PPG-15 stearyl ether, which is relatively more polar. Polydecene gives a low Ostwald ripening rate of 6.4×10^{-30} m^3 s^{-1}, which is one order of magnitude lower than that of PPG-15 stearyl ether (5.5×10^{-29} m^3 s^{-1}).

Table 2.1 Ostwald ripening rates for nanoemulsions based on natural oils.

Oil	Ostwald ripening rate (m^3 s^{-1})
Squalene	2.9×10^{-28}
Squalane	5.2×10^{-30}
Ricinus communis	3.0×10^{-29}
Macadamia ternifolia	4.4×10^{-30}
Buxis chinensis	~ 0

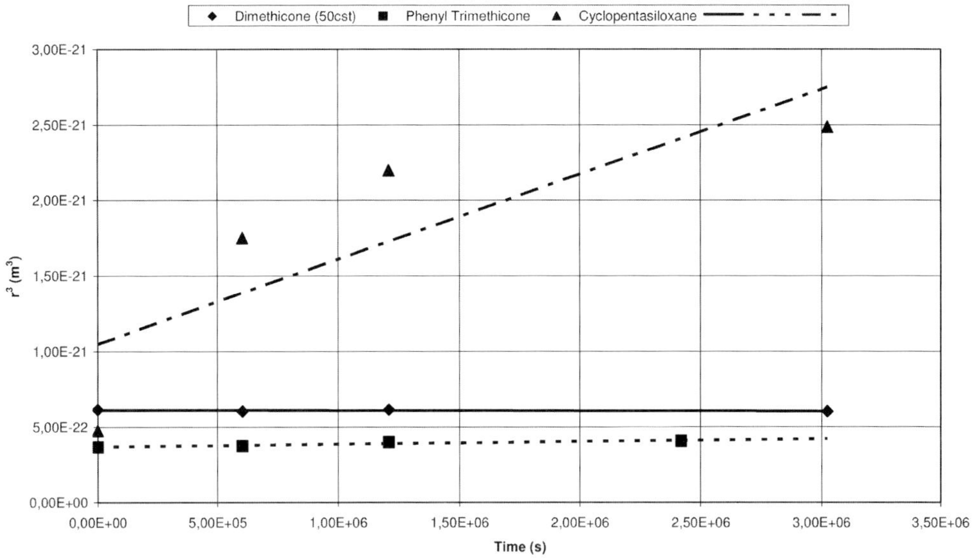

Figure 2.6 r^3 versus t for nanoemulsions based on silicone oils.

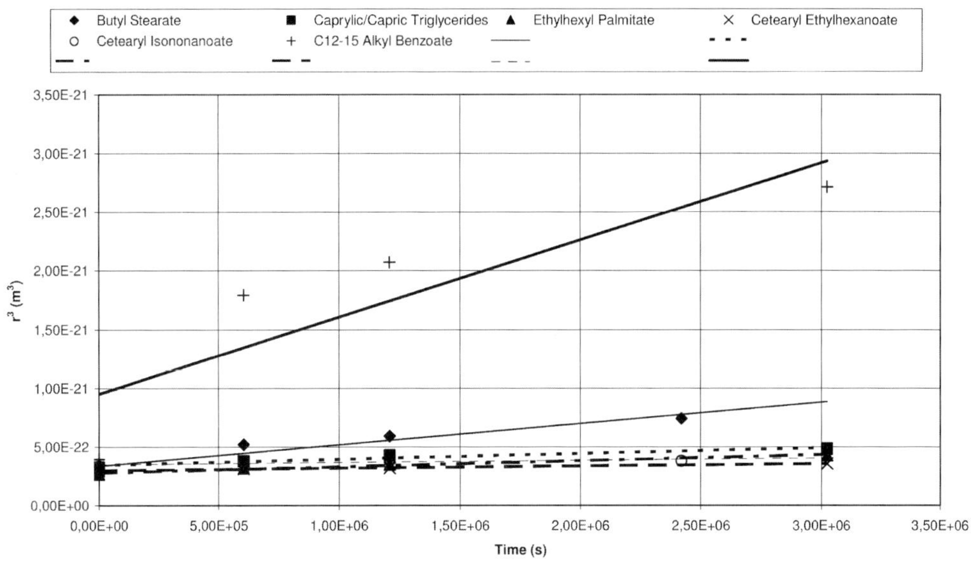

Figure 2.7 r^3 versus t for nanoemulsions based on esters.

Table 2.2 Ostwald ripening rates for nanoemulsions based on esters.

Oil	Ostwald ripening rate ($m^3 s^{-1}$)
Butyl stearate	1.8×10^{-28}
Caprylic capric triglyceride	4.9×10^{-29}
Cetearyl ethylhexanoate	1.9×10^{-29}
Ethylhexyl palmitate	5.1×10^{-29}
Cetearyl isononanoate	1.8×10^{-29}
C_{12-15}-alkyl benzoate	6.6×10^{-28}

The influence of addition of glycerol (which is sometimes added to personal care formulations as a humectant), which can be used to prepare transparent nanoemulsions (by matching the refractive index of the oil and the aqueous phase), on the Ostwald ripening rate is shown in Figure 2.9. With the more insoluble silicone oil, addition of 5% glycerol does not show an increase in the Ostwald ripening rate, whereas for the more soluble isohexadecane oil, glycerol increases the rate.

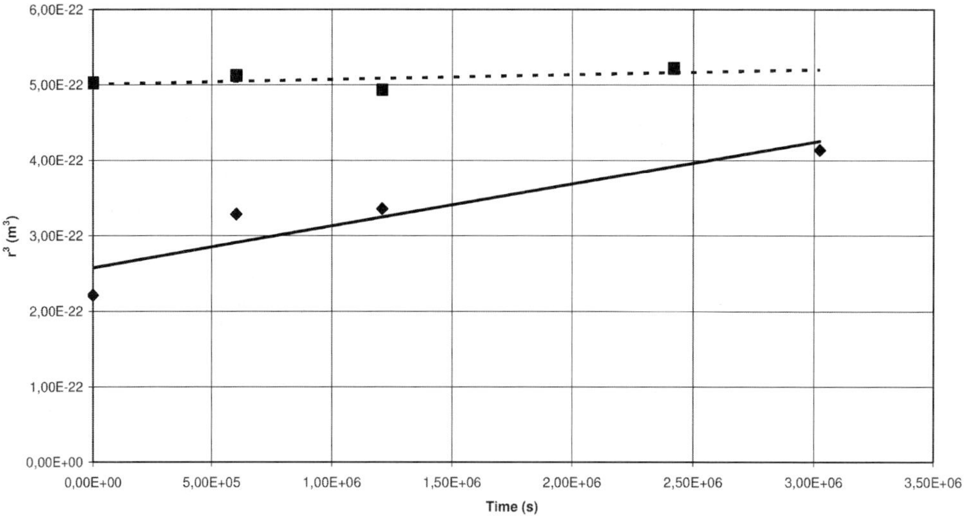

Figure 2.8 r^3 versus t for nanoemulsions based on PPG-15 stearyl ether and polydecene.

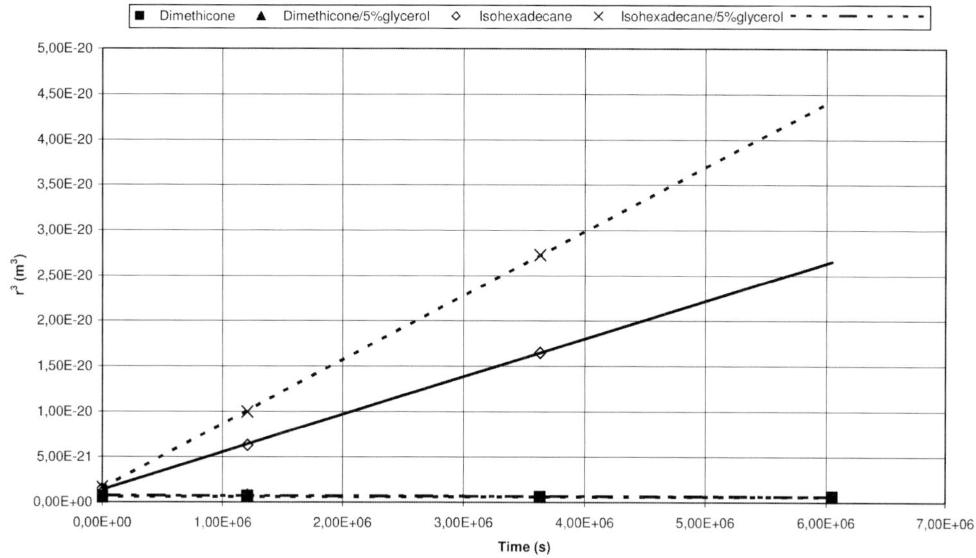

Figure 2.9 Influence of glycerol on the Ostwald ripening rate of nanoemulsions.

2.8 Conclusions

Hydrophobically modified inulin, HMI (INUTEC SP1), reduces the Ostwald ripening rate of nanoemulsions compared with nonionic surfactants such as laureth-4. This is due to the strong adsorption of INUTEC SP1 at the oil/water interface (by multi-point attachment) and enhancement of the Gibbs dilatational elasticity, both reducing the diffusion of oil molecules from the smaller to the larger droplets. The present study also showed a large influence of the nature of the oil phase with the more soluble and more polar oils giving the highest Ostwald ripening rate. However, in all cases, when using INUTEC SP1, the rates are reasonably low, allowing one to use this polymeric surfactant in the formulation of nanoemulsions for personal care applications.

References

1 H. Nakajima, S. Tomomossa, M. Okabe, presented at the First Emulsion Conference, Paris, 1993.

2 H. Nakajima, in *Industrial Applications of Microemulsions*, C. Solans, H. Konieda (eds.), Marcel Dekker, New York, 1997.

3 J. Ugelstadt, M.S. El-Aassar, J.W. Vanderhoff, *J. Polym. Sci.*, **11**, 503 (1973).

4 M. El-Aasser, in *Polymeric Dispersions*, J.M. Asua (ed.), Kluwer, Dordrecht, 1997.

5 A. Forgiarini, J. Esquena, J. Gonzalez, C. Solans, *Prog. Colloid Polym. Sci.*, **115**, 36 (2000).

6 K. Shinoda, H. Kunieda, in *Encyclopedia of Emulsion Technology*, P. Becher (ed.), Marcel Dekker, New York, 1983.

7 S. Benita, M.Y. Levy, *J. Pharm. Sci.*, **82**, 1069 (1993).
8 C.V. Stevens, A. Meriggi, M. Peristerpoulou, P.P. Christov, K. Booten, B. Levecke, A. Vandamme, N. Pittevils, Th.F. Tadros, *Biomacromolecules*, **2**, 1256 (2001).
9 E.L. Hirst, D.I. McGilvary, E.G. Percival, *J. Chem. Soc.*, 1297 (1950).
10 M. Suzuki, in *Science and Technology of Fructans*, M. Suzuki, N.J. Chatterton (eds.), CRC Press, Boca Raton, FL, 1993, p. 21.
11 Th. F Tadros, A. Vandamme, B. Levecke, K. Booten, C.V. Stevens, *Adv. Colloid Interface Sci.*, **108–109**, 207 (2004).
12 P.N. Pusy, in *Industrial Polymers: Characterization by Molecular Weights*, J.H.S. Green, R. Diettz (eds.), Transcripta Books, London, 1973.
13 D.H. Napper, *Polymeric Stabilization of Colloidal Dispersions*, Academic Press, London, 1983.
14 Th.F. Tadros, Polymeric surfactants, in *Principles of Colloid Science and Technology in Cosmetics and Personal Care*, E.D. Goddard, J.V. Gruber (eds.), Marcel Dekker, New York, 1999.
15 W. Thompson (Lord Kelvin), *Philos. Mag.*, **42**, 448 (1871).
16 I.M. Lifshitz, V.V. Slesov, *Sov. Phys. JETP*, **35**, 331 (1959).
17 C. Wagner, *Z. Electrochem.*, **35**, 581 (1961).
18 A.S. Kabalanov, E.D. Shchukin, *Adv. Colloid Interface Sci.*, **38**, 69 (1992).
19 A.S. Kabalanov, *Langmuir*, **10**, 680 (1994).
20 P. Izquirdo, J. Esquena, Th.F. Tadros, C. Dederen, M.J. Gracia, N. Azemar, C. Solans, *Langmuir*, **18**, 26 (2002).
21 P. Walstra, *Chem. Eng. Sci.*, **48**, 333 (1993).

3
Integrating Polymeric Surfactants in Cosmetic Formulations for the Enhancement of Their Performance and Stability

Tharwat F. Tadros, Martine Lemmens, Bart Levecke, and Karl Booten

Abstract

Most personal care formulations are complex multi-phase systems: solid/liquid dispersions (suspensions), liquid/liquid dispersions (emulsions), mixtures of suspensions and emulsions (suspo-emulsions), multiple emulsions (W/O/W or O/W/O), nano-emulsions (covering the size range of 50–200 nm) and micro-emulsions (covering the size range 5–50 nm) [1]. All of these systems are formulated using complex mixtures of surfactants and apart from micro-emulsions they are thermodynamically unstable. On storage of these formulations, some breakdown processes may take place, such as strong flocculation, Ostwald ripening, coalescence and phase-inversion. These problems can be solved to a large extent by using polymeric surfactants of the A–B, A–B–A block, BA*n* or AB*n* graft co-polymers [2, 3]. The B-chain (referred to as the "anchor" chain) is chosen to be highly insoluble in the medium and has a high affinity for the surface. The A-chain (the stabilizing chain) is chosen to be highly soluble in the medium and strongly solvated by its molecules. This provides effective steric stabilization against strong flocculation, Ostwald ripening, coalescence and phase inversion. The objective of this chapter is to show that a newly developed polymeric surfactant, namely hydrophobically modified inulin, HMI (INUTEC® SP1), can be applied to stabilize a wide range of personal care and cosmetic formulations. The A-chain is inulin (a linear polyfructose) on which several alkyl chains are grafted to give an AB*n* graft co-polymer. In addition, this polymeric surfactant imparts good skin-feel to the formulation, showing no stickiness or greasiness. HMI (INUTEC® SP1) can be applied in a wide variety of personal care and cosmetic formulations. It enhances the stabilization of the system against strong flocculation, Ostwald ripening, coalescence and phase-inversion. This is due to the strong adsorption of the polymer surfactant at the oil/water interface (with multi-point attachment with several alkyl groups) and the strong hydration of the linear polyfructose loops and tails (enhanced steric stabilization). With shower gels, which contain oils, addition of INUTEC® SP1 prevents foam-destabilization

Colloids and Interface Science Series, Vol. 4
Colloids in Cosmetics and Personal Care. Edited by Tharwat F. Tadros

ISBN: 978-3-527-31464-5

by precluding the oil droplets from entering into the aqueous foam film between the air-bubbles. In addition, INUTEC® SP1 enhances the performance of formulations on application, such as lack of stickiness and greasiness, thus producing a good skin-feel.

3.1
Introduction

Most personal care formulations are complex multi-phase systems: solid–liquid dispersions (suspensions), liquid–liquid dispersions (emulsions), mixtures of suspensions and emulsions (suspoemulsions), multiple emulsions (W/O/W or O/W/O), nanoemulsions (covering the size range of 50–200 nm) and microemulsions (covering the size range 5–50 nm) [1]. All these systems are formulated using complex mixtures of surfactants and apart from microemulsions they are thermodynamically unstable. On storage of these formulations, some breakdown processes may take place such as strong flocculation, Ostwald ripening, coalescence and phase-inversion. Strong flocculation may occur if there is not a sufficient repulsive energy between the particles or droplets to overcome the van der Waals attraction. This occurs with many surfactant systems that may become displaced from the particle or droplet surface on storage. Ostwald ripening occurs as a result of the higher solubility of the smaller droplets resulting in diffusion of molecules from the small to the large droplets. This results in shift of the particle or droplet size distribution to larger sizes and this can enhance creaming or sedimentation, flocculation and coalescence. With many emulsion systems based on surfactants the liquid film between the droplets may become unstable, resulting in its collapse with ultimately coalescence and oil separation.

The above instability problems can be solved to a large extent by using polymeric surfactants of the A–B, A–B–A block, BA_n or AB_n graft copolymers [2, 3]. The B chain (referred to as the "anchor" chain) is chosen to be highly insoluble in the medium and has a high affinity to the surface. The A chain (the stabilizing chain) is chosen to be highly soluble in the medium and strongly solvated by its molecules. This provides effective steric stabilization against strong flocculation, Ostwald ripening, coalescence and phase inversion. The steric stabilization is the result of two main repulsive energies [4]. The first is unfavorable mixing of the A chains when these are in good solvent conditions. When the particles or droplets approach to a distance h that is smaller than twice the adsorbed layer thickness 2δ, overlap and/or compression of the A chains occurs and this results in an increase in the local segment density in the overlap region. As a result, solvent molecules diffuse to the overlap region, thus separating the particles or droplets. This is referred to as osmotic repulsion or mixing interaction. The second is a decrease in the configurational entropy of the A chains on considerable overlap. This decrease is unfavorable and causes separation of the particles or droplets. This is referred to as entropic or elastic interaction.

The objective of this chapter is to show that a newly developed polymeric surfactant, hydrophobic ally modified insulin (HMI) (INUTEC SP1), can be applied to stabilize a wide range of personal care and cosmetic formulations. The A chain is insulin (linear polyfructose) on which several alkyl chains are grafted to give an AB_n graft copolymer. In addition, this polymeric surfactant imparts good skin feel to the formulation, showing no stickiness or greasiness.

3.2 Materials and Methods

Hydrophobically modified inulin, INUTEC SP1, was supplied by ORAFTI (Belgium). The inulin A backbone (the stabilizing chain) was obtained from native inulin that was extracted from chicory roots. This native inulin consists of polyfructose chains that end with a glucose unit. It has a wide distribution of molecular weights (2–65 fructose units) and it was fractionated to obtain a polyfructose chain with a degree of polymerization > 23. This linear polyfructose A chain was hydrophobically modified by random grafting of several C_{12}-alkyl chains (5) to produce the graft copolymer INUTEC SP1. The polymer gives a clear solution at low concentration (0.1%) and at high concentration it produces a turbid solution, probably due to aggregation of the polymer molecules [6].

Four different personal care formulations have been prepared into which INUTEC SP1 has been incorporated to enhance their stability and performance: massage lotion formulation, hydrating shower cream, soft conditioner and sun spray SPF19: Their compositions are given in Tables 3.1 to 3.4.

Table 3.1 Massage lotion

Phase	Ingredient	Concentration (% w/w)
A	Paraffinium liquidum	42.00
	Helianthus annuus	5.00
	Cetearyl ethylhexanoate	2.50
	Cetearyl isononoate	2.50
	Perfume	0.30
B	Water	36.68
	INUTEC SP1	1.00
	Vitis vinifira (grape) skin extract	0.02
	Preservatives	
	Lactic acid to pH 4–5	
C	C_{10-30}-alkyl acrylate cross-polymer	3.00
	Xanthan gum	3.00
D	10% NaOH to pH 5–6	3.00

Table 3.2 Hydrating shower cream

Phase	Ingredient	Concentration (% w/w)
A	Helianthus annuus (vegetable oil)	9.00
	Questamix (blend skin lipids)	0.10
	Pentaerythritol (substantia oil)	1.00
	Tetracaprylate/caprate	
	Ammonium lauryl sulfate (primary surfactant)	9.00
	Perfume	
B	Water	
	INUTEC SP1 (emulsion stabilizer)	0.20
	Xanthan gum (viscosity modifier)	0.50
	C_{10-30}-alkyl acrylate cross-polymer (viscosity modifier)	0.70
	Preservatives	
C	Cocoamidopropylbetaine (secondary emulsifier)	3.50
D	10% NaOH to pH 4.7–5.2	

The formulations were prepared using standard procedures. The oil phase was added to the aqueous phase while stirring at 10000 rpm using a high-speed stirrer (Ultra-Turrax). Sometimes the thickener was included in the aqueous phase and with a cross-copolymer of polyacrylate, the pH was adjusted with NaOH to produce the microgel. All formulations were kept at ambient temperature and their

Table 3.3 Soft conditioner

Phase	Ingredient	Concentration (% w/w)
A	Cetearyl alcohol (conditioning)	4.00
	Cyclopentamethicone (for shine)	0.60
	Bishydroxyethylbiscetyl (color)	0.05
	Malonamide (maintenance, repairing, strength)	
B	Water	
	Hydroxyethylcellulose (2%), viscosity modifier	25.00
	Polyquaternium-10	0.10
	INUTEC SP1 (emulsion stabilizer)	0.05
	Preservatives	
C	Cetrimonium chloride	1.50
D	Perfume	0.20

Table 3.4 Sun spray SPF19

Phase	Ingredient	Concentration (% w/w)
A	C_{10-30}-alkyl benzoate (dry oil)	6.00
	Jojoba oil pressed (vegetable oil)	2.00
	Isoamyl *p*-methoxycinnamate (UVB filter)	10.00
	Ethylhexyldimethyl-PABA (UVB filter)	7.00
	Cyclopentasiloxane, C_{30-45}-alkyl cetearyl dimethicone cross-polymer (structuring agent)	1.00
	Ethylhexyl palmitate (spreading oil)	1.00
	Sorbitan isostearate (co-emulsifier)	0.50
B	Water	
	Xanthan gum	0.10
	INUTEC SP1 (emulsion stabilizer)	0.75
	Glycerine	3.00
	Preservatives	
C	Aqua, galactobrabinan (stabilizing)	5.00
D	10% NaOH to pH 5–6	

stability was assessed by visual inspection. The lack of any separation after more than 1 year at ambient temperature was taken as proof of stability of the formulations. In addition, the performance of the formulations on application, such as stickiness, greasiness and skin feel, was also assessed.

3.3 Results and Discussion

3.3.1 Massage Lotion

This formulation contains 52.5% oil phase made of five different materials that vary in polarity. Addition of 1% INUTEC SP1 is sufficient to stabilize this formulation against any strong flocculation, coalescence and phase-inversion. The formulation showed no separation for more than 1 year at ambient temperature. This high stability against coalescence is a result of the adsorption and conformation of the polymer at the oil/water interface [7].

The polymer molecule adsorbs with multi-point attachment with several alkyl chains (that can be soluble in the oil phase) leaving loops and tails of the linear polyfructose chain dangling in solution. A schematic representation of the adsorption conformation of the polymer chain at the O/W-interface is given in Figure 3.1.

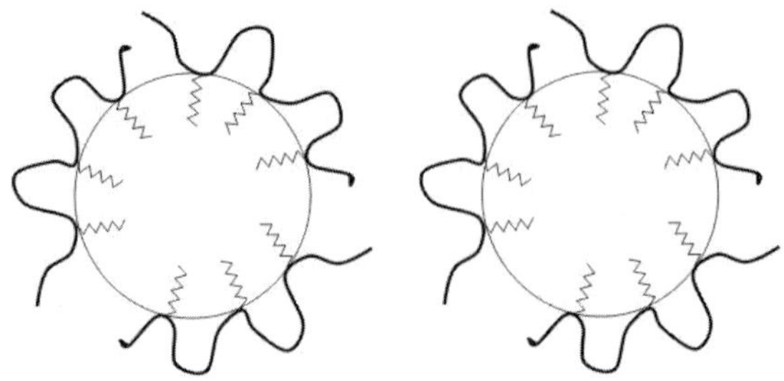

Figure 3.1 Schematic representation of the adsorption and conformation of INUTEC SP1 at the oil/water interface.

The multi-point attachments of the chains at the oil/water interface prevent any desorption on approach of the oil droplets [7, 8]. The strongly hydrated polyfructose loops and tails (which have a thickness in the region of 10 nm) [9] provide strong steric repulsion as a result of the unfavorable mixing of the polyfructose chain and loss in configurational entropy of the chains on considerable overlap [4]. Thus, thinning and disruption of the aqueous film between the oil droplets is prevented, which eliminates coalescence [8]. The polymer surfactant molecule at the oil/water interface enhances the Gibbs dilatational elasticity and this also prevents any coalescence.

Evidence of the high stability of emulsions when using INUTEC SP1 has recently been obtained [10] from disjoining pressure measurements between two oil droplets containing adsorbed polymer surfactant both in water and in high electrolyte solutions. The results showed that by increasing the capillary pressure a stable Newton black film (NBF) is obtained at a film thickness of ~7 nm. The lack of rupture of the NBF up to the highest pressure applied, namely 4.5×10^4 Pa, clearly indicates the high stability of the liquid film in the presence of high NaCl concentrations (2 mol dm^{-3}). This result is consistent with the high emulsion stability obtained at high electrolyte concentrations and high temperature.

To prevent creaming of the emulsion, two rheology modifiers, namely cross-linked polyacrylate and xanthan gum, were added to the aqueous phase. The cross-linked acrylate is neutralized with NaOH and the resulting electrolyte in the system does not affect the stability of the formulation. These rheology modifiers produce a high viscosity at low shear stress or shear rate, thus preventing any creaming or sedimentation [11]. In addition, the "gel" structure of the rheology modifier produces high elasticity, thus preventing any separation of the formulation on storage. However, these "gels" are shear thinning and their viscosity decreases rapidly with increase in shear rate (as produced on application).

An additional advantage of using INUTEC SP1 is the excellent performance on application. The formulation proved to be non-sticky, light and non-greasy and it shows an excellent skin feel. This is not surprising since the polymeric surfactant helps the lubrication of the skin surface.

3.3.2 Hydrating Shower Gel

This formulation contains 11% oil phase and 9% ammonium lauryl sulfate, which is necessary to produce stable foam on application. Several mechanisms have been suggested for explaining the role of surfactants in foam stabilization. One acceptable theory is based on the effect of surfactants on the dilatational elasticity and surface viscosity of the film [12]. The second theory is based on the disjoining pressure concept of Deryaguin [13]. A summary of these theories is given below.

3.3.2.1 Surface Viscosity and Elasticity Theory

The adsorbed surfactant film is assumed to control the mechanical–dynamic properties of the surface layers by virtue of its surface viscosity and elasticity. The above concept may be true for thick films (>100 nm) whereby intermolecular forces are less dominant (i.e. foam stability under dynamic conditions). Surface viscosity reflects the speed of the relaxation process which restores the equilibrium in the system after imposing a stress on it. Surface elasticity is a measure of the energy stored in the surface layer as a result of an external stress. The viscoelastic properties of the surface layer are an important parameter. Some correlations have been found between surface viscosity and elasticity and foam stability, e.g. when adding lauryl alcohol to sodium lauryl sulfate, which tends to increase the surface viscosity and elasticity.

3.3.2.2 The Gibbs–Marangoni Effect Theory

The Gibbs coefficient of elasticity, ε, was introduced as a variable resistance to surface deformation during thinning:

$$\varepsilon = 2\left(\frac{d\gamma}{d \ln A}\right) = -2\left(\frac{d\gamma}{d \ln h}\right) \tag{1}$$

where γ is the interfacial tension, d ln A is the relative change in film area and d ln h is the relative change in lamella thickness. ε is the "film elasticity of compression modulus" or "surface dilatational modulus"; it is a measure of the ability of the film to adjust its surface tension in an instant stress. In general, the higher the value of $\mathring{a}$, the more stable the film is; $\mathring{a}$ depends on surface concentration and film thickness. For a freshly produced film to survive, a minimum $\mathring{a}$ is required. One should also consider diffusion from the bulk solution, i.e. the Marangoni effect. The Marangoni effect tends to oppose any rapid displacement of the surface (Gibbs effect) and may provide a temporary restoring force to "dangerous" thin films. The Gibbs–Marangoni effect explains the maximum foaming behavior at intermediate surfactant concentrations.

3.3.2.3 **Surface Forces Theory (Disjoining Pressure π)** [13]

This theory operates under static (equilibrium) conditions in relatively dilute surfactant solutions ($h < 100$ nm). In addition to the Laplace capillary pressure, three additional forces can operate at surfactant concentrations below the critical micellar comcentration (cmc): electrostatic double layer repulsion $\eth_{el}$, van der Waals attraction $\eth_{vdW}$ and steric (short range) forces $\eth_{st}$:

$$\pi = \pi_{el} + \pi_{vdW} + \pi_{st} \tag{2}$$

At low electrolyte concentrations, double layer repulsion predominates. At high electrolyte concentrations, steric forces predominate.

The presence of oil droplets in a foam is known to destabilize it, as explained by the following mechanism [14]. Undissolved oil droplets form in the surface of the film and this can lead to film rupture. Several examples of oils can destabilize the foam film: alkyl phosphates, diols, fatty acid esters and silicone oils (poldimethylsiloxane).

A widely accepted mechanism for the antifoaming action of oils considers two steps: the oil drops enter the air/water interface and the oil then spreads over the film causing rupture [15–17]. The antifoaming action can be rationalized in terms of the balance between the entering coefficient E and the spreading coefficient S, which are given by the following equations:

$$E = \gamma_{WA} + \gamma_{WO} - \gamma_{OA} \tag{3}$$

$$S = \gamma_{WA} - \gamma_{WO} - \gamma_{OA} \tag{4}$$

where γ_{WA}, γ_{OA} and γ_{WO} are the macroscopic interfacial tensions of the aqueous phase, oil phase and interfacial tension of the oil/water interface, respectively. For antifoaming both E and S should be >0 for entry and spreading.

For most surfactant systems $\gamma_{AW} = 35–45$ mN m^{-1} and $\gamma_{OW} = 5–10$ mN m^{-1} and hence for an oil to act as an antifoaming agent γ_{OA} should be less than 25 mN m^{-1}. This shows why low surface tension silicone oils result in destruction of the foam.

For the above hydrating shower gel without any added INUTEC SP1, oil droplets can enter the liquid film between the air bubbles, adsorbing some of the ammonium lauryl sulfate surfactant, and this causes destabilization of the foam film. In the presence of INUTEC SP1, which is strongly and preferentially adsorbed at the oil/water interface, the oil droplets cannot enter the aqueous film between the air bubbles and hence the film remains stable. A schematic representation of the destabilization and stabilization mechanism of the foam is given in Figure 3.2.

It should be mentioned that the concentration of INUTEC SP1 required to stabilize the system is 0.2% (2% based on the oil phase), which is comparable to that used with the massage lotion. The formulation also has good skin feel on application.

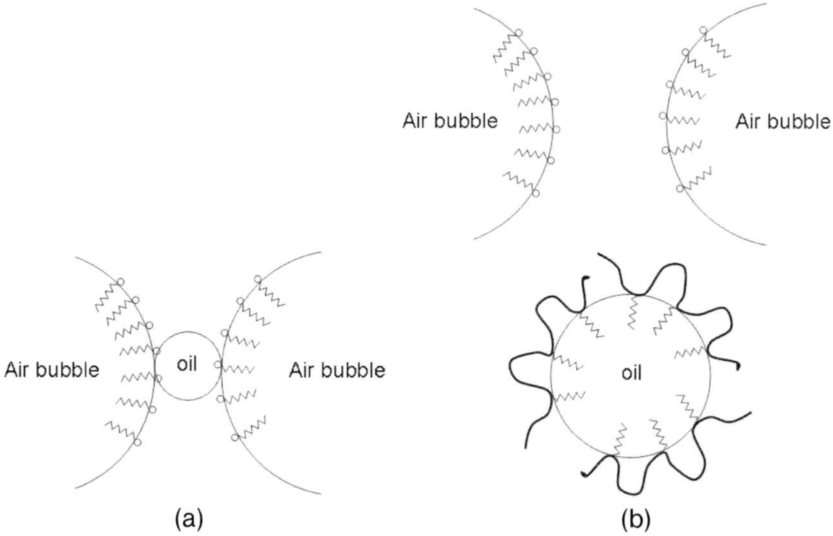

Figure 3.2 (a) Destabilization of the foam film by entering oil droplets. (b) The oil droplets which are stabilized by INUTEC SP1 do not enter the foam film, which remains stable.

3.3.3 Soft Conditioner

This contains 4.6% w/w hydrophobic components (cetearyl alcohol and cyclopentamethicone) and in this case 0.05% INUTEC SP1 was sufficient to stabilize the formulation. This shows that the polymer surfactant is also effective in the stabilization of relatively more polar solids. The formulation also has a smooth skin feel on application.

3.3.4 Sun Spray SPF19

One of the most useful applications of INUTEC SP1 is with sprayable formulations since the polymeric surfactant does not cause any increase in the viscosity of the system on application. The total oil content of this formulation is about 30% w/w and 0.75% INUTEC SP1 was sufficient to stabilize the formulation. This sprayable formulation is non-sticky and gives a nice skin feel.

3.4 Conclusions

Hydrophobically modified inulin (INUTEC SP1) can be applied in a wide variety of personal care and cosmetic formulations. It enhances the stabilization of the system against strong flocculation, Ostwald ripening, coalescence and phase inversion. This is due to the strong adsorption of the polymer surfactant at the oil/water interface (with multi-point attachment with several alkyl groups) and the strong hydration of the linear polyfructose loops and tails (enhanced steric stabilization). The enhanced stabilization could be illustrated using a massage lotion that contained 52.5% oil phase and 1% INUTEC SP1 and this formulation remained stable for more than 1 year. With shower gels, which contain oils, addition of INUTEC SP1 prevents foam destabilization by preventing the oil droplets from entering the aqueous foam film between the air bubbles. INUTEC SP1 could also be used in soft conditioner formulations that contain more polar solids. Due to the low viscosity of formulations containing INUTEC SP1, the polymeric surfactant could also be applied for sprayable formulations. In addition, INUTEC SP1 enhances the performance of formulations on application, such as lack of stickiness and greasiness, thus producing a good skin feel.

References

1 Th.F. Tadros, *Applied Surfactants, Principles and Application*, Wiley-VCH, Weinheim, 2005.

2 Th.F. Tadros, Polymeric Surfactants, in *Principles of Polymer Science and Technology in Cosmetics and Personal Care*, E.D. Goddard, J.V. Gruber (eds.), Marcel Dekker, New York, 1999.

3 Th.F. Tadros, in *Novel Surfactants*, K. Holmberg (ed.), Marcel Dekker, New York, 2003.

4 D.H. Napper, *Polymeric Stabilization of Colloidal Dispersions*, Academic Press, London, 1983.

5 C.V. Stevens, A. Meriggi, M. Peristeropoulou, P.P. Christov, K. Booten, B. Levecke, A. Vandamme, N. Pittevils, Th.F. Tadros, *Biomacromolecules*, **2**, 1256 (2000).

6 J. Nestor, J. Esquena, C. Solans, B. Levecke, K. Booten, Th.F. Tadros, *Langmuir*, **21**, 4837 (2005).

7 Th.F. Tadros, A. Vandamme, K. Booten, B. Levecke, C.V. Stevens, *Adv. Colloid Interface Sci.*, **108–109**, 207 (2004).

8 Th.F. Tadros, A. Vandamme, K. Booten, B. Levecke, C.V. Stevens, *Colloids Surf.*, **250**, 133 (2004).

9 J. Nestor, J. Esquena, C. Solans, P.F. Luckham, B. Levecke, Th.F. Tadros, *J. Colloid Interface Sci.*, **311**, 430 (2007).

10 D. Exerowa, G. Gotchev, T. Kolarev, Khr. Khristov, B. Levecke, Th.F. Tadros, *Langmuir*, **23**, 1711 (2007).

11 Th.F. Tadros, *Adv. Colloid Interface Sci.*, **108–109**, 227 (2004).

12 R.J. Pugh, *Adv. Colloid Interface Sci.*, **64**, 67 (1996).

13 B.V. Deryaguin, *Theory of Stability of Colloids and Thin Films*, Consultants Bureau, New York, 1989.

14 P.R. Garett (ed.), *Defoaming*, Marcel Dekker, New York, 1993.

15 J.V. Robinson, W.W. Woods, *J. Soc. Chem. Ind.*, **1967**, 361 (1948).

16 W.D. Harkins., *J. Phys. Chem.*, **9**, 552 (1941).

17 S. Ross, J.W. McBain, *Ind. Chem. Eng.*, **36**, 560 (1944).

4
Application of Colloid and Interface Science Principles for Optimization of Sunscreen Dispersions

Lorna M. Kessell, Benjamin J. Naden, Ian R. Tooley, and Tharwat F. Tadros

Abstract

Sunscreen dispersions of semiconductor TiO_2 particles require particles in the range 30–50 nm which need to remain stable against aggregation in the formulation and on application. This is essential for the required UV protection. This chapter describes the application of fundamental colloid and interface science principles for the preparation of stable surface-modified TiO_2 dispersions in alkyl benzoate and squalane. For this purpose, poly(hydroxystearic acid) of two molecular weights, 1000 and 2500 (PHS1000 and PHS2500), were used as dispersants. The adsorption isotherms were determined for the two dispersants in both solvents. For comparison, a monomeric dispersant, namely isostearic acid (ISA), was also investigated. In alkyl benzoate, the adsorption of ISA was reversible (of Langmuir type) whereas with PHS1000 and PHS2500 the adsorption was stronger (showing some high affinity character) and a plateau value of 1 mg m^{-2} was obtained for the two polymers. The adsorption of the dispersants from squalane was much stronger, showing a definite high affinity character and there was an increase in the saturation adsorption with increase in molecular weight of the dispersant. The higher adsorption of the dispersant from squalane was due to its poorer solvency for the chain compared with alkyl benzoate. The solvency of the medium for the chain was assessed from solubility parameter calculations. The dispersant demand for producing a colloidally stable dispersion was investigated by measuring the zero shear viscosity of a 40% dispersion as a function of dispersant concentration. The viscosity decreased with increase in dispersant concentration, reaching a minimum at the required concentration for producing a stable dispersion. Viscosity versus solids loading showed a sharp increase at a critical concentration of the solids loading curve. This critical concentration was higher (~50%) for the lower molecular weight PHS1000 compared with the results for the higher molecular weight PHS2500 (~40% for dispersions in alkyl benzoate and ~30% in squalane). The results could be rationalized in terms of the effect of the adsorbed layer thickness on the effective volume fraction of the

Colloids and Interface Science Series, Vol. 4
Colloids in Cosmetics and Personal Care. Edited by Tharwat F. Tadros

ISBN: 978-3-527-31464-5

dispersion. Evidence for the high stability of the dispersions in alkyl benzoate or squalane when using PHS1000 or PHS2500 was obtained from measurement of attenuation of UV/Vis radiation. These results also showed the poor colloid stability when using the monomeric dispersant ISA. The effect of competitive interactions in formulations was investigated by adding the colloidally stable TiO_2 dispersions to a water-in-oil (W/O) emulsion prepared using an A–B–A block copolymer of PHS–PEO–PHS. The results showed that the silica-coated TiO_2 dispersions caused separation of the formulation with water droplet coalescence. This was attributed to the adsorption of the PHS-PEO-PHS emulsifier on the silica-coated particles. This causes depletion of the emulsifier from the water droplets, thus causing coalescence. To compensate for this effect, the emulsifier concentration was increased and this produced a stable formulation. In contrast, the alumina-coated TiO_2 dispersions did not cause any instability when added to the W/O emulsion. This is due to the stronger adsorption of PHS chains on the alumina-coated TiO_2 particles as indicated from the adsorption isotherms.

4.1 Introduction

The increase in skin cancers has heightened public awareness to the damaging effects of the sun and many skin preparations are now available to help protect the skin from UV radiation. The active ingredients employed in these preparations are of two basic types: organics which can absorb UV radiation of specific wavelengths due to their chemical structure and inorganics which both absorb and scatter UV radiation. Inorganics have several benefits over organics in that they are capable of absorbing over a broad spectrum of wavelengths and they are mild and non-irritant. Both of these advantages are becoming increasingly important as the demand for *daily* UV protection against both UVB and UVA radiation increases.

The ability of fine particle inorganics to absorb radiation depends on their refractive index. For inorganic semiconductors such as titanium dioxide and zinc oxide this is a complex number indicating their ability to absorb light. The band gap in these materials is such that UV light up to around 405 nm can be absorbed. They can also scatter light due to their particulate nature and their high refractive indices make them particularly effective scatterers. Both scattering and absorption depend critically on particle size [1]. Particles of around 250 nm, for example, are very effective at scattering visible light and TiO_2 of this particle size is the most widely used white pigment. At smaller particle sizes absorption and scattering maxima shift to the UV region and at 30–50 nm UV attenuation is maximized.

The use of TiO_2 as a UV attenuator in cosmetics was, until recently, largely limited to baby sun protection products due to its poor esthetic properties (viz. scattering of visible wavelengths results in whitening). Recent advances in particle size control and coatings have enabled formulators to use fine particle titanium dioxide and zinc oxide in daily skin care formulations without compromising the cosmetic elegance [2, 3].

The benefits of a pre-dispersion of inorganic sunscreens are widely acknowledged. However, it requires an understanding of the nature of colloidal stabilization in order to optimize this pre-dispersion (for both UV attenuation and stability) and exceed the performance of powder-based formulations. Dispersion rheology and its dependence on interparticle interactions is a key factor in this optimization. Optimization of sunscreen actives, however, does not end there; an appreciation of the end application is crucial to maintaining performance. Formulators need to incorporate the particulate actives into an emulsion, mousse or gel with due regard to esthetics (skin feel and transparency), stability and rheology.

This chapter is aimed at applying colloid and interface science principles for optimization of inorganic sunscreen dispersions. The latter are usually formulated using dispersants that provide effective steric stabilization to avoid flocculation, particularly on application. Maintenance of particle size is essential for effective sunscreens. In addition, these colloidally stable nanoparticles can provide transparency and hence good esthetic characteristics. The chapter starts with a summary of steric stabilization with particular reference to the importance of solvation of the polymer chain by the medium molecules. Results are then presented for the adsorption isotherms of typical dispersants that are used in non-aqueous media. The dispersing power of these polymeric surfactants is assessed using rheological measurements. The UV absorbance of these dispersions is measured to evaluate the effectiveness of the sunscreen dispersions and finally the ability of colloidally stable dispersions to deliver SPF when incorporated into a skincare formulation is investigated.

4.2 Steric Stabilization

Small particles tend to aggregate as a result of the universal van der Waals attraction unless this attraction is screened by an effective repulsion between the particles. The van der Waals attraction energy $G_A(h)$ at close approach depends upon the distance, h, between particles of radius, R, and is characterized by the effective Harnaker constant, A:

$$G_A(h) = -\frac{AR}{12h} \tag{1}$$

The effective Hamaker constant A is given by the following equation:

$$A = ({A_{11}}^{1/2} - {A_{22}}^{1/2})^2 \tag{2}$$

where A_{11} is the Hamaker constant of the particles and A_{22} is that for the medium. For TiO_2, A_{11} is exceptionally high so that in nonaqueous media with relatively low A_{22} the effective Hamaker constant A is high and despite the small size of the particles a dispersant is always needed to achieve colloidal stabilization. This is usually obtained using adsorbed layers of polymers or surfactants. The most effective molecules are the A–B, A–B–A block or BA_n graft polymeric surfactants [4] where B refers to the anchor chain. For a hydrophilic particle this anchor may be a carboxylic acid, an amine or phosphate group or other larger hydrogen bonding type block such as poly(ethylene oxide). The A chains are referred to as the stabilizing chains, which should be highly soluble in the medium and strongly solvated by its molecules. For nonaqueous dispersions the A chains could be poly(propylene oxide), a long-chain alkane, oil-soluble polyester or poly(hydroxystearic acid). A schematic representation of the adsorbed layers and the resultant interaction energy-distance curve is shown in Figure 4.1.

When two particles with an adsorbed layer of hydrodynamic thickness δ approach to a separation distance h that is smaller than 2δ, repulsion occurs as a result of two main effects: (1) unfavorable mixing of the A chains when these are in good solvent condition and (2) reduction in configurational entropy on significant overlap.

Napper [5] derived a form for the so-called steric potential $G(h)$ which arises as polymer layers begin to overlap:

$$G(h) = 2\pi kTR^2\Gamma^2 N_A \left(\frac{{v_p}^2}{V_s}\right)\left(\frac{1}{2}-\chi\right)\left(1-\frac{h}{2\delta}\right)^2 + G_{elastic} \tag{3}$$

where k is Boltzmann's constant, T is the absolute temperature, R is the particle radius, Γ is the amount adsorbed, N_A is Avogadro's number, v is the specific partial volume of the polymer, V_s is the molar volume of the solvent, χ is the Flory–Huggins parameter and δ is the maximum extent of the adsorbed layer.

It is useful to consider the terms in Eq. (3). (1) The adsorbed amount Γ – the higher the value, the greater is the interaction/repulsion. (2) Solvent conditions

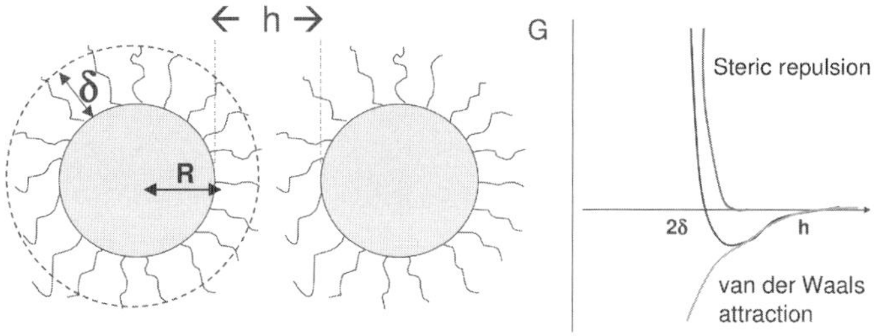

Figure 4.1 Schematic representation of adsorbed polymer layers and resultant interaction energy G on close approach at distance $h < 2R$.

as determined by the value of χ: two very distinct cases emerge. Maximum interaction occurs on overlap of the stabilizing layers when the chains are in good solvent conditions, i.e. $\chi < 0.5$. Osmotic forces cause the solvent to move into the highly concentrated overlap zone, forcing the particles apart. If $\chi = 0.5$, a theta solvent, the steric potential goes to zero and for poor solvent conditions ($\chi > 0.5$) the steric potential becomes negative and the chains will attract, enhancing flocculation. (3) Adsorbed layer thickness δ: the steric interaction starts at $h = 2\delta$ as the chains begin to overlap and increases as the square of the distance. Here the importance is not the size of the steric potential but the distance h at which it begins. (4) The final interaction potential is the superposition of the steric potential and the van der Waals attraction as shown in Figure 4.1.

4.3 Solubility Parameters

The adsorbed layer thickness depends critically on the solvation of the polymer chain and it is therefore important to gain at least a qualitative view as to the relative solubilities of a polymer in different oils employed in dispersion. In this study, solubility parameters were employed to provide that comparison. Generally, the affinity between two materials is considered to be high when the chemical and physical properties of the two materials resemble each other. For example, nonpolar materials can be easily dispersed in nonpolar solvents but hardly dissolved in polar solvents and vice versa.

One of the most useful concepts for assessing the solvation of any polymer by the medium is to use Hildebrand's solubility parameter δ^2, which is related to the heat of vaporization ΔH by the following equation:

$$\delta^2 = \frac{\Delta H - RT}{V_M} \tag{4}$$

where V_M is the molar volume of the solvent.

Hansen [6] first divided Hildebrand's solubility parameter into three terms as follows:

$$\delta^2 = {\delta_d}^2 + {\delta_p}^2 + {\delta_h}^2 \tag{5}$$

where δ_d, δ_p and δ_h correspond to London dispersion effects, polar effects and hydrogen bonding effects, respectively.

Hansen and Beerbower [7] developed this approach further and proposed a stepwise approach such that theoretical solubility parameters can be calculated for any solvent or polymer based on its component groups. In this way we can arrive at theoretical solubility parameters for dispersants and oils. In principle, solvents with a similar solubility parameter to the polymer should also be a good solvent for it (low χ).

4.4
Influence of the Adsorbed Layer Thickness on the Energy–Distance Curve

For sterically stabilized dispersions, the resulting energy–distance curve (Figure 4.1) often shows a shallow minimum G_{min} at a particle–particle separation distance h comparable to twice the adsorbed layer thickness δ. The depth of this minimum depends on the particle size R, Hamaker constant A and adsorbed layer thickness δ. At constant R and A, G_{min} decreases with increase in δ/R. This is illustrated in Figure 4.2.

When δ becomes smaller than 5 nm, G_{min} may become deep enough to cause weak flocculation. This is particularly the case with concentrated dispersions since the entropy loss on flocculation becomes very small and a small G_{min} would be sufficient to cause weak flocculation ($\Delta G_{flocc} < 0$) [8]. This can be explained by considering the free energy of flocculation:

$$\Delta G_{flocc} = \Delta H_{flocc} - T\Delta S_{flocc} \tag{6}$$

Since for concentrated dispersions ΔS_{flocc} is very small, then ΔG_{flocc} depends only on the value of ΔH_{flocc}. This in turn depends on G_{min}, which is negative. In other words, ΔG_{flocc} becomes negative, causing weak flocculation. This will result in a three-dimensional coherent structure with a measurable yield stress [9]. This weak gel can be easily redispersed by gentle shaking or mixing. However, the gel will prevent any separation of the dispersion on storage. Hence we can see that the interaction energies also determine the dispersion rheology.

At high solids content and for dispersions with larger δ/R, viscosity is also increased by steric repulsion. With a dispersion consisting of very small particles, as is the case with UV-attenuating TiO_2, significant rheological effects can be observed even at moderate volume fraction of the dispersion. This is due to the much higher effective volume fraction of the dispersion compared with the core volume fraction due to the adsorbed layer.

Let us consider, for example, a 50% w/w TiO_2 dispersion with a particle radius of 20 nm with a 3000 molecular weight stabilizer giving an adsorbed layer thickness of ~10 nm. The effective volume fraction is given by Eq. (7)

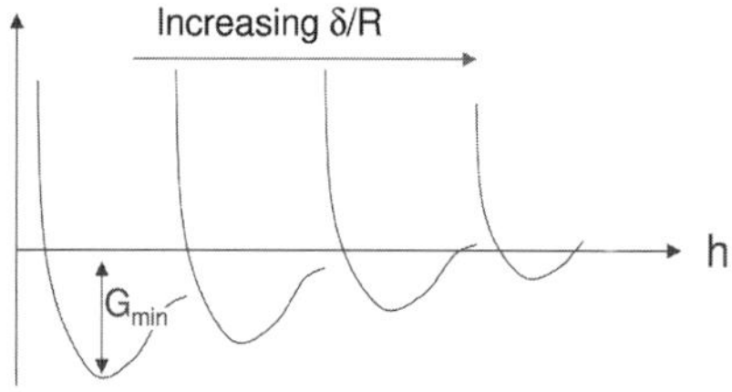

Figure 4.2 Interaction energy depends on the relative layer thickness of adsorbed polymer.

$$\begin{aligned}\phi_{\text{eff}} &= \phi\left(1+\frac{\delta}{R}\right)^3 \\ &= \phi(1+10/20)^3 \\ &\sim 3\phi \end{aligned} \tag{7}$$

The effective volume fraction can be three times that of the core particle volume fraction. For a 50% (w/w) solids TiO_2 dispersion, the core volume fraction ϕ is ~0.25 (taking an average density of $3\ \text{g cm}^{-3}$ for the TiO_2 particles), which means that ϕ_{eff} is about 0.75 which is sufficient to fill the whole dispersion space, producing a highly viscous material. It is important, therefore, to choose the minimum δ for stabilization.

4.5 Criteria for Effective Steric Stabilization and Influence of Other Ingredients in the Formulation

In the case of steric stabilization as employed in these oil dispersions, the important success criteria for well-stabilized but handleable dispersions are (1) complete coverage of the surface – high Γ (adsorbed amount); (2) strong adsorption (or "anchoring") of the chains to the surface; and (3) effective stabilizing chain, chain well solvated, $\chi < 0.5$ and adequate (but not too large) steric barrier δ.

However, this is not the whole story; a stable dispersion does not guarantee a stable and optimized final formulation. TiO_2 particles are always surface modified in a variety of ways in order to improve dispersibility and compatibility with other ingredients. It is important that we understand the impact that these surface treatments may have upon the dispersion and more importantly upon the final formulation. TiO_2 is actually formulated into a suspoemulsion – a suspension in an emulsion. Many additional ingredients are added to ensure cosmetic elegance and function. The emulsifiers used are structurally and functionally not very different to the dispersants used to optimize the fine particle inorganics. Competitive adsorption may occur with some partial desorption of a stabilizer from one or other of the available interfaces. Hence one requires strong adsorption (which should be irreversible) of the polymer to the particle surface.

4.6 Materials and Methods

Dispersions of surface-modified TiO_2 in alkyl benzoate and hexamethyltetracosane (squalane) were prepared at various solids loadings using a polymeric/oligomeric poly(hydroxystearic acid) (PHS) surfactant of molecular weight 2500 (PHS2500) and 1000 (PHS1000). For comparison, results were also obtained using a low molecular weight (monomeric) dispersant, namely isostearic acid

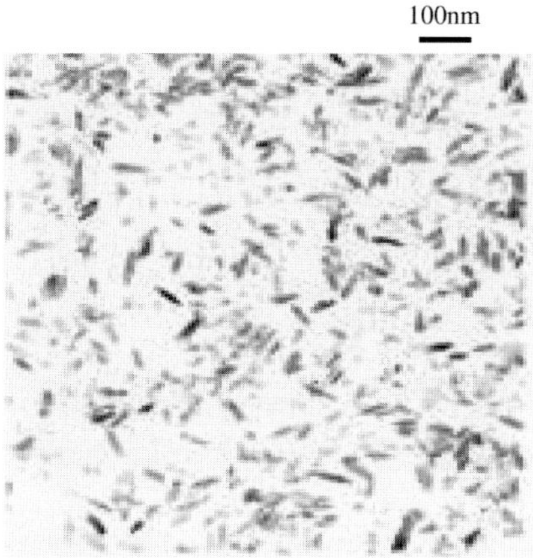

Figure 4.3 Transmission electron micrograph of titanium dioxide particles.

(ISA). The titania particles had been coated with alumina and/or silica. The electron micrograph in Figure 4.3 shows the typical size and shape of these rutile particles. The surface area and particle size of the three powders used are summarized in Table 4.1.

Dispersions of the surface-modified TiO_2 powder, dried at 110 °C, were prepared by milling (using a horizontal bead mill) in polymer solutions of different concentrations for 15 min and were then allowed to equilibrate for more than 16 h at room temperature before making measurements.

Adsorption isotherms were obtained by preparing dispersions of 30% w/w TiO_2 at different polymer concentrations (C_0, mg L^{-1}). The particles and adsorbed dispersant were removed by centrifugation at 20 000 rpm (~48 000 g) for 4 h, leaving a clear supernatant. The concentration of the polymer in the supernatant was determined by acid value titration. Isotherms were calculated by mass balance to determine the amount of polymer adsorbed at the particle surface (Γ, mg m^{-2}) of a known mass of particulate material (m, g) relative to that equilibrated in solution (C_e, mg L^{-1}).

Table 4.1 Surface-modified TiO_2 powders.

Powder	Coating	Surface area[a] (m^2 g^{-1})	Particle size[b] (nm)
A	Alumina/silica	95	40–60
B	Alumina/stearic acid	70	30–40
C	Silica/stearic acid	65	30–40

a) BET N_2 method.
b) Equivalent sphere diameter, X-ray disc centrifuge.

The surface area of the particles (A_S, $m^2\ g^{-1}$) was determined by the BET nitrogen adsorption method. Dispersions of various solids loading were obtained by milling at progressively increasing TiO_2 concentration at an optimum dispersant/solids ratio. The dispersion stability was evaluated by viscosity measurement and by attenuation of UV/Vis radiation. The viscosity of the dispersions was measured by subjecting the dispersions to an increasing shear stress, from 0.03 to 200 Pa over 3 min at 25 °C using a Bohlin CVO rheometer. It was found that the dispersions exhibited shear thinning behavior and the zero shear viscosity, identified from the plateau region at low shear stress (where viscosity was apparently independent of the applied shear stress), was used to provide an indication of the equilibrium energy of interaction that had developed between the particles.

UV/Vis attenuation was determined by measuring transmittance of radiation between 250 and 550 nm. Samples were prepared by dilution with a 1% w/v solution of dispersant in cyclohexane to approximately 20 mg L^{-1} and placed in a 1-cm pathlength cuvette in a UV/Vis spectrophotometer. The sample solution extinction ε ($L\ g^{-1}\ cm^{-1}$) was calculated from Beer's law:

$$\varepsilon = \frac{A}{cl} \tag{8}$$

where A is absorbance, c is the concentration of attenuating species ($g\ L^{-1}$) and l is pathlength (cm).

The dispersions of powders B and C were finally incorporated into typical water-in-oil sunscreen formulations at 5% solids with an additional 2% of organic active (butylmethoxydibenzoylmethane) and assessed for efficacy, SPF (sun protection factor) and stability (visual observation, viscosity). SPF measurements were made on an Optometrics SPF-290 analyzer fitted with an integrating sphere, using the method of Diffey and Robson [10].

4.7 Results

4.7.1 Adsorption Isotherms

Figure 4.4 shows the adsorption isotherms of ISA, PHS1000 and PHS2500 on TiO_2 (powder A) in Figure 4.4a alkyl benzoate and in Figure 4.4b squalane. The adsorption of the low molecular weight ISA from alkyl benzoate is of low affinity (Langmuir type), indicating reversible adsorption (possibly physisorption). In contrast, the adsorption isotherms for PHS1000 and PHS2500 are of the high-affinity type, indicating irreversible adsorption and possible chemisorption due to acid–base interaction. From squalane, all adsorption isotherms are of high-affinity type and they show higher adsorption values compared with the results

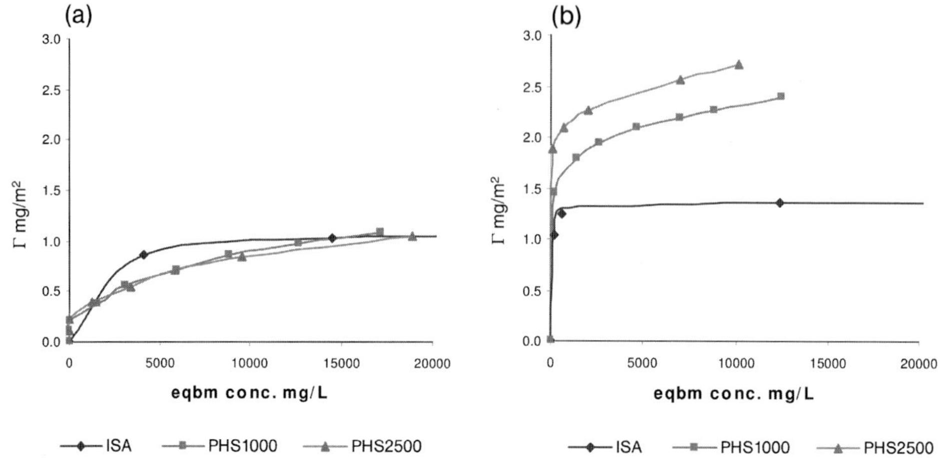

Figure 4.4 Adsorption isotherms in (a) alkyl benzoate and (b) squalane.

using alkyl benzoate. This reflects the difference in solvency of the dispersant by the medium, as will be discussed below.

4.7.2
Dispersant Demand

Figure 4.5 shows the variation of zero shear viscosity with dispersant loading percentage on solid for a 40% dispersion. It can be seen that the zero shear viscosity decreases very rapidly with increase in dispersant loading and eventually the vis-

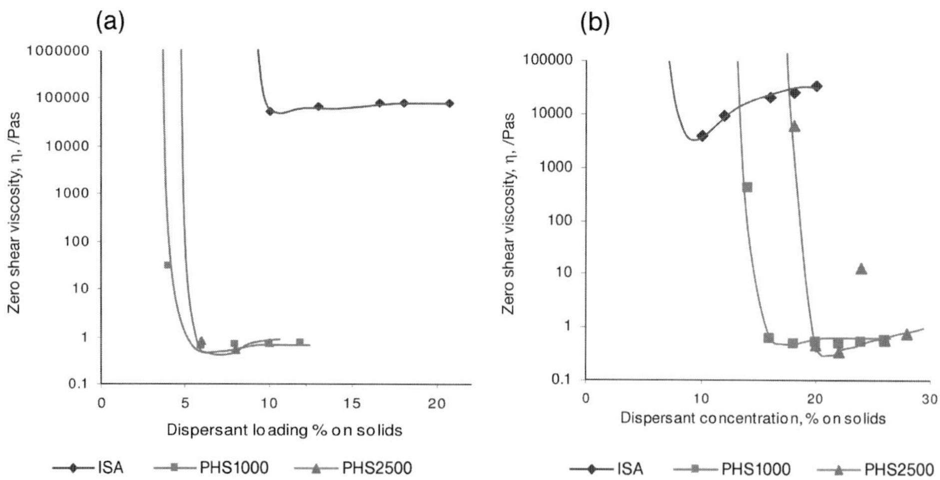

Figure 4.5 Dispersant demand curve in (a) alkyl benzoate and (b) squalane.

cosity reaches a minimum at an optimum loading that depends on the solvent used and the nature of the dispersant. With the molecular dispersant ISA, the minimum viscosity that could be reached at high dispersant loading was very high (several orders of magnitude more than the optimized dispersions), indicating poor dispersion of the powder in both solvents. Even reducing the solids content of TiO_2 to 30% did not result in a low-viscosity dispersion. With PHS1000 and PHS2500, a low minimum viscosity could be reached at 8–10% dispersant loading in alkyl benzoate and 18–20% dispersant loading in squalane. In the latter case, the dispersant loading required for reaching a viscosity minimum is higher for the higher molecular weight PHS.

4.7.3 Quality of Dispersion UV-Vis Attenuation

At very low dispersant concentration, a high solids dispersion can be achieved by simple mixing, but the particles are aggregated, as demonstrated by the UV/Vis curves (Figure 4.6).These large aggregates are not effective as UV attenuators. As the PHS dispersant level is increased, UV attenuation is improved and above 8 wt.% dispersant on particulate mass, optimized attenuation properties (high UV, low visible attenuation) are achieved (for PHS1000 in alkyl benzoate). However, milling is also required to break down the aggregates into their constituent nanoparticles and a simple mixture which is unmilled has poor UV attenuation even at 14% dispersant loading.

The UV/Vis curves obtained when monomeric isostearic acid was incorporated as a dispersant (Figure 4.7) indicate that these molecules do not provide a sufficient barrier to aggregation, resulting in relatively poor attenuation properties (low UV, high visible attenuation).

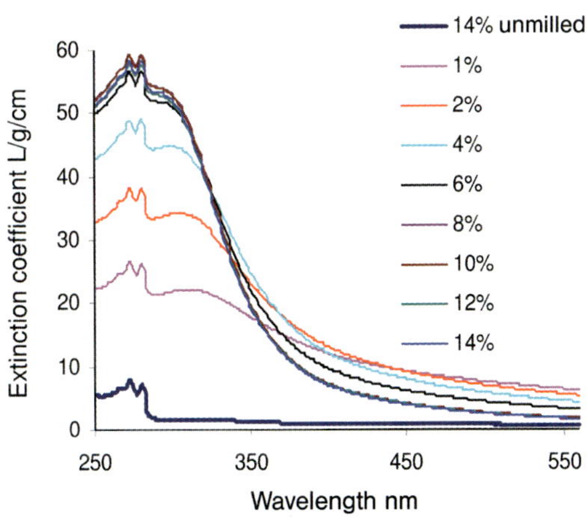

Figure 4.6 UV/Vis attenuation for milled dispersions with 1–14% PHS1000 dispersant and unmilled at 14% dispersant on solids.

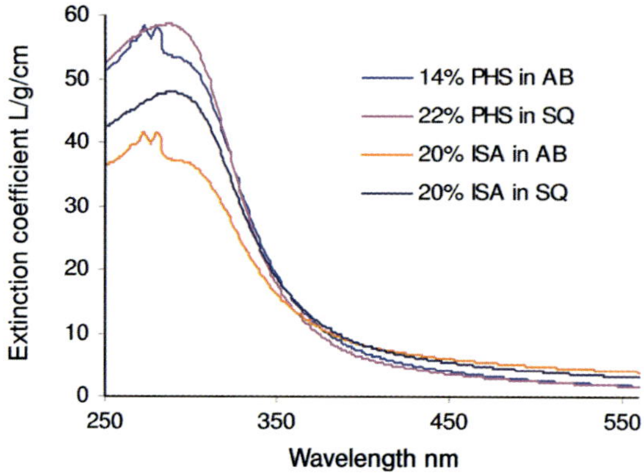

Figure 4.7 UV/Vis attenuation for dispersions in squalane (SQ) and alkyl benzoate (AB) using 20% isostearic acid (ISA) as dispersant compared with optimized PHS1000 dispersions in the same oils.

4.7.4
Solids Loading

The steric layer thickness δ could be varied by altering the dispersion medium and hence the solvency of the polymer chain. This had a significant effect on dispersion rheology. Solids loading curves (Figure 4.8) demonstrate the differences in effective volume fraction due to the adsorbed layer [Eq. (7)].

In the poorer solvent case (squalane), the effective volume fraction and adsorbed layer thickness showed a strong dependence on molecular weight, with solids loading becoming severely limited above 35% for the higher molecular

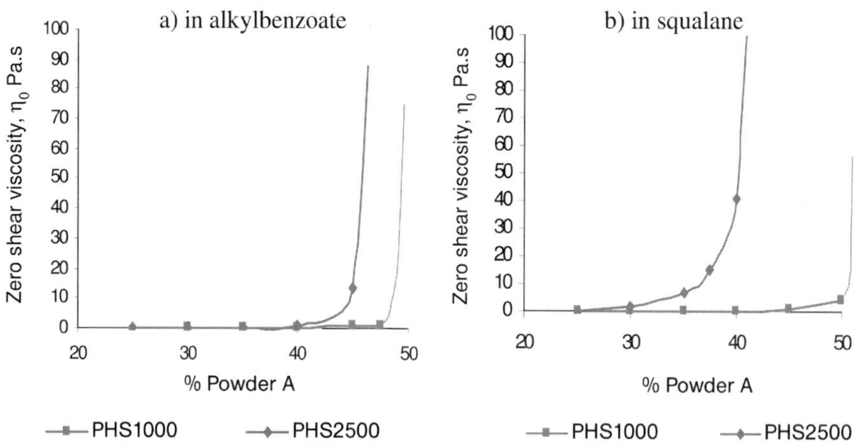

Figure 4.8 Zero shear viscosity dependence on solids loading in (a) alkyl benzoate and (b) squalane.

weight whereas ~50% could be reached for the lower molecular weight polymer. In alkyl benzoate, no strong dependence was seen, with both systems achieving more than 45% solids. Solids weight fractions above 50% resulted in very high viscosity dispersions in both solvents.

4.7.5 SPF Performance in Emulsion Preparations

The same procedure as described above permitted optimized dispersion of equivalent particles with alumina and silica inorganic coatings (powders B and C). Both particles additionally had the same level of organic (stearate) modification. These optimized dispersions were incorporated into water-in-oil formulations and their stability/efficacy was monitored by visual observation and SPF measurements (Table 4.2).

Table 4.2 Sunscreen emulsion formulations from dispersions of powders B and C.

Emulsion	Visual observation	SPF	Emulsifier level (5)
Powder B emulsion 1	Good homogeneous emulsion	29	2.0
Powder C emulsion 1	Separation, inhomogeneous	11	2.0
Powder C emulsion 2	Good homogeneous emulsion	24	3.5

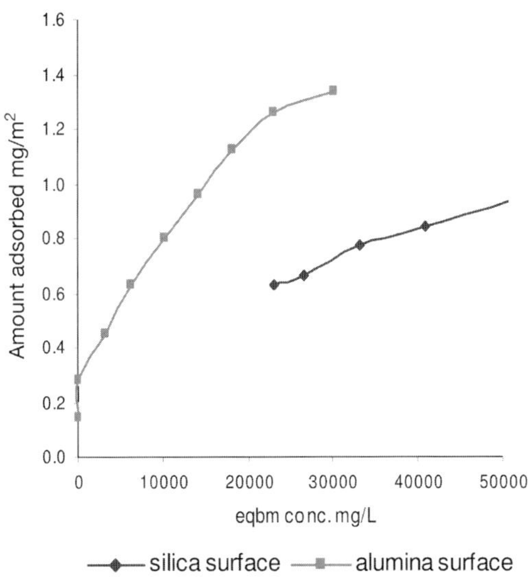

Figure 4.9 Adsorption isotherms for PHS2500 on powder B (alumina surface) and powder C (silica surface).

The formulation was destabilized by the addition of the powder C dispersion and poor efficacy was achieved despite an optimized dispersion before formulation. When the emulsifier concentration was increased from 2 to 3.5% (emulsion 2), the formulation became stable and efficacy was restored.

The anchor of the chain to the surface (described qualitatively through χ_s) is very specific and this could be illustrated by silica-coated particles, which showed lower adsorption of the PHS (Figure 4.9).

In addition, when a quantity of emulsifier was added to an optimized dispersion of powder C (silica surface), the acid value of the equilibrium solution was seen to rise, indicating some displacement of the PHS2500 by the emulsifier.

4.8 Discussion

The dispersant demand curves (Figure 4.5a and b) and solids loading curves (Figure 4.8a and b) show that one can reach a stable dispersion using PHS1000 or PHS2500 both in alkyl benzoate and in squalane. These can be understood in terms of the stabilization produced when using these polymeric dispersants. Addition of sufficient dispersant allows coverage of the surface and results in a steric barrier (Figure 4.1) preventing aggregation due to van der Waals attraction. Both molecular weight oligomers were able to achieve stable dispersions. The much smaller molecular weight "monomer" isostearic acid, however, is insufficient to provide this steric barrier and dispersions were aggregated, leading to high viscosities, even at 30% solids. UV/Vis curves confirm that these dispersions are not fully dispersed since their full UV potential is not realized (Figure 4.7). Even at 20% isostearic acid the dispersions are seen to give a lower E_{max} and increased scattering at visible wavelengths, indicating a partially aggregated system.

The differences between alkyl benzoate and squalane observed in the optimum dispersant concentration required for maximum stability can be understood by examining the adsorption isotherms in Figure 4a and b. The nature of the steric barrier depends on the solvency of the medium for the chain and is characterized by the Flory–Huggins interaction parameter χ. Information on the value of χ for the two solvents can be obtained from solubility parameter calculations [Eq. (5)]. The results of these calculations are given in Table 4.3 for PHS, alkyl benzoate and squalane.

Table 4.3 Hansen and Beerbower solubility parameters for the polymer and both solvents.

Polymer/solvent	δ_T	δ_d	δ_p	δ_h	$\Delta\delta_T$
PHS	19.00	18.13	0.86	5.60	
Alkyl benzoate	19.64	19.13	1.73	4.12	1.99
Squalane	15.88	15.88	0	0	6.1

It can be seen that both PHS and alkyl benzoate have polar and hydrogen bonding contributions to the solubility parameter δ_T. In contrast, squalane, which is nonpolar, has only a dispersion component to δ_T. The difference in the total solubility parameter $\Delta\delta_T$ value is much smaller for alkyl benzoate than squalane. Hence one can expect that alkyl benzoate is a better solvent than squalane for PHS. This explains the higher adsorption amounts of the dispersants in squalane compared with alkyl benzoate (Figure 4.4). PHS finds adsorption at the particle surface energetically more favorable than remaining in solution. The adsorption values at the plateau for PHS in squalane (>2 mg m^{-2} for PHS1000 and >2.5 mg m^{-2} for PHS2500) is more than twice the value obtained in alkyl benzoate (1 mg m^{-2} for both PHS1000 and PHS2500). It should be mentioned, however, that both alkyl benzoate and squalane will have χ values less than 0.5, i.e. good solvent conditions and a positive steric potential. This is consistent with the high dispersion stability produced in both solvents. However, the relative difference in solvency for PHS between alkyl benzoate and squalane is expected to have a significant effect on the conformation of the adsorbed layer. In squalane, a poorer solvent for PHS, the polymer chain is denser than the polymer layer in alkyl benzoate. In the latter case, a diffuse layer that is typical for polymers in good solvents is produced. This is illustrated in Figure 4.10a. which shows a greater hydrodynamic layer thickness for the higher molecular weight PHS2500. A schematic representation of the adsorbed layers in squalane is shown in Figure

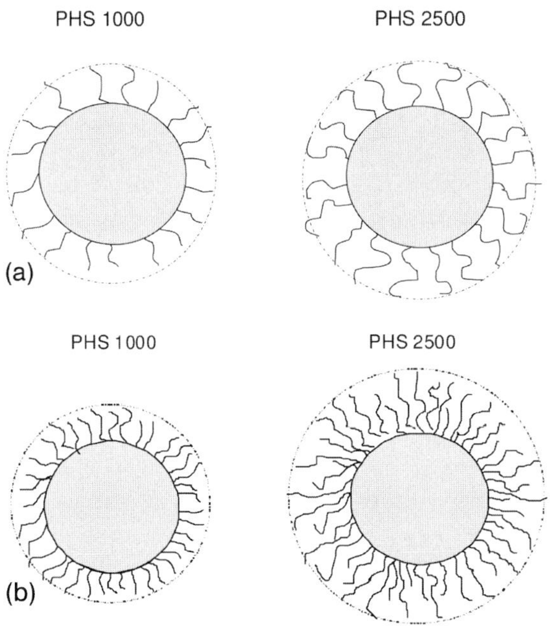

Figure 4.10 (a) Well-solvated polymer results in diffuse adsorbed layers (alkyl benzoate). (b) Polymers are not well solvated and form dense adsorbed layers (squalane).

4.10b, which also shows a greater thickness for the higher molecular weight PHS2500.

In squalane, the dispersant adopts a close-packed conformation with little solvation and large amounts are required to reach full surface coverage ($\Gamma > 2$ mg m^{-2}). It seems also that in squalane there is much more dependence of the amount of adsorption on the molecular weight of PHS than in the case of alkyl benzoate. It is likely that with the high molecular weight PHS2500 in squalane the adsorbed layer thickness can reach higher values compared with the results in alkyl benzoate. This greater layer thickness increases the effective volume fraction and this restricts the total solids that can be dispersed. This is clearly shown from the results in Figure 4.8, which shows a rapid increase in zero shear viscosity at a solids loading >35%. With the lower molecular weight PHS1000, with a smaller adsorbed layer thickness, the effective volume fraction is lower and a high solids loading (~50%) can be reached. The solids loading that can be reached in alkyl benzoate when using PHS2500 is higher (~40%) than that obtained in squalane. This implies that the adsorbed layer thickness of PHS2500 is smaller in alkyl benzene compared with the value in squalane, as shown schematically in Figure 4.10. The solids loading with PHS1000 in alkyl benzoate is similar to that in squalane, indicating a similar adsorbed layer thickness in both cases.

The solids loading curves demonstrate that with an extended layer such as that obtained with the higher molecular weight (PHS2500), the maximum solids loading becomes severely limited as the effective volume fraction [Eq. (5)] is increased.

In squalane, the monomeric dispersant isostearic acid shows a high-affinity adsorption isotherm with a plateau adsorption of 1 mg m^{-2}, but this provides an insufficient steric barrier (δ/R too small, Figure 4.2) to ensure colloidal stability.

4.8.1 Competitive Interactions in Formulations

On addition of the sunscreen dispersion to an emulsion to produce the final formulation, one has to consider the competitive adsorption of the dispersant/emulsifier system. In this case the strength of adsorption of the dispersant to the surface-modified TiO_2 particles must be considered. As shown in Figure 4.9, the silica-coated particles (C) show lower PHS2500 adsorption than the alumina-coated particles (B). However, the dispersant demand for the two powders to obtain a colloidally stable dispersion was similar in both cases (12–14% PHS2500). This appears at first sight to indicate similar stabilities. However, when added to a water-in-oil emulsion prepared using an A–B–A block copolymer of PHS–PEO–PHS as emulsifier, the system based on the silica-coated particles (C) became unstable, showing separation and coalescence of the water droplets. The SPF performance also dropped drastically from 29 to 11. In contrast, the system based on alumina-coated particles (B) remained stable, showing no separation, as illustrated in Table 4.2. These results are consistent with the stronger adsorption (higher χ_s) of PHS2500 on the alumina-coated particles. With the silica-coated particles, it is likely that the PHS–PEO–PHS block copolymer becomes adsorbed on the particles, thus depleting the emulsion interface from the polymeric emulsifier, and this is the cause of coalescence. It is well known that molecules based on PEO can adsorb on silica surfaces [11]. By addition of more emulsifier (increasing its concentration from 2 to 3.5%), the formulation remained stable, as is illustrated in Table 4.2.

This final set of results demonstrates how a change in surface coating can alter the adsorption strength, which can have consequences for the final formulation. The same optimization process as used for powder A enabled stable dispersions to be formed from powders B and C. Dispersant demand curves showed optimized dispersion rheology at similar added dispersant levels of 12–14% PHS2500. To the dispersion scientist these appeared to be stable TiO_2 dispersions. However when the optimized dispersions were formulated into the external phase of a water-in-oil emulsion, differences were observed and alterations in formulation were required to ensure emulsion stability and performance.

4.9 Conclusion

The application of colloid and interface science principles give a sound basis on which to carry out true optimization of consumer-acceptable sunscreen formulations based on particulate TiO_2. It was found that both dispersion stability and dispersion rheology depended on the adsorbed amount Γ and steric layer thickness δ (which in turn depends on oligomer molecular weight M_n and solvency χ), but that in order to optimize formulation, the adsorption strength χ_s must also be considered. The nature of the interaction between particles, dispersant, emulsifiers and thickeners must be considered with regard to competitive adsorption and/or interfacial stability if a formulation is to deliver its required protection when spread on the skin.

References

1 J.L. Robb, L.A. Simpson, D.F. Tunstall, Scattering and absorption of UV radiation by sunscreens containing fine particle and pigmentary titanium dioxide. *Drug Cosmet. Ind.*, March 1994, pp. 32–39.

2 J.P. Hewitt, A moment of clarity. *Soap Perfum. Cosmet.* **75**(3), 47–50 (2002).

3 G.P. Dransfield, S. Cutter, P.L. Lyth, Particulate metal oxide. *PCT Patent Application* WO 02/00797, 2001.

4 G.J. Fleer, M.A. Cohen-Stuart, J.M.H.M. Scheutjens, T. Cosgrove, B. Vincent, *Polymers at Interfaces*, Chapman and Hall, London, 1993.

5 F.H. Napper, *Polymeric Stabilization of Colloidal Dispersions*, Academic Press, London, 1983.

6 C.M. Hansen, *J. Paint Technol.*, **39**, 104–117, 505–514 (1967).

7 C.M. Hansen, A. Beerbower, in: Handbook of Solubility Parameters and Other Cohesion Parameters (ed. A.F.M. Barton). CRC Press, Boca Raton, FL, 1983.

8 Th.F. Tadros, P. Izquierdo, J. Esquena, C. Solans. Formulation and stability of

nanoemulsions, *Adv. Colloid Interface Sci.*, **108–109**, 303–318 (2004).

9 L.M. Kessell, B.J. Naden, Th.F. Tadros, *Attractive and repulsive gels from inorganic sunscreen actives, Poster 213* in *Proceedings of the IFSCC 23rd Congress*, October 2004.

10 B.L. Diffey, J. Robson, *J. Soc. Cosmet. Chem.*, **40**, 127–133 (1989).

11 J.A. Shar, T.M. Obey, T. Cosgrove, *Colloids Surf. A*, **150**, 15–23 (1999).

5
Use of Associative Thickeners as Rheology Modifiers for Surfactant Systems

Tharwat F. Tadros and Steven Housley

Abstract

This chapter describes a comparison between surfactant systems "thickened" with NaCl or an associated thickener based on PEG-150 distearate (Promidium LTS). It starts with a discussion of the use of surfactant systems as rheology modifiers with emphasis on the effect of addition of salt which induces the formation of thread-like structures that can produce "gels" by overlap of the chains. Rheology modifiers based on associative thickeners produce "gels" by association of the hydrophobic chains producing micelle-like structures. The gel produced is affected by addition of surfactants and, at some surfactant concentration, the micelles can "bridge" the polymer chains. Viscoelastic measurements were carried out using a typical shampoo surfactant system consisting of a mixture of anionic and amphoteric surfactants. This surfactant system was thickened by addition of either NaCl or Promidium TLS. Results were obtained as a function of applied stress (at constant frequency of 1 Hz) and frequency (whereby the stress was fixed in the linear viscoelastic region). The results showed a distinct difference between the two thickened systems. That based on NaCl was much more elastic than viscous, whereas the system thickened with Promidium TLS was more viscous than elastic. Oscillatory measurements show a much higher relaxation time for the NaCl-thickened system compared with that thickened using Promidium TLS. This explains the better performance on surfactant systems thickened with Promidium TLS, which did not show the "stringy" character observed with systems thickened by addition of NaCl.

5.1
Introduction

Most aqueous-based cleansing formulations (shampoo, shower gel, body and facial washes) are based on concentrated surfactant systems [1, 2]. Mostly these surfactant systems are blends of anionics such as alkyl sulfates, with amphoterics

Colloids and Interface Science Series, Vol. 4
Colloids in Cosmetics and Personal Care. Edited by Tharwat F. Tadros

ISBN: 978-3-527-31464-5

such as betaines, with an overall concentration of approximately 10% active surfactant. Electrolytes (typically sodium chloride) are usually added to these surfactant blends to increase the viscosity for optimum application to skin and hair. The addition of salt causes a change in the micellar structure from spherical to rod-shaped units [3]. This is caused by a change in the critical packing parameter, P, from $\leq \frac{1}{3}$ to $\leq \frac{1}{2}$ by screening the charge on the surfactant's polar head group [1, 2]. These rod-shaped micelles can produce a "gel" network by interaction between one another, becoming "worm-like" in structure [3]. However, addition of salt may not produce the optimum rheological characteristics (viscoelasticity) for ease of application and sometimes formulations appear stringy and rubbery, giving poor sensorial effects (both visual and through touch). The aim of this chapter is to demonstrate that by using an associative thickener such as a blend of PEG-150 distearate and PPG-2 hydroxyethyl cocamide, marketed as Promidium LTS (Croda), one can achieve the right viscoelastic behavior with good sensory characteristics.

The chapter starts with a discussion of the use of surfactant systems as rheology modifiers. This is followed by a section on associative thickeners as rheology modifiers with particular reference to their interaction with surfactant micelles.

5.2 Surfactant Systems as Rheology Modifiers

In dilute solutions, surfactants tend to form spherical micelles with aggregation numbers in the range 50–100 units. These micellar solutions are isotropic with low viscosity.

At much higher surfactant concentrations (>30% depending on the surfactant nature), they produce liquid crystalline phases of the hexagonal (H_1) and lamellar (L_α) phases, which are anisotropic with much higher viscosities. A schematic representation of the hexagonal and lamellar phases is shown in Figures 5.1 and 5.2. These liquid crystalline phases, which are viscoelastic, can be used as rheol-

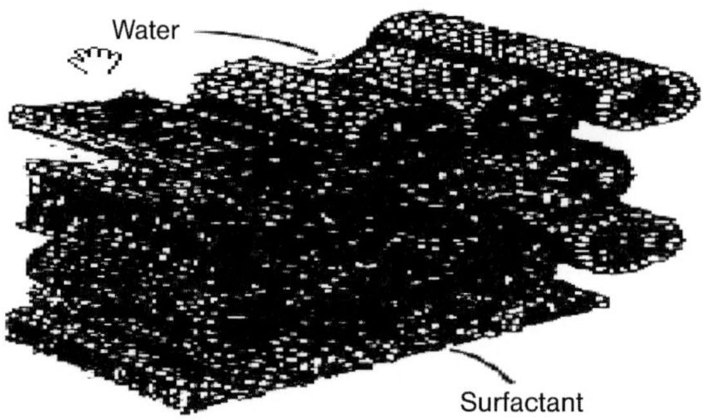

Figure 5.1 Schematic representation of the hexagonal phase.

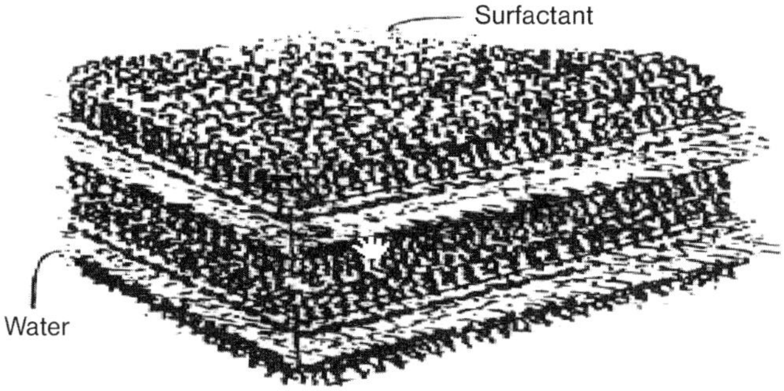

Figure 5.2 Schematic representation of the lamellar phase.

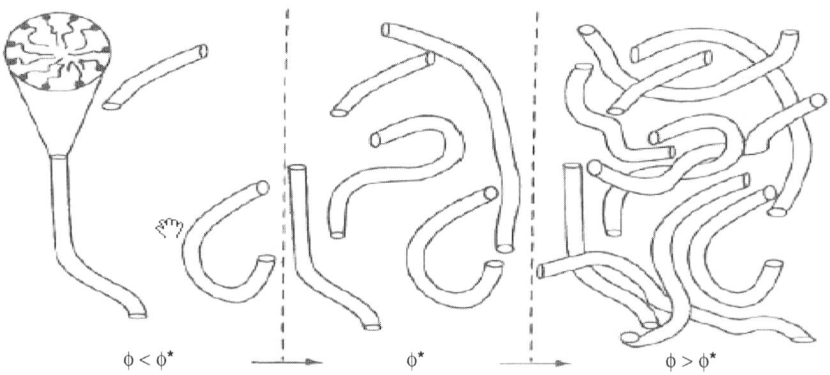

Figure 5.3 Schematic representation of overlap of thread-like micelles.

ogy modifiers. However, for practical applications such as in shampoos, such very high surfactant concentrations are undesirable. One way to increase the viscosity of a surfactant solution at lower concentrations is to add an electrolyte that causes a change from spherical to cylindrical micelles which can grow in length, and at above a critical surfactant volume fraction ϕ^* these worm-like micelles begin to overlap, forming a "gel", as illustrated in Figure 5.3.

5.3 Associative Thickeners as Rheology Modifiers

Associative thickeners are hydrophobically modified polymer molecules whereby alkyl chains (C_{12}–C_{16}) are either randomly grafted on a hydrophilic polymer molecule such as hydroxyethylcellulose (HEC) or simply grafted at both ends of the hydrophilic chain. An example of hydrophobically modified HEC is Natrosol Plus (Hercules), which contains 3–4 C_{16} chain randomly grafted on to HEC. An

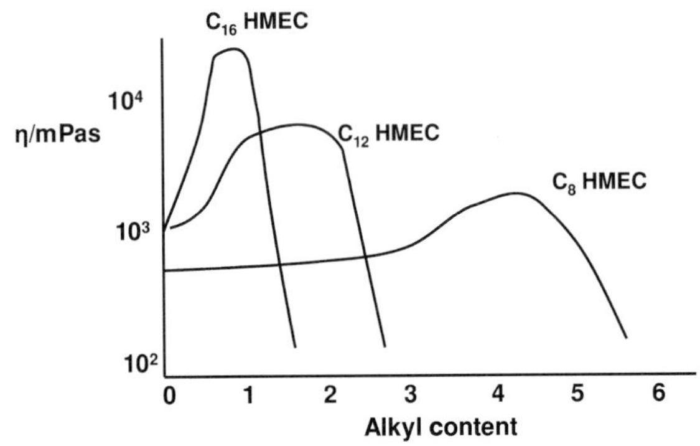

Figure 5.4 Variation of viscosity of 1% HMHEC versus alkyl group content of the polymer.

example of a polymer that contains two alkyl chains at both ends of the molecule is HEUR (Rohm and Haas), which is made of poly(ethylene oxide) (PEO) that is capped at both ends with linear C_{18} hydrocarbon chain. These molecules are similar to PEG-150 distearate (one of the components of Promidium LTS) that is used in the present study.

The above hydrophobically modified polymers form gels when dissolved in water. Gel formation can occur at relatively lower polymer concentrations compared with the unmodified molecule.

The most likely explanation of gel formation is due to hydrophobic bonding (association) between the alkyl chains in the molecule. This effectively causes an apparent increase in the molecular weight. These associative structures are similar to micelles, except that the aggregation numbers are much smaller. Figure 5.4 shows the variation of viscosity (measured using a Brookfield at 30 rpm as a function of the alkyl content (C_8, C_{12} and C_{16}) for hydrophobically modified HEC (i.e. HMHEC). The viscosity reaches a maximum at a given alkyl group content that decreases with increase in the alkyl chain length. The viscosity maximum increases with increase in the alkyl chain length.

Associative thickeners also show interaction with surfactant micelles that are present in the formulation. The viscosity of the associative thickeners shows a maximum at a given surfactant concentration that depends on the nature of surfactant. This is shown schematically in Figure 5.5.

The increase in viscosity is attributed to the hydrophobic interaction between the alkyl chains on the backbone of the polymer and the surfactant micelles. A schematic diagram showing the interaction between HM polymers and surfactant micelles is shown in Figure 5.6.

At higher surfactant concentration, the "bridges" between the HM polymer molecules and the micelles are broken (free micelles) and η decreases.

The viscosity of hydrophobically modified polymers shows a rapid increase at a critical concentration, which may be defined as the critical aggregation concentration (CAC) as illustrated in Figure 5.7 for HMHEC (WSP-D45 from Hercules).

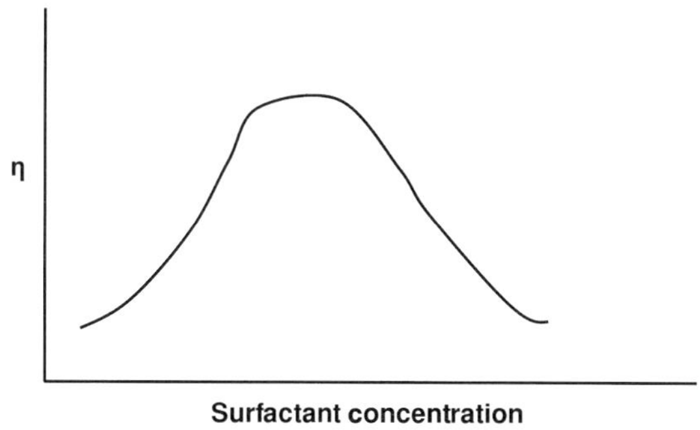

Figure 5.5 Schematic plot of viscosity of HM polymer with surfactant concentration.

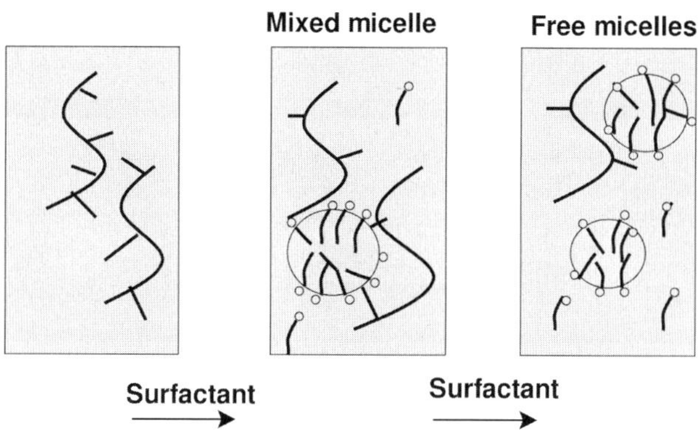

Figure 5.6 Schematic representation of the interaction of polymers with surfactants.

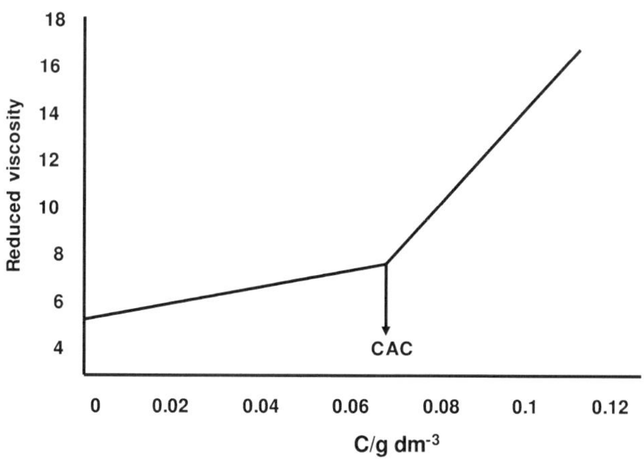

Figure 5.7 Variation of reduced viscosity with HMHEC concentration.

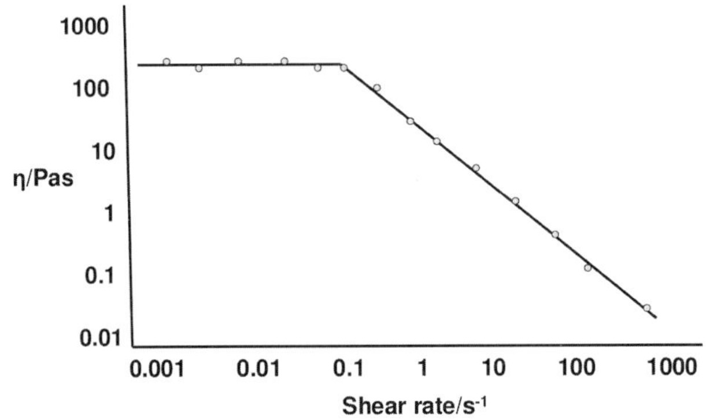

Figure 5.8 Variation of viscosity with shear rate for HMEC WSP-47 at 0.75 g per 100 cm^3.

The assumption is made that the CAC is equal to the coil overlap concentration C^*. From a knowledge of C^* and the intrinsic viscosity $[\eta]$ one can obtain the number of chains in each aggregate. For the above example $[\eta] = 4.7$ and $C^*[\eta] = 1$, giving an aggregation number of ~4.

At C^* the polymer solution shows non-Newtonian flow (shear thinning behavior) and it shows a high viscosity at low shear rates. This is illustrated in Figure 5.8, which shows the variation of apparent viscosity with shear rate (using a constant stress rheometer).

Below ~0.1 s^{-1}, a plateau viscosity value $\eta(0)$ (referred to as residual or zero shear viscosity) is reached (~200 Pa s). With increase in polymer concentration above C^*, the zero shear viscosity increases with increase in polymer concentration. This is illustrated in Figure 5.9.

The above hydrophobically modified polymers are viscoelastic. This is illustrated in Figure 5.10 for a solution 5.25% of C_{18} end-capped PEO with $M =$

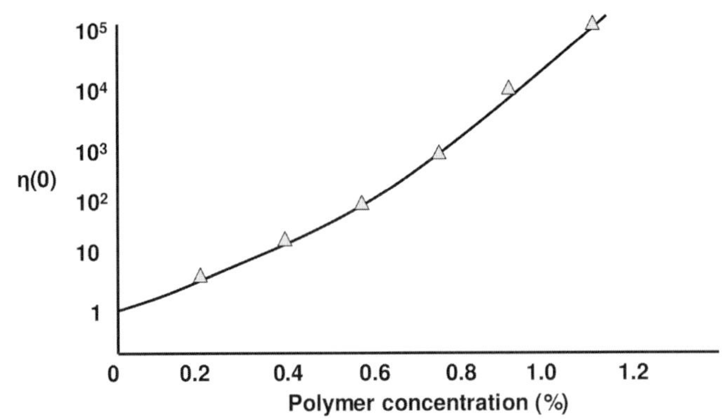

Figure 5.9 Variation of $\eta(0)$ with polymer concentration.

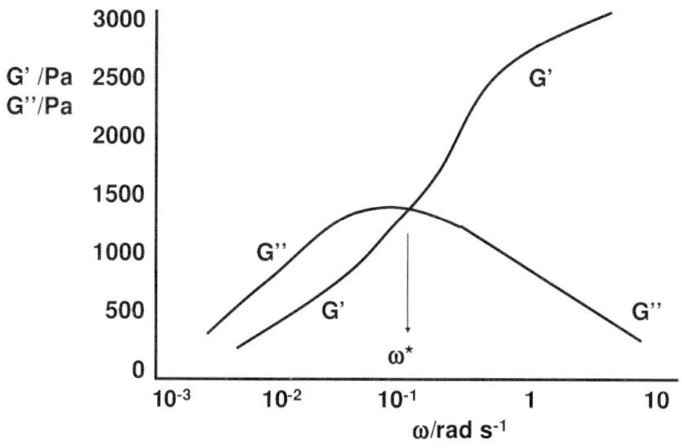

Figure 5.10 Variation of G' and G'' with frequency for 5.24 HM PEO.

35 000, which shows the variation of the storage modulus G' and loss modulus G'' with frequency ω (rad s^{-1}).

G' increases with increase in frequency and ultimately it reaches a plateau value at high frequency. G'' (which is higher than G' in the low-frequency regime) increases with increase in frequency, reaches a maximum at a characteristic frequency ω^* (at which $G' = G''$) and then decreases to near zero in the high-frequency regime. This variation of G' and G'' with ω is typical for a system that shows Maxwell behavior.

From the cross-over point ω^* (at which $G' = G''$), one can obtain the relaxation time τ of the polymer in solution:

$$\tau = \frac{1}{\omega^*} \tag{1}$$

For the above polymer, $\tau = 8$ s.

The above gels (sometimes referred to as rheology modifiers) are used in many surfactant formulations to produce the right consistency.

The high-frequency modulus, sometimes referred to as the network modulus, can be used to obtain the number of "links" in the gel network structure. Using the theory of rubber elasticity, the network modulus G_N is related to the number of elastically effective links N and a factor A that depends on the junction functionality:

$$G_N = ANkT \tag{2}$$

where k is Boltzmann's constant and T is the absolute temperature.

For an end-capped PEO (i.e. HEUR), the junctions should be multifunctional ($A = 1$). For tetrafunctional junctions $A = \frac{1}{2}$.

5.4 Materials and Methods

The following is a simplified shampoo/body wash formulation and was used throughout this work: 7% active sodium laureth sulfate (2 mol of ethylene oxide) (SLES), 3% cocamidopropylbetaine (CAPB) and 1% preservative (Germaben II). The rheology of this aqueous surfactant solution was modified in increments by addition of either NaCl or Promidium LTS (PEG-150 distearate and PPG-2 hydroxyethylcocamide). For simplicity, the trade name Promidium LTS will be used throughout this chapter.

The viscoelastic behavior was investigated using dynamic (oscillatory) measurements. For this purpose, a Bohlin CVO rheometer (Malvern Instruments, UK) was used, the samples being measured using a cone and plate geometry (4° cone angle/40 mm diameter). All measurements were carried out at 25 °C.

In oscillatory measurements, a sinusoidal stress or strain is applied on the cone or plate and the resulting strain or stress through the sample is measured simultaneously [4]. From the time shift of stress and strain, Δt, and the frequency of oscillation, ω (rad s^{-1}), one can obtain the phase angle shift, δ:

$$\delta = \Delta t \omega \tag{3}$$

From the stress and strain amplitudes, σ_0 and γ_0, respectively, and the phase angle shift, δ, one can obtain the following rheological parameters:

$$\text{complex modulus, } |G^*| = \sigma_0/\gamma_0 \tag{4}$$

$$\text{storage modulus (elastic component), } G' = |G^*| \cos\delta \tag{5}$$

$$\text{loss modulus (viscous component), } G'' = |G^*| \sin\delta \tag{6}$$

$$\tan\delta = G''/G' \tag{7}$$

$$\text{dynamic viscosity, } \eta' = G''/\omega \tag{8}$$

In dynamic measurements, one usually fixes the frequency, for example at 1 Hz, and G' and G'' are measured as a function of strain (or stress) amplitude. This allows one to obtain the linear viscoelastic region whereby G' and G'' are independent of the strain (or stress) amplitude. In a second experiment, the strain (or stress) is fixed within the determined linear region and G' and G'' are then measured as a function of frequency, ω.

For a viscoelastic system, $G'' > G'$ when the frequency is below a certain characteristic frequency, ω^*. Above this characteristic frequency, ω^*, the reverse is true and $G' > G''$. The cross-over point at which $G' = G''$ (and $\tan\delta = 1$) gives the characteristic frequency, ω^*, which allows the relaxation time, t^*, of the formulation to be calculated:

$$t^* = 1/\omega^* \tag{9}$$

5.5 Results

Figure 5.11 gives typical stress sweep results obtained at 1 Hz for the surfactant base thickened with (a) 1.6% NaCl and (b) 1.75% Promidium LTS. In both cases G' and G'' remained constant up to a critical stress, above which both G' and G'' start to decrease with decrease in applied stress. The region below the critical stress at which G' and G'' remain constant with increase in stress is denoted the linear viscoelastic region. It should be mentioned that the surfactant system based on NaCl gives a lower critical stress compared with the system thickened with Promidium LTS. This reflects the difference in "gel" structure between the two systems. It is likely that the system thickened with Promidium TLS gives a more coherent region (with a longer linear viscoelastic region) compared with the system based on NaCl.

It can be also seen from the results in Figure 5.11 that the surfactant base thickened with Promidium LTS (Figure 5.11b) is far more viscous than elastic ($G'' \gg G'$) compared with the same base thickened with NaCl (Figure 5.11a) where $G' > G''$. In fact, regardless of the quantity of Promidium LTS used (and

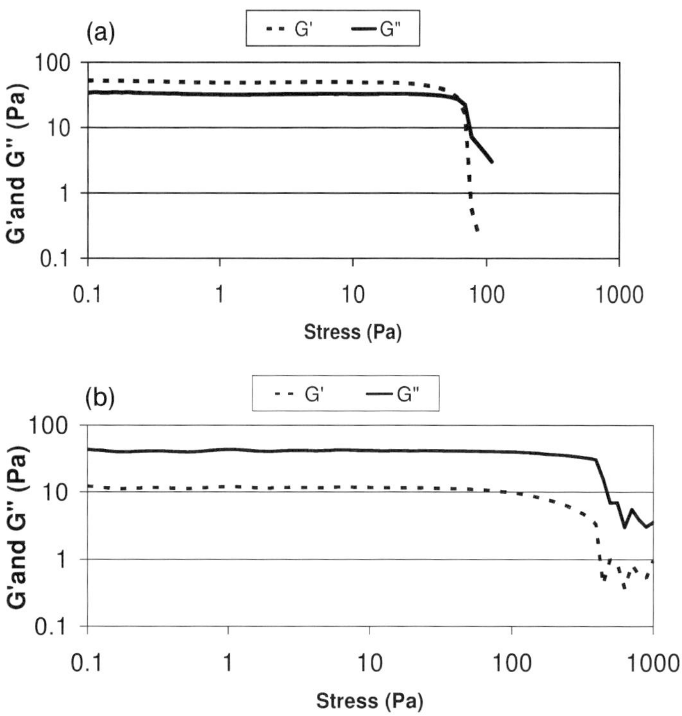

Figure 5.11 Typical stress sweep results (1 Hz) for surfactant blends thickened with (a) 1.6% NaCl and (b) Promidium LTS.

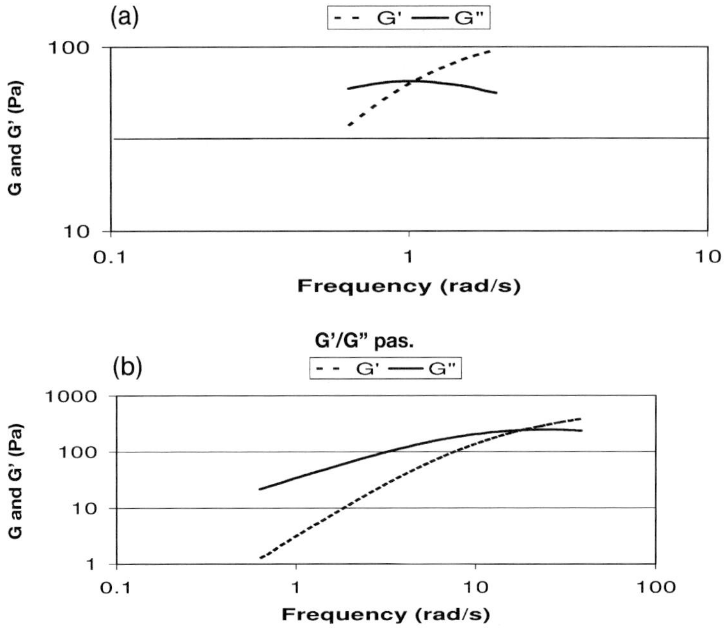

Figure 5.12 Typical frequency sweeps for surfactant base thickened with (a) 2.5% NaCl and (b) 2.5% Promidium LTS.

hence the final viscosity of the formulation), the thickened surfactant always remains viscous dominant, even at extremely high viscosity. In the case of surfactant thickened with NaCl, at some critical concentration (in this case close to 1.6%) the base becomes elastic dominant. This can be seen even at reasonably low viscosity.

Once the linear viscoelastic region was known, it was possible to measure the effect of frequency on these surfactant bases. As an example, typical frequency sweeps for surfactant bases thickened with 2.5% NaCl and 2.5% Promidium LTS are given in Figure 5.12.

It can be seen from Figure 5.12 that the cross-over point (at which $G' = G''$) occurs at much higher frequency for the surfactant base thickened with Promidium LTS compared with the same base thickened with salt. This implies that the relaxation time for the base thickened with Promidium LTS is much smaller than the values for the salt-thickened system. A plot of relaxation time versus both NaCl and Promidium LTS concentration is given in Figure 5.13.

At high frequencies (corresponding to short time scales) the response is more elastic than viscous ($G' > G''$) for surfactants thickened with both NaCl and Promidium LTS. The high-frequency modulus values are significantly higher for the bases thickened with Promidium LTS than those thickened with NaCl. However, G'' is beginning to plateau at 1.4% NaCl whereas the Promidium LTS G'' values are continuing to rise (over the whole concentration range). Again, this implies that independent of Promidium LTS concentration and the structure this gives,

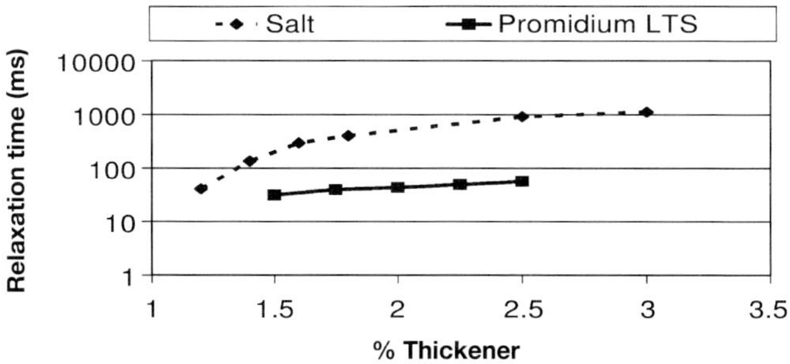

Figure 5.13 Relaxation time versus NaCl and Promidium LTS concentration.

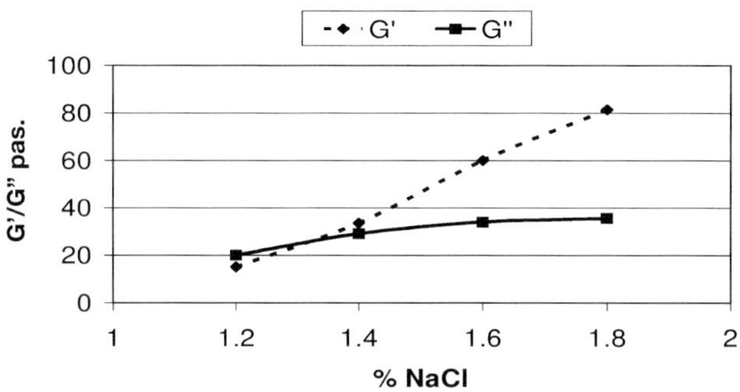

Figure 5.14 Variation of G' and G'' (at a frequency higher than the G'/G'' cross-over point) for surfactants thickened with NaCl.

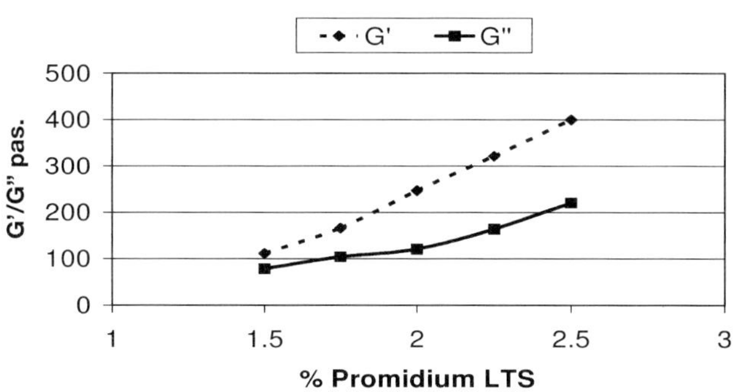

Figure 5.15 Variation of G' and G'' (at a frequency higher than the G'/G'' cross-over point) for surfactant base thickened with Promidium LTS.

the surfactant bases thickened with this associative thickener remain viscous in behavior. Those thickened with salt become predominantly elastic.

Figures 5.14 and 5.15 show the variation of G' and G'' of a high-frequency oscillation that is above the G'/G'' cross-over point. This is shown as a function of both NaCl and Promidium LTS concentration. For NaCl the frequency is 10 rad s^{-1}, for LTS the frequency is 50 rad s^{-1}.

5.6 Discussion

The increase in the viscosity or elasticity of surfactant blends thickened with NaCl is due to the change in micellar structure from spherical to rod-shaped micelles. This can be understood from consideration of critical packing parameter, P, for surfactant molecules [5]:

$$P = \frac{v}{l_c a} \tag{10}$$

where v is the average value of the volume of the hydrocarbon chains with an extended length, l_c, and a is the cross-sectional area of the head group. For ionic surfactants, in the absence of added electrolyte, $P \le \frac{1}{3}$ and spherical micelles are produced. This results in a low-viscosity solution. On addition of electrolyte, the charge on the polar head group of the surfactant is screened, thus reducing a and P can subsequently reach values of $\frac{1}{2}$. This results in the formation of rod-shaped micelles, as discussed before, which in turn produce entangled "worm-like" structures, hence producing a "gel" structure with higher viscosity and elasticity. However, we have seen that such structures give a more elastic than viscous response even at low frequency (long time scales), i.e. 1 Hz. This could be comparable to the time scales used in pouring, pumping and spreading during in-use application. The cross-over point for such electrolyte-thickened systems occurs at much lower frequencies, giving long relaxation times. With increasing NaCl concentration the relaxation time increases, reaching very high values. For example, at 2.5% NaCl, the cross-over point occurs at 1 rad s^{-1} (0.16 Hz) giving a relaxation time of 1 s.

An alternative and more elegant way of thickening surfactant cleansing products is to use associative thickeners. The associative thickener studied here consists of a hydrophilic chain of 150 ethylene oxide units (PEG-150), with two stearate chains attached, one at each end of the hydrophilic chain. As discussed before, this produces "micelle-like" structures [6]. These structures are seen to have much shorter relaxation times compared with surfactants thickened with salt (Figure 5.3). The relaxation times are more than one order of magnitude lower than the bases thickened with salt. The cross-over point for the formulations thickened using Promidium LTS occurs at much higher frequency than those thickened by addition of NaCl. This means that at low frequency the system thickened with Promidium LTS is more viscous than elastic, independent of stress amplitude (Figure 5.1). This will contribute a great degree to the sensory characteristics of the shampoo, shower gel or facial cleanser both in terms of feel during application and also visually, i.e. a lack of stringiness and stickiness.

5.7 Conclusion

Many personal cleansing products (shampoos and body washes) which contain a high concentration of anionic and amphoteric surfactants are thickened by addition of electrolytes, e.g. NaCl. By screening the ionic charge on the surfactant head group, the critical packing parameter is increased from $\leq\frac{1}{3}$ to $\leq\frac{1}{2}$. This results in the formation of rod-shaped micelles forming three-dimensional worm-like structures. This results in an increase in the viscosity and elasticity of the system. However, these salt-thickened shampoo and body washes appear stringy and sticky, giving poor sensory attributes. These problems are solved by replacing the salt with an associative thickener such as Promidium LTS. The latter consists of a hydrophilic PEG chain with a stearate group at each end. In solution these associative thickeners form "micelle-like" structures by hydrophobic bonding between the stearate chains. These thickened systems have much shorter relaxation times than salt-thickened systems. At low frequency (1 Hz), the system is predominantly more viscous than elastic. This eliminates any stringiness and stickiness, providing better esthetics and skin feel.

References

1 Th. F. Tadros, *Applied Surfactants*, Wiley-VCH, Weinheim, 2005.
2 K. Holmberg, B. Jonnson, B. Kronberg, B. Lindman, *Surfactants and Polymers in Aqueous Solution*, Wiley-VCH, Weinheim, 2003.
3 K. Penfield, *IFSCC Mag.*, **8**, 115 (2005).
4 Th. F. Tadros, *Adv. Colloids Interfaces*, **68**, 91 (1996).
5 J. N. Israelachvili, *Intermolecular and Surface Forces*, Academic Press, London, 1985.
6 E. D. Goddard, J. V. Gruber (eds.), *Principles of Polymer Science and Technology in Personal Care*, Marcel Dekker, New York, 1999.

6
Cosmetic Emulsions Based on Surfactant Liquid Crystalline Phases: Structure, Rheology and Sensory Evaluation

Tharwat F. Tadros, Sandra Léonard, Cornelis Verboom, Vincent Wortel, Marie-Claire Taelman, and Frederico Roschzttardtz

Abstract

A brief description of the various liquid crystalline phases (hexagonal, cubic and lamellar) that are produced in concentrated surfactant solutions is given. The driving force for the formation of each type is given in terms of the critical packing parameter *P*. The most important type in emulsion systems is the lamellar phase that consists of several bilayers of surfactants that can "wrap" around the emulsion droplets, thus producing an energy barrier that prevents coalescence of the droplets. These lamellar phases can extend from the droplet surface to the bulk liquid, producing a "three-dimensional" gel network structure that prevents creaming or sedimentation. This type of liquid crystalline phase is referred to as "oleosomes". The lamellar liquid crystalline structure may simply extend in the aqueous phase and entrap the oil droplets, forming what is referred to as "hydrosomes". Using oscillatory and creep measurements, it is possible to discriminate between the oleosomes and the hydrosomes. An attempt was made to correlate the emulsion structure and rheology with sensory attributes.

6.1
Introduction

In dilute solutions, surfactants tend to form spherical micelles with aggregation number in the range 50–100 units [1, 2]. These micellar solutions are isotropic (L_1 phase) with low viscosity. At higher surfactant concentrations, the spherical micelles grow to form cylindrical micelles, which show flow birefringence (a single phase). At much higher surfactant concentrations, a series of mesophases, referred to as liquid crystalline phases [3], appear whose structure depends on the nature and concentration of the surfactant. At such high surfactant concentrations, the solution shows a dramatic increase in viscosity and birefringence with marked changes in self-assembly. These liquid crystalline structures can be

Colloids and Interface Science Series, Vol. 4
Colloids in Cosmetics and Personal Care. Edited by Tharwat F. Tadros

ISBN: 978-3-527-31464-5

identified using polarizing microscopy, X-ray diffraction, NMR spectroscopy and various rheological techniques [1–5].

Two main phase structures can be identified: (1) structure built of limited or discrete self-assemblies (spherical, prolate or cylindrical), and (2) infinite or unlimited self-assemblies whereby the aggregates are connected over macroscopic distances in one, two or three dimensions. The hexagonal phase (see below) is an example of one-dimensional continuity, the lamellar phase is an example of two-dimensional continuity and the bicontinuous cubic phase is an example of three-dimensional continuity.

In this paper, we will give a brief description of the structure of the above three liquid crystalline phases and the driving force for their formation. This is followed by a description of the application of lamellar liquid crystalline structures in emulsions. Two main types of systems can be produced, depending on the nature of the surfactants and oil. Two main systems could be produced, namely oleosomes and hydrosomes, and their structures are represented schematically. The role of liquid crystalline phases in emulsion stabilization is briefly described. An attempt to correlate the structure of oleosomes and hydrosomes with their rheological characteristics and sensory properties is given. In this study, a group of 34 formulations containing hydrosomes and 17 oleosome-based formulations were investigated using rheological analysis, and their sensory attributes were assessed using an expert panel.

6.2
Structure of Liquid Crystalline Phases

Liquid crystalline structures behave as fluids and are usually highly viscous. X-ray studies of these phases [3] show relatively sharp lines that resemble those produced by crystals and hence they are more ordered than ordinary liquids. Rheologically liquid crystalline phases show viscoelasticity (both viscous and elastic response) that is characteristic of liquid and solid behavior [4, 5]. Three main liquid crystalline phases can be identified and these are briefly described below.

- *Hexagonal phase (H_1)*: this phase is built of (infinitely) long cylinders in a hexagonal pattern, with each micelle surrounded by six other micelles (Figure 6.1a). The radius of the circular cross-section (which may be somewhat deformed) is close to the surfactant molecule length.
- *Micellar cubic phase (I_1)*: this phase is made of a regular packing of small micelles, which have properties similar to those of micelles in the solution phase. However, the micelles are short prolates (with axial ratio of 1–2) (Figure 6.1b) rather than spheres, since this allows better packing. The micellar cubic phase is highly viscous.

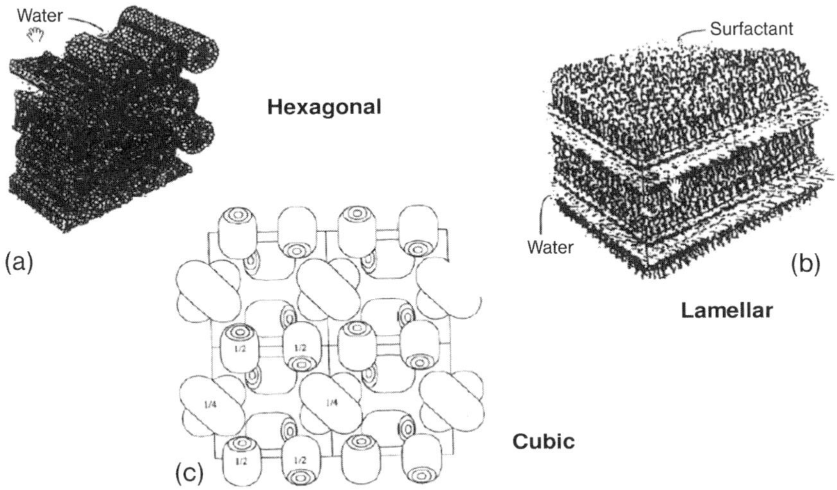

Figure 6.1 Schematic representation of the liquid crystalline phases: (a) hexagonal phase, (b) cubic phase, (c) lamellar phase.

- *Lamellar phase* (L_α): this phase is built of bilayers of surfactant molecules alternating with water layers (Figure 6.1c). The thickness of the bilayer is less than twice the surfactant molecular length. The thickness of the water layers can vary over wide ranges depending on the nature of the surfactant. The surfactant bilayer can range from being stiff and planar to very flexible and undulating.

The hexagonal and lamellar phases are anisotropic and show specific textures under a polarizing microscope. The hexagonal phase shows a "fan-like" texture, whereas the lamellar phase shows "oily streaks" with the appearance of some "Maltese crosses". The cubic phase is isotropic and hence it does not show any texture under a polarizing microscope.

6.3 Driving Force for the Formation of Liquid Crystalline Phases

One of the simplest methods of predicting the shape of an aggregated structure is based on the critical packing parameter (P) concept. Consider a spherical micelle with radius r and aggregation number n; the volume of the micelle is given by

$$\frac{4}{3}\pi r^3 = nv \quad (1)$$

where v is the volume of a surfactant molecule. The area of the micelle is given by

$$4\pi r^2 = na_0 \tag{2}$$

where a_0 is the area per surfactant head group. Combining Eqs. (1) and (2):

$$a_0 = \frac{3v}{r} \tag{3}$$

The cross-sectional area of the hydrocarbon chain a is given by the ratio of its volume to its extended length, l_c:

$$a = \frac{v}{l_c} \tag{4}$$

From Eqs. (3) and (4):

$$P = \frac{a}{a_0} = \frac{1}{3}\frac{r}{l_c} \tag{5}$$

Since $r < l_c$, then $P \leq 1/3$.

For a cylindrical micelle with length d and radius r:

$$\text{Volume of the micelle} = \pi r^2 d = nv \tag{6}$$

$$\text{Area of the micelle} = 2\pi rd = na_0 \tag{7}$$

Combining Eqs. (5) and (6):

$$a_0 = \frac{2v}{r} \tag{8}$$

$$a = \frac{v}{l_c} \tag{9}$$

$$P = \frac{a}{a_0} = \frac{1}{2}\frac{r}{l_c} \tag{10}$$

Since $r < l_c$, then $1/3 < P \leq 1/2$.

For vesicles (liposomes) $1 > P \geq 2/3$ and for lamellar micelles $P \approx 1$.

Using the above concept, one can predict the shape of the micelle. For example, for a nonionic surfactant $C_{12}E_6$ (where E represents an ethylene oxide unit) with a large head group area in dilute solution, the preferred shape will be a spherical micelle ($P \leq 1/3$). As the volume fraction of the surfactant is increased, repulsion between the micelles tends to space them out, forming a cubic array of micelles. With further increase in the volume fraction of the surfactant, the free energy of the system can be minimized by changing to a packing geometry of cylindrical units. The formation of the lamellar phase is the result of relieving the "strain" of increasing the volume fraction even further. This argument explains the sequence $L_1 \rightarrow$ hexagonal $\rightarrow$ lamellar for the $C_{12}E_6$–water system.

6.4 Formulation of Liquid Crystalline Phases

The formulation of liquid crystalline phases is based on the application of the above concepts. However, one must take into account the penetration of the oil between the hydrocarbon tails (which affects the volume and hence a of the chain) and also hydration of the head group, which affects a_o.

The most useful liquid crystalline phases are those of the lamellar structure, which can bend around the droplets, producing an energy barrier against coalescence and Ostwald ripening. As mentioned above, these lamellar liquid crystals can also extend in the bulk phase, forming a "gel" network that prevents creaming or sedimentation. These liquid crystalline phases also provide the optimum consistency for sensorial application. Due to the high water content of the liquid crystalline structure (water incorporated between several bilayers), it can also provide increased skin hydration.

The key to producing lamellar liquid crystals is to use mixtures of surfactants with different P values (different HLB numbers) whose composition can be adjusted to produce the right units.

Using the above concepts, we have developed two different types of liquid crystals in oil-in-water (O/W) emulsions: oleosomes and hydrosomes. These structures were obtained by using several surfactant mixtures whose concentration ratio and total concentration were carefully adjusted to produce the desired effect. These systems are described below.

6.4.1 Oleosomes

These are multilayers of lamellar liquid crystals surrounding the oil droplets that become randomly distributed as they progress into the continuous phase. The rest of the liquid crystals produce the "gel" phase that is viscoelastic. The oleosomes are produced using a mixture of Brij 72 (Steareth-2), Brij 721 (Steareth-21), a fatty alcohol and a minimum of a specific emollient. The nature of the emollient is crucial; it should be a medium to polar oil such as Arlamol E (PPG-15 stearyl ether) or Estol 3609 (triethylhexanoin). Very polar oils such as Prisorine 2034 (propylene glycol isostearate) and Prisonine 2040 (glyceryl isostearate) disturb the oleosome structure. Nonpolar oils such as paraffinic oils inhibit the formation of oleosomes.

The oleosomes are anisotropic and they can be identified using polarizing microscopy. Figure 6.2 shows a schematic diagram of the oleosomes.

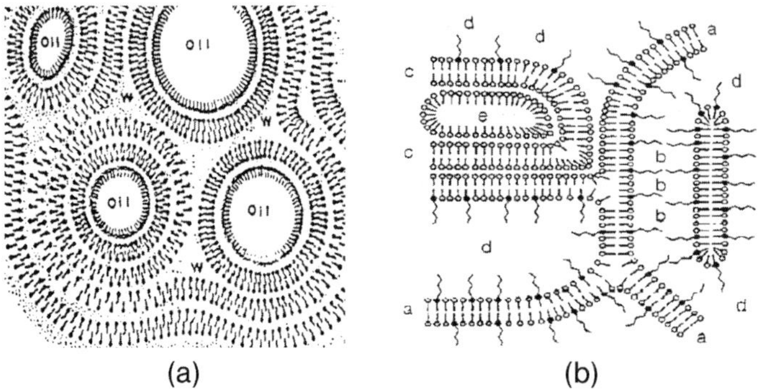

Figure 6.2 Schematic representation of (a) oleosomes and (b) hydrosomes.

6.4.2
Hydrosomes

In this case a "gel" network is produced in the aqueous phase by the lamellar liquid crystals. The surfactant mixture is dispersed in water at high temperature (80 °C) and this creates the lamellar phase, which becomes swollen with water between the bilayers. The oil is then emulsified and the droplets become entrapped in the "holes" of the "gel" network. The viscoelastic nature of the "gel" prevents close approach of the oil droplets. The hydrosomes can be obtained using Arlatone 2121 (sorbitan stearate and sucrose cocoate) or Arlatone LC (sorbitan stearate and sorbityl laurate). A schematic representation of hydrosomes is shown in Figure 6.2.

6.5
Emulsion Stabilization Using Lamellar Liquid Crystals

The lamellar liquid crystals produce several bilayers that "wrap" the droplets. This produces an energy barrier preventing coalescence. This is similar to the process of steric stabilization produced by polymeric surfactants [6]. As a result of the presence of these multilayers, the potential drop between two droplets is shifted to longer distances, thus preventing any coalescence [7]. For coalescence to occur, these multilayers have to be removed "two-by-two" and this produces an effective barrier against emulsion coalescence. The liquid crystalline structure can also prevent Ostwald ripening by providing a high elasticity at the O/W interface.

One of the most useful techniques to study liquid crystalline structures is dynamic (oscillatory) measurements. The storage modulus G' (the elastic component) and the loss modulus G'' (the viscous component) are measured as a function of strain amplitude at a constant frequency of 1 Hz.

With Arlatone LC (sorbitan stearate and sorbityl stearate) at 5%, G' and G'' remain constant up to a strain amplitude of 0.015 (long linear viscoelastic region). This is consistent with the formation of a coherent "gel" structure that is important for application and stabilization of the emulsion. In contrast, if the sorbityl laurate is removed from the system, i.e. using sorbitan stearate alone at the same concentration (5%), G' starts to decrease rapidly with increase in applied strain. In this case no liquid crystalline structure is produced and only reversed micelles (L_2 phase) are formed. With the latter system, emulsion stabilization is not possible.

6.6 Materials and Methods

The surfactants used for the preparation of the oleosome-based emulsions are ethoxylated stearyl alcohol in combination with an emollient (such as isohexadecane or PPG-15 stearyl ether). For the hydrosome-based emulsions a blend of sorbitan stearate and sucrose cocoate (or sorbityl laurate) was used. The oleosome emulsions were prepared by the direct emulsification technique with the emulsifiers dissolved in the oil phase. For hydrosome emulsions the gel network of the surfactant system was first prepared by heating and swelling the mixture in the water phase followed by addition of the oil while stirring.

The rheological measurements were carried out using a Physica USD 200 universal dynamic spectrometer (Paar Physica, Germany) and a cone–plate geometry device (50 mm radius, 2° angle). Two rheological tests were performed: constant stress (creep test) and dynamic measurements (frequency sweep test) (Figure 6.3).

In the creep measurements, a constant stress was applied on the system and the deformation (strain) γ was followed as a function of time for 2 min. The compliance J calculated is simply the strain divided by the applied stress for each

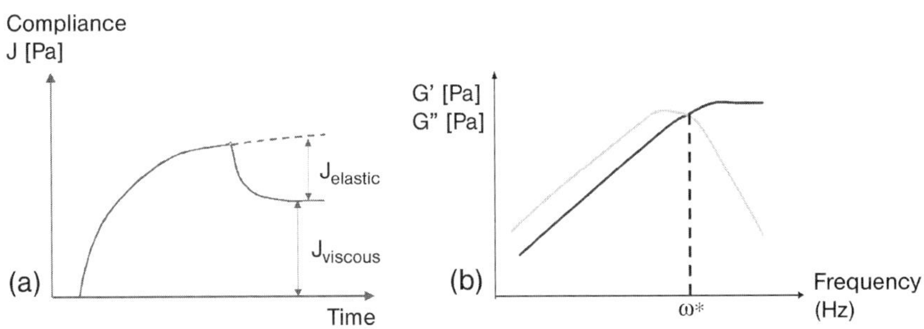

Figure 6.3 Illustration of (a) a typical creep curve and (b) a typical frequency sweep curve. Variation of G' (elastic componenet) and G'' (viscous componenet) with frequency. ω^* is the characteristic frequency at which $G' = G''$ (ω^* is the reciprocal of the relaxation time of the sample).

measured point. After this time, the stress was removed, keeping its value at zero and the strain was followed for another 2 min to obtain the recovery curve of the sample. The total compliance could be resolved into an elastic component J_e and a viscous component J_v.

In the frequency sweep measurements the strain is kept constant at a value in the linear viscoelastic region whereas the frequency is changed from 10 to 0.01 Hz. The frequency sweep gives information about the *gel strength* and time dependence of the structure present in the formulation. A large slope of the G' curve indicates low strength and a significant dependence of viscoelastic behavior on time; in contrast, small slope indicates a high-strength microstructure and its low dependence on time.

From these measurements, the following rheological parameters could be established:

- The critical stress, σ_{crit}, obtained from creep measurements. This value could be considered as the limit of the viscoelastic domain above which the viscosity of the system starts to decrease with further increase in the stress (flowing). This is designated crit_strs.
- The elastic modulus, G' (the storage component of the complex modulus), obtained from the frequency sweep at a frequency of 1 Hz and at a strain value in the linear region. This will be designated em_1Hz.
- The slope of the G' versus frequency (in the range 10^{-2}–1 Hz) obtained during the dynamic test. This will be designated slem_fsw.

All samples were evaluated using the Spectrum Descriptive Analysis™ method [8]. This sensory technique relies on obtaining accurate numbers by a well-trained sensory test panel. This panel consists of approximately 10–15 members and each panelist evaluates each product once using well-defined attributes with a fixed meaning. The 21 attributes can be subdivided into several groups: appearance, pick-up, rub-out, immediate after-feel and after-feel after 20 min. This study mainly focuses on cohesiveness (pick-up), wax (rub-out and after-feel), grease (after-feel) and integrity of shape (appearance), because these are obviously most related to the structure of the emulsions. These attributes can be defined as follows:

- *Cohesiveness*: This sensory attribute is evaluated during the pick-up phase and is evaluated by compressing the product slowly between index finger and thumb, after which the fingers are separated. The amount that the sample strings rather than breaks when fingers are separated is defined as cohesiveness. A stringy product has a high cohesiveness number.

- *Wax (rub-out and after-feel)*: This parameter represents the amount of wax perceived during rub-out and after-feel.

- *Grease (rub-out)*: This attributes evaluated the amount of grease perceived during rub-out.

- *Integrity of shape*: This parameter characterized one criterion of the appearance of the product and is evaluated by putting a nickel-sized portion on a Petri dish using a spiral motion (edges to center). The panel evaluates the degree of product which holds its shape.

A SIMCA (Soft Independent Modeling of Class Analogy) [9] was performed on hydrosome and oleosome samples. This is a statistical method based on construction of mathematical descriptions of clusters of data. This reduced the dimensionality of the data and increased the quality of the information.

6.7 Results and Discussion

6.7.1 Emulsion Structure and Rheology

Figure 6.4 shows the SIMCA results for oleosomes and hydrosomes based on rheological attributes. A discriminating power is plotted for three attributes, namely slem_fsw, em_1Hz and crit_strs. The discrimination power plot of the three rheological variables indicates the ability of that variable to discriminate hydrosome emulsions from oleosome samples. A discrimination power >3 indicates that the variables can be considered as important to distinguish the two emulsion structures. The plot in Figure 6.4, however, shows a discrimination

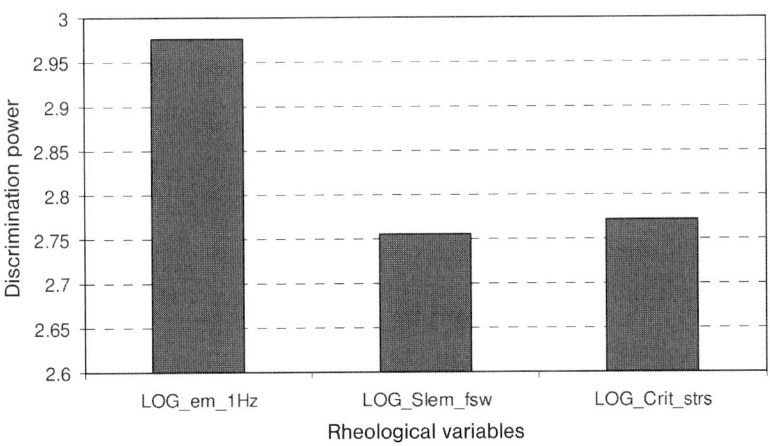

Figure 6.4 Discrimination power plot of three rheological variables showing the ability to discriminate oleosomes from hydrosomes based on emulsion structure. The higher the discrimination power, the greater is the ability of the variable to discriminate.

power between 2.75 and 3, indicating that these variables do not have a strong ability to discriminate the two structures from one another reasonably.

Figure 6.5 shows a plot of crit_strs versus slem_fsw for both hydrosome- and oleosome-based formulations. Comparing the slem_fsw, slope of G', of both emulsion structures, hydrosomes show a lower slope value than oleosomes at

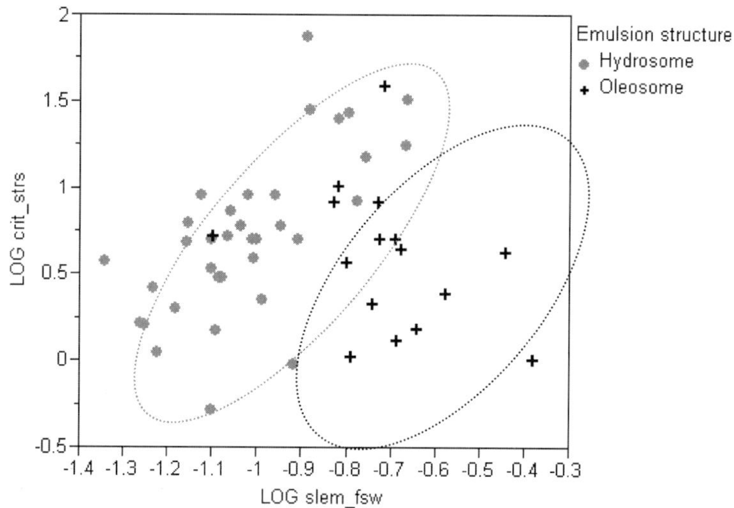

Figure 6.5 Scatter plot of logarithm-based crit_strs versus slem_fsw for hydrosomes and oleosomes.

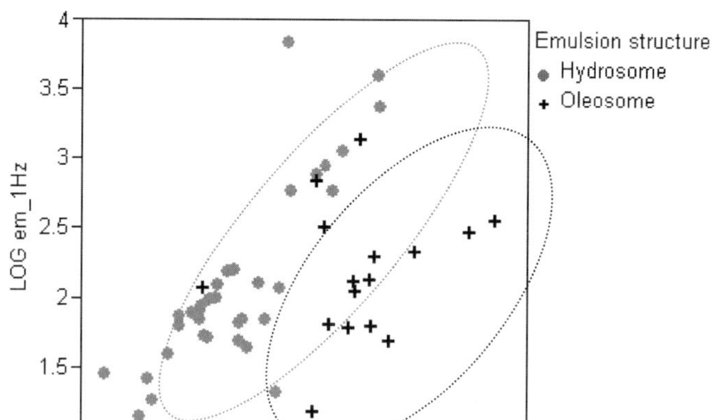

Figure 6.6 Scatter plot of logarithm-based em_1Hz versus slem_fsw for hydrosomes and oleosomes.

the same crit_strs. This trend is consistent with the three-dimensional gel network of the hydrosomes (with a higher number of contact points) compared with the multilayer structure of oleosomes. This difference in structure is also visible in terms of dynamic of restructuring, where hydrosomes present shorter relaxation times than oleosomes, thus giving a lower dependence on time for the hydrosome structures.

Another observation can be made considering both structures with similar slope of G', slem_fsw, where hydrosomes show higher values of storage modulus at 1 Hz, em_1Hz, than those obtained from oleosomes (Figure 6.6). This is also consistent with the more coherent gel network for the hydrosomes compared with the oleosomes.

6.7.2 Emulsion Structure and Sensory Attributes

Figure 6.7 shows the SIMCA results for oleosomes and hydrosomes based on sensory attributes. The discriminating power is plotted for five attributes, namely cohesiveness (COHES), waxy rub-out (WAXro), waxiness after-feel [WAX(%)], grease after-feel [GRS(%)] and integrity of shape (INToSHP).

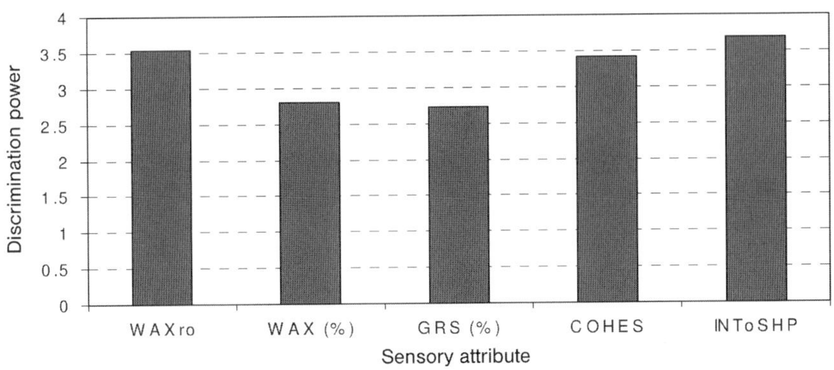

Figure 6.7 Discrimination power plot of five sensory attributes showing the ability to discriminate oleosomes from hydrosomes.

6.7.3 Emulsion Structure, Rheology and Sensory Attributes

The plot in Figure 6.8 represents the relation between rheology, sensory attributes and emulsion structure. It shows the correlation between cohesiveness (COHES) and slope elastic modulus (slem_fsw), although not strong but significant at a probability level of >99%. This plot confirms a class difference between the hydrosome and oleosome emulsion structures.

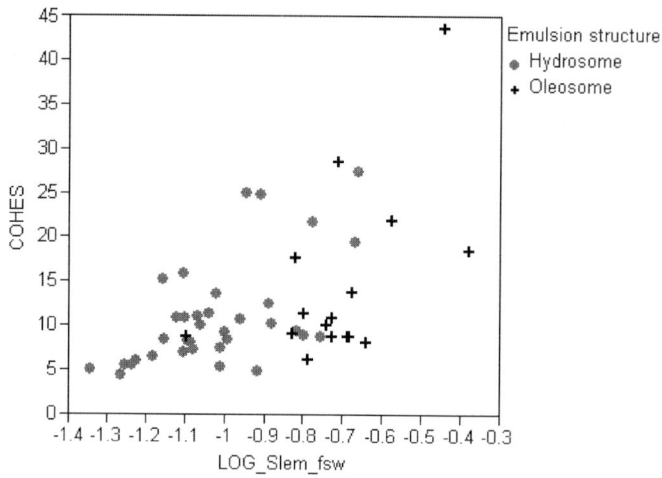

Figure 6.8 Scatter plot between cohesiveness and log slem_fsw for hydrosomes and oleosomes.

6.8
Conclusion

This study shows that in principle it is possible to distinguish hydrosome- from oleosome-based emulsions using basic rheological measurements and sensorial evaluations.

It may be concluded that in principle one can relate structure to rheology and sensory evaluation for the present systems of hydrosomes and oleosomes (Figure 6.9) [10].

In future work it will be required to establish the exact structures that are present in these complex emulsion systems, and this will require good measurements using freeze fracture and electron microscopy. It may also be possible to extend the rheological measurements to include results that are obtained under conditions whereby this structure is broken down, and the time scale required for the recovery.

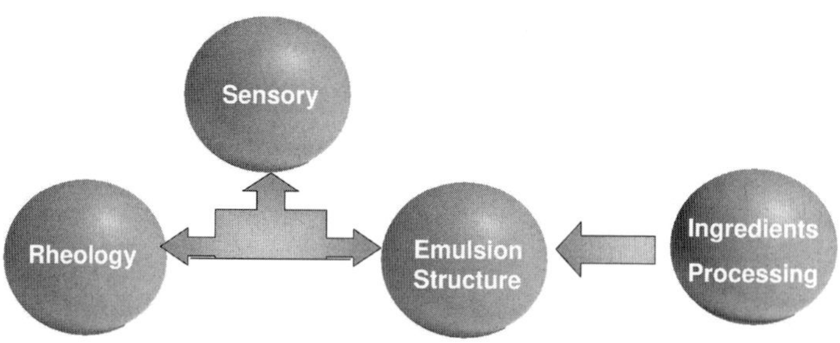

Figure 6.9 Model linking emulsion structure with sensory attributes and rheology [10].

References

1 K. Holmberg, B. Jonsson, B. Kronberg, B. Lindman, *Surfactants and Polymers in Aqueous Solution*, 2nd edn., Wiley, New York, 2002.
2 Th.F. Tadros, *Applied Surfactants: Principles and Applications*, Wiley-VCH, Weinheim, 2005.
3 R.G. Laughlin, *The Aqueous Phase Behavior of Surfactants*, Academic Press, London, 1994.
4 G.T. Dimitrova, Th.F. Tadros, P.F. Luckham, *Langmuir*, **11**, 1101–1111 (1995).
5 G.T. Dimitrova, Th.F. Tadros, P.F. Luckham, M. Kipps, *Langmuir*, **12**, 315–318 (1996).
6 Th.F. Tadros, in *Principles of Polymer Science and Technology in Cosmetics and Personal Care*, E.D. Goddard, J.V. Gruber (eds.), Marcel Dekker, New York, 1999, Chapter 3, pp. 73–112.
7 S. Frieberg, P.O. Jansson, E. Cederberg, *J. Colloid Interface Sci.*, **55**, 614 (1976).
8 M. Meilgaard, G.V. Civille, B.T. Carr, *Sensory Evaluation Techniques*, CRC Press, Boca Raton, FL, 1991.
9 B.G.M. Vandeginste, *Handbook of Chemometrics and Qualimetrics*, Elsevier, Amsterdam, 1998.
10 V. Wortel, C. Verboom, M.-C. Taelman, S. Léonard, J.W. Wiechers, Th.F. Tadros, *Linking sensory and rheology characteristics – a first step to understand the influence of emulsion structure on sensory characteristics*, presented at IFSCC, 2004.

7
Personal Care Emulsions Based on Surfactant–Biopolymer Mixtures: Correlation of Rheological Parameters with Sensory Attributes

Tharwat F. Tadros, Sandra Léonard, Cornelis Verboom, Vincent Wortel, Marie-Claire Taelman, and Frederico Roschzttardtz

Abstract

The stability of oil-in-water emulsions using a mixture of surfactants and biopolymers was been investigated using rheological techniques. The biopolymer was a mixture of Konjac mannan and xanthan gums (KX). Two types of emulsifiers were used, namely a mixture of alcohol ethoxylates or of sucrose esters. This produced two surfactant–biopolymer mixtures, namely Arlatone V100 and Arlatone V175. Constant stress measurements of the individual components of the biopolymer mixture and their combinations showed a much higher zero shear viscosity for the mixture when compared with that of the individual components and this clearly showed the synergetic effect of Konjac and xanthan gums. This is attributed to the interaction between the two polysaccharide molecules. Rheological measurements for the surfactant–biopolymer mixtures, Arlatone V100 and Arlatone V175, showed a reduction in the zero shear viscosity when compared with KX alone. This indicates that the surfactants reduce the interaction between the two polysaccharides. Rheological investigations were carried out for emulsions prepared using Arlatone V100 or Arlatone V175 at various intervals of time and the results showed high stability of these emulsions both at room temperature and higher temperatures. These emulsions showed no separation (no creaming) as a result of the presence of a "gel" network in the continuous phase. However, the emulsions are shear thinning and their viscosity reached low values at high shear rates. Several emulsions were prepared and their sensory attributes were determined using expert panels. The results obtained were assessed using statistical analysis. The sensory attributes of several emulsions based on these "surfactant–biopolymer" mixtures were compared with those obtained using classical surfactants and hydrocolloids (such as carbomer). Generally, the emulsions based on the emulsifier–biopolymer mixtures showed higher spreadability, higher wetness, lower firmness, lower greasiness, lower thickness and lower integrity of shape compared with the other emulsions. This was mainly due to the lower viscosity at high shear rate and the lower (but coherent) "gel" structure,

Colloids and Interface Science Series, Vol. 4
Colloids in Cosmetics and Personal Care. Edited by Tharwat F. Tadros

ISBN: 978-3-527-31464-5

which can be easily broken under shear. With the emulsions containing thickeners such as carbomer, a higher cohesive energy density is obtained and the gel structure cannot be easily broken under shear. These results clearly indicate the advantage of using the surfactant–biopolymer mixtures for the formulation of personal care emulsions.

7.1 Introduction

Oil-in-water (O/W) emulsions that are commonly used in many personal care formulations are usually formulated using nonionic surfactants of the alcohol ethoxylate or sucrose ester types. However, these systems may suffer from some instability problems [1, 2]: (1) creaming or sedimentation since the droplet size distribution (usually in the range 1–5 μm) is outside the range where Brownian diffusion overcomes the gravity; (2) flocculation arising from the lack of sufficient repulsion to overcome the van der Waals attraction; this is particularly the case when the poly(ethylene oxide) chain is not sufficiently large and/or in poor solvent conditions; (3) Ostwald ripening that arises from the difference in solubility between small and large droplets; on storage, particularly at high temperature, the droplet size distribution shifts to larger sizes and this accelerates creaming or sedimentation and flocculation; and (4) coalescence that arises from the thinning and disruption of the liquid film between the droplets; this process may arise from the lack of sufficient interfacial elasticity and low viscosity of the liquid film.

To overcome the above instability problems, we have recently developed an emulsifier–hydrocolloid system using a mixture of a nonionic surfactant and two hydrocolloids, namely Konjac mannan (K) and xanthan (X) gums. K is a β-1,4-linked glucomannan with branches consisting of about 16 sugar units linked t °C-3 of the glucose and mannose at approximately every 10 residues along the chain [3]. Native K is acetylated and it does not gel in water. However, on deacetylation in the presence of alkali, a thermally reversible gel is produced [4]. xanthan gum is a charged polysaccharide consisting of a β-1,4-linked glucopyranose backbone with a trisaccharide side-chain linked to every second glucose residue. The side-chain consists of two mannose units separated by a glucuronic acid residue. xanthan gum does not gel at any concentration but it undergoes a temperature-induced conformational transition from an ordered helical structure (where the side-chains are folded in and associated with the backbone) to a disordered structure (where the side-chains project away from the backbone). The transition temperature depends on the ionic strength and nature of the electrolyte and also the pH [5]. Mixtures of Konjac mannan and xanthan (KX) gums form thermally reversible gels which most workers agree are due to molecular association [6, 7]. Dea et al. [8] proposed that association takes place between the ordered xanthan helix and unsubstituted regions of the galactomannan backbone. The interaction between Konjac mannan and xanthan gums has been investigated by Annable et al. [9] using differential scanning calorimetry (DSC), electron spin resonance (ESR) and viscoelastic measurements.

In this chapter, we will demonstrate the synergy between the two hydrocolloids which results in the formation of a robust "gel" structure that can prevent any creaming or sedimentation. In addition, the possible adsorption of the gums at the O/W interface will enhance stabilization against coalescence. The preparation of the O/W emulsion using a special procedure will be described. These emulsions were assessed using microscopic and rheological measurements. Using creep and oscillatory measurements, a comparison will be made between the present hydrocolloid system and those used in other personal care formulations, e.g. hydrophobically modified polyacrylates, cross-linked polyacrylates and xanthan gum alone. Several formulations have been prepared using the above system and several emollients. Also, more complex emulsions containing emollient blends, waxes and pigments were also prepared. All these emulsions were investigated using rheological analysis and their sensory attributes were assessed using an expert panel. The O/W emulsions used in cosmetic formulations need to have the required sensory profile for application. In this chapter, an attempt is made to correlate some of the sensory attributes with some of the rheological parameters.

7.2 Materials and Methods

7.2.1 Materials

Several oils have been used for preparation of the emulsions. Initially a four-component oil mix was used to evaluate the stability of the emulsion system. This oil mix consisted of Arlamol HD (isohexadecane) 55.5% w/w, Estol 3603 (caprylic/capric triglyceride) 22.3% w/w (both supplied by Croda, Wilton, UK), Avocado oil (*Persea gratissama*) 11.1%, supplied by Mosselman, Belgium) and Florasun (*Helianthus annus*) 90 11.1% (supplied by Florateck, USA). The thickener systems consists of a mixture of Konjac (mannan) and xanthan gums, referred to as KX. Two emulsifier–biopolymer systems were used: mixture A consisted of Steareth-100, Steareth-2, glyceryl stearate citrate, sucrose, mannan and xanthan gum and mixture B consisted of sucrose palmitate, glyceryl stearate, glyceryl stearate citrate, sucrose, mannan and xanthan gum.

Keltrol F (Kelco, USA), Carbopol 2001 (carbomer) and Pemulin TR2 (Acrylat/C_{10-30}-alkyl acrylate cross-polymer) (Noveon, USA) were used as received. A preservative, Nipaguard BPX (Clarient, Germany) was used in all solutions.

7.2.2 Preparation of Powder Dispersions

Dispersions of KX, Arlatone V100 or Arlatone V175 were prepared using several procedures. In the first method (A), the powder was added slowly to water at room temperature, while stirring at 600 rpm (using an RW 20 IK-A-Werk Janke and Kunkel) until all the powder was dispersed. One should make sure that a vortex is produced in the liquid for incorporating the powder in order to prevent any air entrapment. The preservative was then added while stirring and homogenization at 9500 rpm (using an Ultraturrax) was carried out to produce the gel; stirring was then continued for at least 5 min at 600 rpm until a smooth dispersion was obtained. In the second method (B), the powder of KX, Arlatone V100 or Arlatone V175 was added part by part while stirring at 600 rpm until complete dispersion of the powder occurred and then the temperature of the dispersion was increased to 80 °C while stirring for 15–20 min and then the dispersion was left to cool to room temperature while stirring at 600 rpm. Alternatively, the same procedure as B was followed except that the dispersion was homogenized for 2 min at 80 °C (method C). The fourth procedure (D) consisted of powder dispersion in water at 80 °C with homogenization. The best procedure for powder dispersion methods A or C as shown by the rheological results (see below). Dispersion of the powder at high temperature was undesirable since "lumps" were formed and these were difficult to redisperse. If the dispersion is carried out at room temperature, it is preferable to disperse the powder in glycerol (one part Araltone and three parts glycerol), followed by homogenization according to methods A and C.

The dispersion of all other powders, namely Keltrol F, Carbopol 2001 and Permulen TR 2, was carried out at room temperature and at 600 rpm until all the powder was completely dispersed.

7.2.3
Preparation of the Emulsion

The powder of Arlatone V100 or V175 was dispersed in water at 20 °C. Oil was then added while stirring at 600 rpm followed by homogenization for 2 min at 9500 rpm. Alternatively, the powder was dispersed in water at 60 °C and the oil that was kept at 20 °C was added followed by homogenization as described before (the homogenization temperature in this case was 50 °C). Alternatively the temperature of the aqueous phase was increased to 80 °C, while the oil was still kept at 20 °C and the homogenization temperature reached 60 °C. Another procedure was adopted whereby the temperature of both the aqueous and oil phases was increased to 80 °C. The best procedure was found to be that with homogenization at 60 °C. In this case, the powder dispersion that was carried out at 20 °C followed by heating to 80 °C ensured the gel formation on cooling. Emulsions with an oil volume fraction of 0.2 could be prepared using 0.5% Arlatone V100 or V175. Alternatively, an emulsion with an oil volume fraction of 0.6 could be prepared using 2% emulsifier. Recent experiments showed that 1% emulsifier was sufficient to make an emulsion with volume fraction of 0.6. This emulsion was then diluted with water to give an oil volume fraction of 0.2 while keeping the emulsifier concentration at 0.5%. Both emulsions were of the same quality as assessed by droplet size analysis.

For comparison, other emulsions were also prepared using classical surfactant mixtures with a classical thickener (called in this chapter "classical surfactant–thickener" systems): emulsifier 1 = sorbitan ester and carbomer blend; emulsifier 2 = glyceryl stearate–soap and stearic acid blend. For the "classical surfactant–thickener" systems, the emulsifiers–oil blend was added to the heated aqueous phase while stirring and followed by the homogenization step.

7.2.4
Rheological Measurements

A Physica USD 200 universal dynamic spectrometer (Paar Physica, Germany) was used for the rheological measurements. Three types of measurements were carried out, namely flow–viscosity curves, constant stress measurements (creep tests) and the dynamic (oscillatory) technique [1].

A plate–plate geometry was used with a 1-mm gap. The temperature was controlled using a Peltier plate and a solvent trap was used to prevent evaporation. In the creep measurements, a constant stress was applied on the system and the deformation (strain) γ was followed as a function of time for 2 min. The compliance J is simply the strain divided by the applied stress. After this time, the stress was removed and the strain (which reversed sign) was followed for another 2 min to obtain the recovery of the sample. The total compliance could be resolved into an elastic component J_e and a viscous component J_v.

In the oscillatory technique, two types of measurements were carried out: (1) amplitude sweep at a constant frequency of 1 Hz, which allows one to obtain the linear viscoelastic region where the moduli are independent of the applied strain; and (2) frequency sweep, whereby the strain is kept constant at a value in the linear viscoelastic region whereas the frequency is changed from 10 to 0.01 Hz.

For a viscoelastic system, the stress and strain amplitudes are shifted by a time Δt and this allows one to obtain the phase angle shift δ ($\delta = \Delta t\omega$, where ω is the frequency in rad s^{-1}). From the amplitudes of stress and strain τ_0 and γ_0 and δ, one can obtain the complex modulus G^* the storage modulus G' (the elastic component) and the loss modulus G'' (the viscous component),

$$|G^*| = \frac{\tau_0}{\gamma_0} \tag{1}$$

$$G' = |G^*| \cos \delta \tag{2}$$

$$G'' = |G^*| \sin \delta \tag{3}$$

$$\tan \delta = \frac{G''}{G'} \tag{4}$$

G' is a measure of energy stored elastically during a cycle of oscillation, whereas G'' is a measure of the energy dissipated as viscous flow during oscillation. The ratio of G'' to G' is $\tan \delta$.

7.2.5 Principal Component Analysis (PCA)

The basic principle of PCA is to reduce the number of dimensions by identifying linear relationships between the variables. This technique involves a mathematical procedure that transforms a number of (possibly) correlated variables into a (smaller) number of uncorrelated variables called *principal components*. The first principal component accounts for as much of the variability in the data as possible and each succeeding component accounts for as much of the remaining variability as possible. A loading plot provides information about the variables and how these are related. A score plot reveals how the formulations are arranged in principal space, revealing which are similar or dissimilar towards other formulations. Projecting the loadings in the score plot provides information on which (group of) formulations have comparable variable scores and which do not [10].

Hierarchical cluster analysis [11] is a statistical tool that is used to classify formulations, characterized by the values of a set of variables (sensory attributes), into groups. It can be seen as an alternative to PCA for describing the structure of a data table. A scatter plot is used to summarize the results of hierarchical cluster analysis; the identified rheological variable or sensory attributes of formulations are plotted against different classes of emulsions systems.

All samples were evaluated using the spectrum descriptive analysis method [4]. This sensory technique relies on obtaining accurate numbers by a well-trained sensory test panel. This panel consists of approximately 10 members and each panelist evaluates each product three times using well-defined attributes with a fixed meaning. The 21 attributes can be subdivided into several groups: appearance, pick-up, rub-out, immediate after-feel and after-feel after 20 min.

7.3 Results

7.3.1 Rheological Results for Xanthan Gum and KX Solutions

Table 7.1 shows the variation of G', G'', low shear rate viscosity and critical stress (obtained from creep measurements) as a function of xanthan gum (Keltrol) concentration, and Table 7.2 show the results for KX. With xanthan gum, there is a gradual increase in all rheological parameters with increase in polymer concentration (in the range 0.5–1% studied). With KX, the same trend is observed within the concentration range 0.05–0.1%, which is clearly an order of magnitude lower concentration than that used for xanthan gum alone. It is also clear from the results in Tables 7.1 and 7.2 that the low shear rate viscosity values for KX are more than one order of magnitude higher than the values for xanthan gum alone, even though the latter solution was an order of magnitude higher in concentration than the corresponding KX solutions. G' for 0.1% KX is compara-

Table 7.1 Summary of the rheological results for xanthan gum (Keltrol) at 25 °C.

Xanthan gum (%)	G' (Pa)	G'' (Pa)	Low shear viscosity (Pa s) Shear rate $\sim 7 \times 10^{-3}\ s^{-1}$	Critical stress (Pa)
0.5	5	3	61	0.5
0.6	7	3.5	111	0.8
0.7	9	4	140	0.9
0.8	9	4	244	1.0
0.9	10	5	239	1.5
1.0	14	6	228	1.9

Table 7.2 Summary of the rheological results for KX at 25 °C.

KX (%)	G' (Pa)	G'' (Pa)	Low shear viscosity (Pa s) Shear rate $\sim 6 \times 10^{-4}\ s^{-1}$	Critical stress (Pa)
0.05	2	1.5	3874	2.4
0.06	3	1.5	5540	2.5
0.07	4	2	5950	3.1
0.08	7	2	4250	4.2
0.09	8	2	9723	8.5
0.10	10	2	6907	5.9

ble to the value obtained for 0.9% xanthan gum, indicating the higher elasticity of the KX solutions. In addition, the KX solutions also have a relatively higher elastic to viscous components (G/G'') when compared with the solutions of xanthan gum alone.

The above constant stress (creep measurements) using solutions of KX and X alone showed a clear synergy between K and X. The residual viscosity of KX solutions (in the range 0.05–0.1%) is several orders of magnitude higher than that of X solutions (in the range 0.5–1%). In addition, the elastic modulus, G', of 0.1% KX solution is close to that of 0.9% X solution. This synergy is due to the interaction between the ordered xanthan chains and sequences along the mannan backbone where the galactose residues are positioned on one side [5–8].

7.3.2 Rheological Investigation of Stabilizing Systems

Creep curves for 1.0% Arlatone V175, Arlatone V100 and KX solutions prepared using the cold or cold/hot procedure showed the typical behavior of a viscoelastic system. In all cases, the strain increases very slowly with time when the stress is below a certain critical value and when the stress is removed appreciable recovery of the strain is obtained. Above this critical stress, the strain shows a rapid increase with time and when the stress is removed only partial recovery of the strain is obtained. The results show that the stabilizer system prepared using the cold procedure gives higher critical stress than that obtained using the cold/hot procedure. The viscosity at low shear rate values can be obtained from the slope of the line before the critical stress is reached. The slope of the curve of strain versus time gives the shear rate and the viscosity is simply equal to the stress applied divided by this shear rate. The zero shear viscosity for the stabilizer system is higher for the cold procedure than for the cold/hot procedure. For example, at 1% Arlatone V175, the zero shear viscosities are 1362 and 855 Pa s for the cold and cold/hot procedures, respectively.

Frequency sweep results for 0.5% Arlatone V175 also showed that the storage modulus G' is much higher than the loss modulus G'' in the frequency range 10^{-2}–1 Hz. Within that frequency range, $\tan\delta$ is <1. This is typical of a system that is more elastic than viscous.

The above rheological investigations can be applied to test the method of preparation of the dispersions. Variation of the low shear viscosity versus percentage Arlatone V100 obtained by two procedures, namely dispersing the powder in cold water and dispersing the powder in cold water followed by heating the solution, showed that the second procedure gives much higher viscosities than the values obtained when the stabilizer is dispersed in cold water. Similar results were obtained for KX. In contrast, the results for Arlatone V175 showed lower zero shear viscosity when the sample was prepared cold/hot compared with the results for a sample prepared using the cold procedure. It should be mentioned that the viscosity of the stabilizers Arlatone V100 and Arlatone V175 was lower than that of the thickener KX. This implied that addition of surfactant to the thickener reduces the synergy between xanthan and Konjac gums.

The results for Pemulin TR2 showed much lower viscosities at low shear rates (<500 Pa s) at higher concentrations (>0.25%) than the Arlatone stabilizers and comparable viscosities could only be reached at much higher Pemulin TR2 concentrations (1%). With xanthan gum, the viscosities were much lower (61 Pa s at 0.5%) compared with the values obtained when using KX.

With Carbopol that has been neutralized, high viscosities at low shear rates were produced at low Carbopol concentrations (>50 000 Pa s at 0.2%). This high viscosity is due to the formation of "microgel" particles when the Carbopol is neutralized. Dissociation of the COOH groups results in the formation of charged "gel" particles, which swell as a result of the extended double layers produced.

7.3.3 Rheological Investigations of Emulsions

7.3.3.1 Influence of Arlatone Concentration

The results for creep measurements for an O/W emulsion (using the four-component mix) with a volume fraction φ of 0.2 as a function of Arlatone con-

Table 7.3 Summary of the creep measurements for emulsions ($\varphi = 0.2$) stabilized with Arlatone V100 as a function of stabilizer concentration and storage time.

Arlatone V100 (%)	Low shear viscosity (Pa s) Shear rate $\sim 10^{-3}$ s^{-1}			Critical stress (Pa)		
	1 day	1 month	2 months	1 day	1 month	2 months
0.5	907	806	830	1.1	0.9	1.1
0.6	1347	863	853	1.3	0.9	0.9
0.7	1675	1460	613	1.7	2.0	1.1
0.8	1500	1413	1314	2.0	2.2	1.6
0.9	1855	1265	1243	2.5	1.9	1.75
1.0	2494	1415	1947	2.5	2.25	2.25

centration (0.5–1%) are summarized in Tables 7.3 and 7.4. As mentioned before, the low shear viscosity was obtained from the slope of the strain versus time creep curve before the critical stress value. The critical stress was taken to be the value above which the strain shows a rapid increase with time.

The results in Tables 7.3 and 7.4 show in general an increase in the low shear viscosity and critical stress with increase in stabilizer concentration. However, in some cases such a trend is not always followed and the results sometimes show an increases followed by a decrease as the stabilizer concentration is increased. The storage results did not always show a regular trend, although in some cases, particularly with Arlatone V175 at stabilizer concentrations above 0.8%, there was

Table 7.4 Summary of the creep measurements for emulsions ($\varphi = 0.2$) stabilized with Araltone V175 as a function of stabilizer concentration and storage time.

Arlatone V175 (%)	Low shear viscosity (Pa s) Shear rate $\sim 10^{-3}$ s^{-1}			Critical stress (Pa)		
	1 day	1 month	2 months	1 day	1 month	2 months
0.5	1899	609	167	1.5	0.8	0.3
0.6	1272	614	1110	1.2	0.8	1.2
0.7	1333	1076	1487	1.7	1.2	1.8
0.8	2550	2447	2296	1.6	2.3	2.3
0.9	1584	1640	2376	2.4	2.4	3.0
1.0	1168	1016	2547	2.0	1.25	2.0

Table 7.5 Summary of the results of oscillatory measurements for an emulsion ($\varphi = 0.2$) stabilized with Arlatone V100 as a function of stabilizer concentration and storage time.

Arlatone V100 (%)	G' (Pa)			G'' (Pa)		
	1 day	1 month	2 months	1 day	1 month	2 months
0.5	3.8	5.6	6.0	1.4	1.5	1.5
0.6	5.0	7.0	7.0	1.5	1.8	1.8
0.7	7.0	10.0	8.5	1.8	2.0	2.0
0.8	8.0	9.0	10.0	2.0	2.0	2.0
0.9	10.5	10.5	10.5	2.5	2.5	3.0
1.0	10.5	10.5	10.5	2.5	2.5	2.5

an increase in the rheological parameters with increase in storage time, which may indicate some flocculation of the samples on storage.

A summary of the results for the oscillatory measurements is given in Tables 7.5 and 7.6. In all cases the systems were more elastic than viscous, i.e. $G' \gg G''$, and the tables show the variation of G' and G'' with Arlatone concentration and storage time.

The critical strain, the value above which G' and G'' show variation with strain amplitude was also measured as a function of stabilizer concentration and storage time. In most cases the critical strain was in the region of 1–2% and it did not show much change with stabilizer concentration and/or storage time. The values of the critical strain are reasonably high and they indicate a coherent

Table 7.6 Summary of the results of the oscillatory measurements for an emulsion ($\varphi = 0.2$) stabilized with Arlatone V175 as a function of stabilizer concentration and storage time.

Arlatone V175 (%)	G' (Pa)			G'' (Pa)		
	1 day	1 month	2 months	1 day	1 month	2 months
0.5	4.0	4.0	4.0	1.0	1.5	1.5
0.6	6.2	10.0	9.0	1.8	2.8	1.8
0.7	8.0	10.0	8.9	2.0	2.5	2.0
0.8	8.0	11.0	20.0	2.0	2.8	2.5
0.9	10.5	20.0	20.0	3.0	4.0	3.0
1.0	10.0	15.0	20.0	2.5	3.0	3.0

viscoelastic structure of the emulsion that is stabilized with Arlatone V100 or Arlatone V175.

The results of oscillatory measurements show a regular trend of an increase in G' and G'' with increase in stabilizer concentration. On storage, these moduli values did not show significant changes up to 2 months. The only exception is the results for Arlatone V175, which showed an increase on storage, indicating some flocculation of the emulsion.

7.3.3.2 Influence of Oil Volume Fraction

Emulsions with oil volume fractions between 0.2 and 0.6 were prepared by keeping the surfactant to oil ratio constant (3.6% based on the oil phase). Table 7.7 gives a summary of the creep results for emulsions stabilized using Arlatone V100 and Table 7.8 those for Arlatone V175.

Table 7.7 Summary of the creep measurements for emulsions stabilized with Arlatone V100 as a function of oil volume fraction φ and storage time.

φ	Low shear viscosity (Pa s) Shear rate 10^{-4}–10^{-3} s^{-1}			Critical stress (Pa)		
	1 day	1 month	2 months	1 day	1 month	2 months
0.2	2000	3500	3404	1.5	2.1	2.25
0.3	1642	2582	1170	1.5	1.5	1.25
0.4	2020	2336	6008	1.5	1.25	1.25
0.5	2266	3500	4544	3.0	3.5	1.5
0.6	17997	54428	22356	3.0	2.0	4.0

Table 7.8 Summary of the creeps measurements for emulsions stabilized with Arlatone V175 as a function of oil volume fraction φ and storage times.

φ	Low shear viscosity (Pa s) Shear rate 10^{-4}–10^{-3} s^{-1}			Critical stress (Pa)		
	1 day	1 month	2 months	1 day	1 month	2 months
0.2	1670	1025	1060	1.8	1.4	1.0
0.3	909	954	909	1.0	1.1	1.1
0.4	1965	2230	7871	1.0	1.3	3.5
0.5	4586	13366	11675	4.5	4.5	5.5
0.6	29302	33101	15089	8.0	7.0	6.0

The results show a small change in low shear viscosity and critical stress when the oil volume fraction is increased from 0.2 to 0.4. However, when φ exceeds 0.4 there is a rapid increase in low shear viscosity and critical stress with further increase in the oil volume fraction. The increase at high oil volume fractions is indicative of flocculation of the emulsion.

The results for the oscillatory measurements are summarized in Tables 7.9 and 7.10, which show the variation of G' (measured in the linear viscoelastic region and at a frequency of 1 Hz) and critical strain with storage time at various oil volume fractions.

The results of oscillatory measurements show a systematic increase in the value of G' with increase in the oil volume fraction. This trend is expected since the value of G' depends on the number of contact points and their strength of interaction, both of which increase with increase in the oil volume fraction.

The results also show a significant increase in the value of G' with increase in storage time when the oil volume fraction exceeds 0.4. This indicates flocculation

Table 7.9 Summary of the results of oscillatory measurements for emulsions stabilized with Arlatone V100 as a function of oil volume fraction φ and storage times.

φ	G' (Pa)			Critical strain (%)		
	1 day	1 month	2 months	1 day	1 month	2 months
0.2	8	9	9	1	1	1
0.3	10.5	16	20	1	1	1
0.4	20	35	45	0.5	0.5	0.4
0.5	45	60	75	1	0.5	0.5
0.6	120	130	150	0.6	1.0	0.5

Table 7.10 Summary of the results of oscillatory measurements for emulsions stabilized with Arlatone V175 as a function of oil volume fraction φ and storage times.

φ	G' (Pa)			Critical strain (%)		
	1 day	1 month	2 months	1 day	1 month	2 months
0.2	8	8	8	1	1	4
0.3	15	15	16	1	1	0.4
0.4	25	26	40	1	1	0.5
0.5	40	75	90	1	0.3	1.5
0.6	110	118	220	2	1	3

of the emulsion on storage at such high oil volume fraction. The critical strain did not show a regular change with either oil volume fraction or storage time. However, accurate determination of the critical strain is difficult since the exact location of γ_{cr} (the point at which G' shows a reduction with increase in applied strain) is difficult to estimate. In general, the oscillatory measurements are consistent with the results obtained using creep measurements (Tables 7.5 and 7.6) and they demonstrate clearly the effect of increasing the oil volume fraction on the emulsion rheology and its long-term stability.

7.3.3.3 Influence of Temperature on the Rheology of KX, Arlatone V100, Arlatone V175 and the Emulsions Prepared Using the Stabilizers

The results for the variation of storage modulus (G') and viscous modulus (G'') with temperature for an aqueous solution containing 0.1% KX showed a shallow maximum at 35 °C, after which it decreased with increase in temperature in the measurement range 35–50 °C. However, G'' seemed to decrease when the temperature of the thickener was increased above 30 °C.

With 1% Arlatone V100, G' remained virtually constant up to 35 °C, after which it showed a gradual reduction with further increase in temperature. The same trend was also obtained for G''. Arlatone V175 (1%) seemed to show a rapid reduction in G' and G'' when the temperature was increased from 25 to 30 °C, after which no change in G' was obtained in the temperature range 30–50 °C. In contrast, G'' showed a gradual decrease with increase in temperature in this range.

The results for the emulsions ($\varphi = 0.2$) based on 1% Arlatone V100 and 1% Arlatone V175 showed a remarkable independence of G' on temperature within the range 25–50 °C for both emulsions. G'' showed a small reduction when the temperature was increased above 30 °C.

7.3.4 PCA Results

Figure 7.1 shows a PCA scores loading plot for the sensory attributes for formulations based on emulsifier–biopolymer systems combined with single emollients (■), classical surfactant–thickener systems (▲) and emulsifier–biopolymer systems associated with emollient–wax mixtures (●).

It can be seen that most formulations based on emulsifier–biopolymer mixture are clustered in one area of the score plot. The only exception is the sample F464 (●) that contains cetearyl alcohol, which is now placed on the right part of the score plot.

All other formulations are spread across the whole score plot. In general, the formulations based on the emulsifier–biopolymer mixtures show high spreadabililty, high wetness, low firmness, low grease rub-out, low thickness and low integrity of shape compared with the other formulations.

For comparison, Figure 7.2 shows a PCA scores loading plot for the rheological parameters. In this case the formulations based on the emulsifier–biopolymer mixtures (■) show a more distinct behavior compared with classical surfactant–thickener formulations (▲).

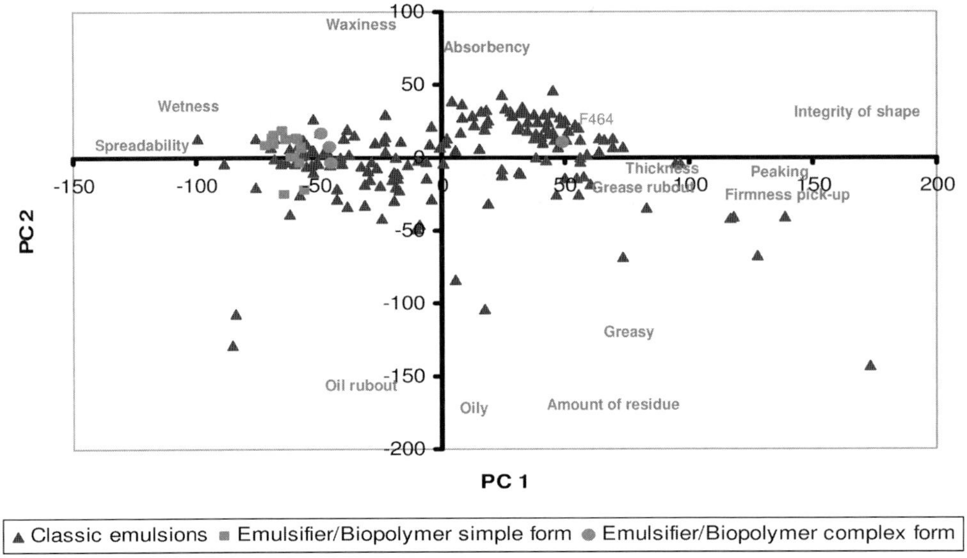

Figure 7.1 PCA score and loading plot for the sensory attributes.

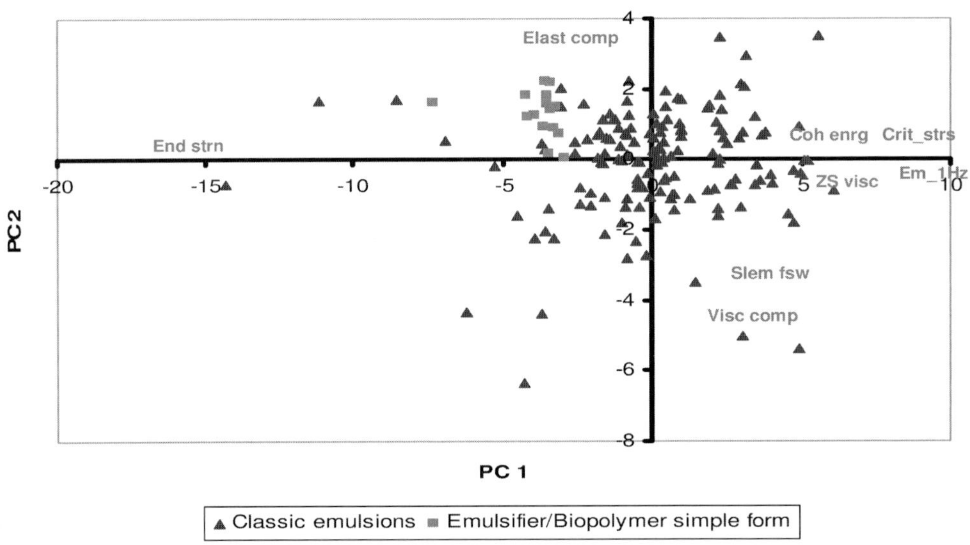

Figure 7.2 PCA score and loading plot for the rheological parameters.

In general, the formulations based on the emulsifier–biopolymer mixtures show a lower residual viscosity (ZS_vsc), a lower elastic modulus at a frequency of 1 Hz (Em_1Hz) and a higher end strain (End_strn) compared with the classical formulations. This explains the high spreadability of the emulsifier–biopolymer formulations (which have lower residual viscosity) and the relatively weak but coherent "gel" structure (with a low modulus, higher end strain and the low dependence of the modulus on frequency). This does not imply the system will have a higher cohesive energy (Coh_enrg) since the bond strengths are not very high; in other words; this gel structure can be easily broken down under shear. With the samples based on classical emulsifiers and classical thickeners such as carbomer, they give a higher modulus but a lower end strain. In this case the cohesive energy density of the structure is higher and the gel structure can not be easily broken under shear and this result in lower spreadability compared with the formulations based on the emulsifier–biopolymer mixtures.

Although the PCA score plots as depicted in Figures 7.1 and 7.2 have shown that emulsifier–biopolymer-based formulations can be grouped, most of the classical surfactant–thickener formulations have wide varieties of emollients in comparison with emulsifier–biopolymer-based formulations. As a consequence, the comparison between classical surfactant–thickener formulations is less well balanced towards emulsifier–biopolymer-based formulations. To make the comparison better balanced, we therefore compare emulsifier–biopolymer with emulsifier 1- and emulsifier 2-based formulations, all from the same set of emollients.

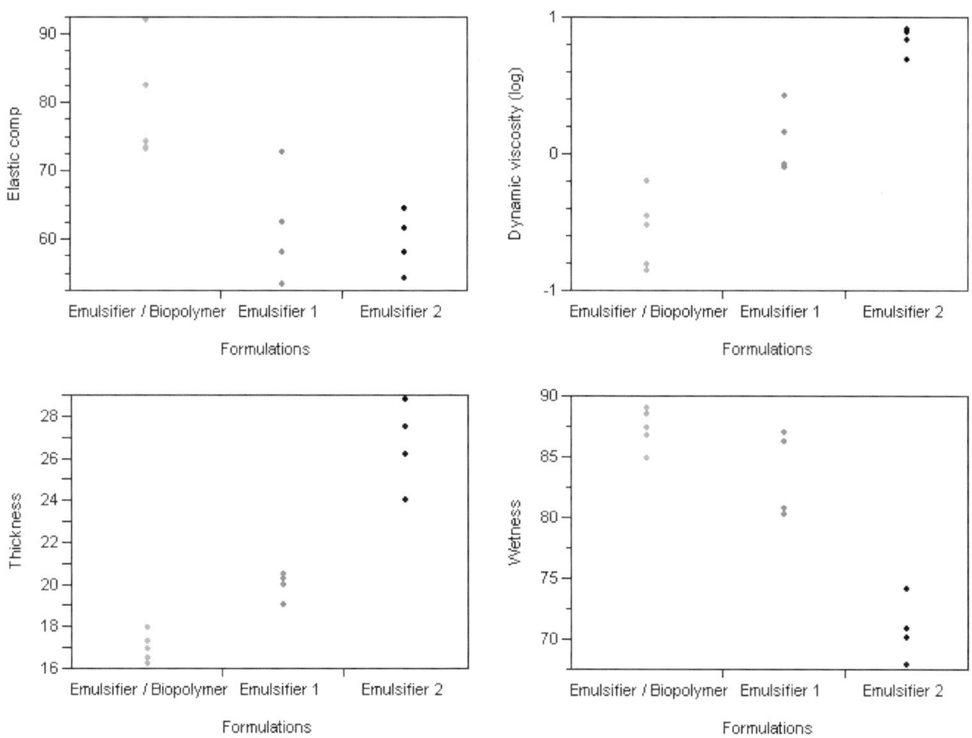

Figure 7.3 Scatter plots (results of summarized cluster analysis) of some of the rheological and sensorial properties for the different emulsifier systems.

Figure 7.3 shows four scatter plots as summarized results of cluster analysis between emulsifier–biopolymer, classical emulsifier 1 and 2 from the same group of emollients. The three different emulsifier systems with same group of emollients were compared for the sensory attributes (wetness–thickness) and rheological properties (elastic component–dynamic viscosity).

Based on Figure 7.3, there are clear clusters between classical emulsifier 2 and emulsifier–biopolymer for both sensory attributes and rheological variables. The cluster formation of classical emulsifier 1 and emulsifier–biopolymer is less convincing for wetness sensorial attribute, but the latter formulation generally scores higher wetness.

7.4
Discussion

The emulsifier–stabilizer systems used in the present study show a number of interesting features. The emulsifiers used in Arlatone V100 and Arlatone V175 are both nonionic and they are commonly used in many personal care formulations. Both emulsifiers are expected to lower significantly the interfacial tension between the oil and water, thus aiding the emulsification process. For an O/W emulsion with an oil volume fraction φ of 0.2, the optimum concentration of Arlatone is 0.8%. This amount is sufficient to cover the interface completely with emulsifier molecules, thus reducing the interfacial tension to a minimum, and this helps the emulsification process. This low stabilizer concentration (against creaming) must be due the synergy obtained when using the two gums. Using either gums at such low concentration would certainly be insufficient for reduction of creaming and/or separation. There is ample evidence in the literature concerning the interaction between the two polysaccharide molecules and a summary of the arguments presented is given below.

As mentioned in the Introduction, Konjac mannan gum is essentially a noncharged polysaccharide, whereas xanthan gum is a charged polysaccharide. The mixture forms thermally reversible gels, which most workers agree to be due to molecular association [6, 7]. However, the nature of the interaction has been a matter of great debate [9, 13]. As mentioned in the Introduction, Dea et al. [8] proposed that association occurs between the ordered xanthan helix and unsubstituted regions of the galactomannan backbone. However, McCleary [14] modified this model and suggested that interaction could occur between the ordered xanthan chains and sequences along the mannan backbone where the galactose residues are positioned only on one side. X-ray diffraction studies [15] confirmed the molecular binding. However, since gels are only formed at temperatures above that expected for the xanthan conformational changes, one may conclude that the xanthan interaction must occur in the disordered form. Williams and coworkers [16, 17] proposed that xanthan–K interaction could occur with xanthan in both the ordered and disordered form depending on the ionic strength of the solution. For example, in pure water the interaction appears to occur a few degrees above the conformational transition (>82 °C). In 0.04 mol dm^{-3} NaCl, the interaction occurred at much lower temperature (~42 °C). The gels formed in pure water are stronger than those in electrolyte solutions. Using the spin label technique (ESR), Annable et al. [9] were able to follow the molecular motion of the polymer chains in solution, which allowed them to follow the conformational change and K association as a function of temperature, and they correlated these processes with gelation. The ESR spectra of spin-labeled xanthan alone in water showed an isotropic behavior at high temperature and, since the spin label is attached to the xanthan side-chain rather than the backbone, it indicates a high degree of xanthan side-chain mobility. At lower temperatures, composite spectra containing a significant proportion of an anisotropic component were observed. This indicated substantial loss of the side-chain mobility, which now associates with the cellulosic backbone forming the ordered structure. By resolving the spectra, it was shown that the anisotropic component p increased rapidly when the temperature was decreased below ~80 °C, reaching a maximum at ~50°. This means that the xanthan molecules become ordered within this temperature range

For mixtures of xanthan and K, the anisotropic component p started to increase at slightly higher temperature compared with solutions of xanthan alone, indicating that the xanthan ordering process shifted to higher temperatures when K was added. Using spin-labeled K and non-spin-labeled xanthan (in a 1:1 mixture), it was shown that p starts to increase below ~65 °C, which corresponds closely to the temperature noted for the onset of gelation. These results indicate that association only occurs after xanthan chain ordering. K seems to have little effect on the ordering of xanthan chains. The proportion of K chain segments which associate with the xanthan molecules decreases as the proportion of K in the mixture increases. This implied that at higher K concentrations, all of the available interaction sites are occupied and that the "excess" K molecules my not interact and remain free in solution.

In the presence of electrolytes, it was shown [9] that the interaction also occurred after ordering of the xanthan chains. The xanthan coil → helix transition occurs at ~80 °C in the presence of monovalent cations and above 100 °C in the presence of divalent cations. However, gelation occurs at 53 °C or less, depending on the electrolyte present. This means that a reduced interaction occurs in the presence of electrolytes, implying the presence of fewer interaction sites.

Based on the above results, Annable et al. [9] proposed a model for the interaction between K and xanthan gums. On cooling, the xanthan molecules are thermodynamically driven to adopt a predominantly ordered structure. The side-chains which are directed away from the xanthan cellulosic main-chain of the disordered xanthan molecules (single coils or expanded dimers) begin to associate with the backbone, thus reducing polymer–solvent contacts and give rise to chain stiffening. In the absence of K this leads to either double helix formation or xanthan self-association. In the presence of K, ordered or disordered sequences within the xanthan molecules prefer to interact with K chains rather than other xanthan chains. One of the main reasons for this interaction is the uncharged chain of K, in contrast to xanthan chains, which are highly charged. Addition of electrolytes serves to promote xanthan–xanthan association at the expense of xanthan–K association due to a charge screening effect. The cations that are most effective in promoting xanthan ordering and hence xanthan–xanthan interaction give rise to the weakest interaction with K, as noted from the lower value of the elastic modulus G' and gelation temperature. The rheological results are also consistent with this concept. Whereas G' for xanthan in solution alone is enhanced by addition of monovalent cations and to a larger extent by divalent cations, the exact reverse is true for xanthan–K mixtures. This arises from the fact that addition of electrolytes enhances xanthan ordering and self-association at the expense of xanthan–K interaction.

From the above discussion, it is now clear that mixtures of K and xanthan gums (KX) should result in synergy in the rheological behavior compared with the two gums alone. This is clearly demonstrated by the results shown in Tables 7.1 and 7.2. For example, the low shear viscosity of 0.1% KX is 6149 Pa s, whereas that of 1% xanthan gum (which is an order of magnitude higher in concentration) is 228 Pa s (i.e. more than one order of magnitude lower). This very high viscosity at low shear rates [sometimes referred to as residual or zero shear viscosity, $\eta(0)$] explains the absence of creaming of emulsions stabilized with Arlatone V100 or V175, when the concentration of KX in the continuous phase is only 0.1%.

It should be mentioned that addition of surfactants to KX, i.e. when using Arlatone V100 or Arlatone V175, resulted in a decrease in the viscosity of KX. This seems to due to the interaction of the surfactants with the gums, resulting in a lowering of their interaction.

The above reduction in creaming rate with increase in $\eta(0)$ have been discussed in detail [18]. One should consider the stress exerted by a droplet σ_p in the continuous phase, which is simply the ratio of the gravity force exerted on the droplet to the area of the droplet, i.e.

$$\sigma_p = \frac{\frac{4}{3}\pi R^3 \Delta\rho g}{4\pi R^2} = \frac{R\Delta\rho g}{3} \tag{5}$$

where R is the droplet radius, $\Delta\rho$ is the density difference between oil and continuous phase and g is the acceleration due to gravity.

For a droplet with $R = 10$ µm and density difference of 0.2 g cm^{-3}, σ_p is equal to 6.5×10^{-3} Pa. This means that the viscosity of the emulsion needs to be measured at low stresses or low shear rates, i.e. the residual or zero shear viscosity $\eta(0)$, to predict the creaming and separation of the emulsion. The results using KX have shown that $\eta(0)$ at low shear rates (of the order of 6×10^{-4} s^{-1}) is >6000 Pa s. Emulsions prepared using 1% Arlatone V100 or Arlatone V175 also gave values of $\eta(0) > 1000$ Pa s. These high residual viscosities are sufficient to eliminate any creaming of the emulsions, as found experimentally over a long period (several month of storage).

It should be mentioned that the emulsions prepared using Arlatone V100 or Arlatone V175 give a different consistency compared with emulsions prepared using Carbopol ETD 2001. The critical stress obtained when using 0.1% KX was significantly higher than that obtained at the same concentration when using Carbopol (5.9 versus 0.5 Pa, respectively). This high critical stress indicates that the "gels" produced when using KX are more "coherent" than those using Carbopol. This may have some implication on the long-term physical stability and also the "skin feel" of the emulsions prepared using the present system.

Another important stabilizing mechanism when using Arlatone V100 or V175 is the absence of coalescence as detected using optical microscopy. The lack of coalescence in these emulsions is due to two main effects: (1) steric stabilization produced when using mixtures of nonionic surfactants [19] and (2) possible adsorption of the gums at the O/W interface, thus producing a viscoelastic film at the interface, which prevents any thinning and disruption of the aqueous film between emulsion droplets.

Emulsions prepared using the emulsifier–biopolymer mixtures showed high residual viscosities compared with emulsions based on the same emulsifier system but in the absence of KX. These results explain the high stability against creaming/sedimentation of the emulsions. In addition, the elastic modulus of these emulsions is much higher than the viscous modulus and it shows little dependence on temperature in the range 25–50 °C [8]. This also explains the high stability of the emulsions on storage. As we shall see later, the residual viscosities obtained using the emulsifier–biopolymer mixtures are lower than those obtained using classical surfactant–thickener systems, and this will have a major impact on sensory attributes.

7.5 Conclusions

Using a combination of nonionic surfactants and two polysaccharides, namely Konjac mannan and xanthan gums (referred to as KX), one can produce very stable emulsions against any creaming or coalescence. The stabilizing mechanism of KX is due to the interaction between Konjac mannan and xanthan gums, resulting in a synergistic effect. The residual or zero shear viscosity of KX is much higher than that of xanthan gum alone and this explains the absence of creaming. Emulsions prepared using Arlatone V100 or V175 also show a residual or zero shear viscosity >1000 Pa s, which is sufficient to eliminate any creaming and coalescence of the emulsion. Any flocculation of the emulsion on storage is fairly weak and the samples can be simply redispersed by gentle shaking. The mixture of surfactants and KX eliminates coalescence by effective steric repulsion produced by the nonionic surfactants and also due to the possible co-adsorption of the gums at the interface which produce a viscoelastic film, thus preventing any thinning and disruption of the liquid film between the droplets to the interaction between the Konjac and xanthan gum. The emulsions prepared using the surfactant–biopolymer mixtures were very stable, showing no separation after several months of storage. This was due to the high residual viscosity obtained and the lack of dependence of the storage modulus on temperature (in the range 25–50 °C).

The sensory attributes of several emulsions based on these surfactant–biopolymer mixtures were compared with those obtained using classical surfactants and hydrocolloids (such as carbomer). Generally, the emulsions based on the emulsifier–biopolymer mixtures showed higher spreadability, higher wetness, lower firmness, lower greasiness, lower thickness and lower integrity of shape than the other emulsions. This was mainly due to the lower viscosity at high shear rate and the lower (but coherent) "gel" structure, which can be easily broken under shear. With the emulsions containing thickeners such as carbomer, a higher cohesive energy density is obtained and the gel structure cannot be easily broken under shear. These results clearly indicate the advantage of using the surfactant–biopolymer mixtures for the formulation of personal care emulsions.

References

1 Th.F. Tadros, B. Vincent, in *Encyclopedia of Emulsion Technology*, P. Becher (ed.), Marcel Dekker, New York, Vol. **I**, 1983.

2 B.P. Binks, *Modern Aspects of Emulsion Science*, Royal Society of Chemistry, Cambridge, 1998.

3 K. Nishinari, P.A. Williams, G.O. Phillips, *Food Hydrocolloids*, **6**, 199 (1992).

4 K. Maekaji, *Agric. Biol. Chem.*, **38**, 315 (1974).

5 E.D. Goddard, J.V. Gruber (eds.), *Principles of Polymer Science and Technology in Cosmetics and Personal Care.*

6 K.P. Shatwell, I.W. Sutherland, S.B. Ross-Murphy, I.C.M. Dea, *Carbohydr. Polym.*, **14**, 131 (1991).

7 F.M. Goycoolea, T.J. Foster, R.K. Richardson, E. Morris, M.J. Gidley, in *Gums and Satbilisers for the Food Industry*, G.O. Phillips, P.A. Williams, D.J. Wedlock (eds.), Oxford University Press, Oxford, 1994, pp. 333–344.

8 I.C.M. Dea, E.R. Morris, D.A. Rees, E.J. Walsh, *Carbohydr. Res.*, **57**, 249 (1977).

9 P. Annable, P.A. Williams, K. Nishinari, *Macromolecules*, **27**, 4204 (1994).

10 T. Naes, H. Martens, *Multivariant Calibration*, Wiley, Chichester, 1989.

11 M Meilgaard, GV Civille, BT Carr, *Sensory Evaluation Techniques*, CRC Press, Boca Raton, FL, 1991.

12 M Meilgaard, GV Civille, BT Carr, *Sensory Evaluation Techniques*, CRC Press, Boca Raton, FL, 1991.

13 P.A. Williams, M. Hicky, D. Mitchell, *Cosmet. Toiletries*, **118**, 51 (2003).

14 B.V. McCleary, *Carbohydr. Res.*, **71**, 205 (1979).

15 G.J. Brownsey, P. Cairns, M.J. Miles, V.J. Morris, *Carbohydr. Res.*, **176**, 329 (1988).

16 P.A. Williams, D.H. Day, M.J. Langdon, G.O. Phillips, K. Nishinari, *Food Hydrocolloids*, **4**, 4891 (1991).

17 P.A. Williams, S.M. Clegg, D.H. Day, G.O. Phillips, in *Food Polymers, Gels and Colloids*, E. Dickinson (ed.), RSC Publication No. 82, Royal Society of Chemistry, Cambridge, 1991, p. **339**.

18 Th.F. Tadros, *Adv. Colloid Interface Sci.*, **108–109**, 227 (2004).

19 D.H. Napper, *Polymeric Stabilisation of Colloidal Dispersions*, Academic Press, London, 1983.

8
Correlation of "Body Butter" Texture and Structure of Cosmetic Emulsions with Their Rheological Characteristics

Tharwat F. Tadros, Sandra Léonard, Cornelis Verboom, Vincent Wortel, Marie-Claire Taelman, and Frederico Roschzttardtz

Abstract

The rheological behavior of several body butter formulations was investigated. Three different types of measurements were carried: (1) steady state by carrying out measurements at various shear rates and the variation of stress and viscosity with applied shear rate was established; (2) constant stress (creep) measurements whereby a constant stress σ was applied on the sample and the variation of strain γ (or compliance $J = \gamma/\sigma$) was followed as a function of time; various creep curves were obtained at increasing applied stress; (3) dynamic (oscillatory) measurements whereby a sinusoidal strain is applied on the sample and the stress is simultaneously measured. From the stress and strain amplitudes σ_0 and γ_0 and the phase angle shift δ, the complex modulus G^*, the storage modulus G' (the elastic component), and the loss modulus G'' (the viscous component) were obtained as a function of strain amplitude γ_0 and frequency. The steady-state results showed that with some body butter formulations, a well-defined maximum (stress overshoot) is reached, after which the stress reaches a steady value. This was attributed to the structure of the system, which consists of oil droplets dispersed in water that contains waxes that form a "three-dimensional" gel network. Alternatively, for W/O body butter the water droplets are dispersed in an oil continuous phase that also contains waxes which produce a gel structure. On application of the formulation, the system requires some pressure before the gel structure starts to be "broken down" and once this pressure has been overcome the formulation begins to spread and it does so readily. Constant stress (creep) measurements showed that the O/W body butters have a higher elastic modulus than viscous components compared with W/O body butters. In addition, the critical stress above which considerable flow occurs was also higher for the O/W body butters compared with the W/O systems. The same behavior was also obtained when using oscillatory measurements, which showed higher elastic moduli for the O/W body butters than the W/O systems. Spectrum Descriptive Analysis™ and principal component analysis (PCA) methods showed

Colloids and Interface Science Series, Vol. 4
Colloids in Cosmetics and Personal Care. Edited by Tharwat F. Tadros

ISBN: 978-3-527-31464-5

that the body butter formulations can be easily distinguished from classical formulations using rheology. In addition, rheology could also distinguish between O/W body butters and W/O systems. In particular, the elastic modulus G' could distinguish unambiguously between body butter formulations and classical emulsions. PCA analysis also revealed that body butter formulations have high firmness, stickiness, integrity of shape, grease and low wetness and spreadability compared with the classical formulations.

8.1 Introduction

Recently, several cosmetic formulations have been prepared with a consistency similar to that of butter. These systems consist of oil-in-water (O/W) or water-in oil (W/O) emulsions to which waxes are added to give a consistency similar to that of butter.

The rheological characteristics of body butters are very important during application; for example, on compression of the sample there will be an instantaneous elastic response followed by a retarded response. On removing this stress partial recovery occurs and the structure rebuilds very slowly. A useful method to study the rheology of these systems is to use constant stress (creep) and oscillatory measurements. These techniques allow one to study the viscoelastic behavior of these systems [1, 2]. Below a critical stress the system behaves as a viscoelastic solid. For example, when a stress (below the critical stress) is applied to the system, rapid elastic deformation occurs and the initial low deformation (strain) remains virtually constant (giving a near zero shear rate). When the stress is removed, almost complete recovery occurs and the strain decreases gradually to zero. However, above a critical stress that depends on the structure of the body butter, the system behaves as a viscoelastic liquid. In this case, the strain increases rapidly to a certain value (giving an instantaneous elastic response), after which it increases slowly with time until a steady state (with constant shear rate). When the stress is removed, only partial recovery occurs and the strain reaches a limiting value after some period. Oscillatory measurements can be applied to study the rheology of body butters. Two types of measurements are carried out. First, the frequency of oscillation is kept constant (say at 1 Hz) and the amplitude of the strain is gradually increased. Below a critical strain, the system shows linear viscoelastic behavior, whereby the elastic modulus remains constant and above this critical strain the elastic modulus starts to decrease (nonlinear behavior). This critical strain may be identified with the value above which the structure begins to "break down" and flow starts to occur. In a second experiment, the strain is kept constant in the linear viscoelastic region and the elastic and viscous moduli are measured as a function of frequency. Below a certain frequency (denoted characteristic frequency) the viscous modulus is higher than the elastic modulus. This region corresponds to a long time experiment, as is the case, for example, on application of the body butter to the skin surface. Above the characteristic frequency, the elastic modulus becomes higher than the viscous modulus. This region corresponds to short time scales, i.e. before application of much stress to the sample.

Due to the above rheology profiles, "body butters" give a rich, nourishing appearance with particular sensory skin feel properties such as firmness and spreading during melting of the emulsion structure on the skin. In principle, one can relate these effects to the rheological characteristics of body butters.

The main objective of this chapter is to correlate the rheological and sensorial characteristics of body butters with their structure. The latter consists of oil or water droplets that are dispersed in a continuous phase of water or oil. The rheology of the O/W or W/O emulsions (to give a consistency similar to that of butter) is controlled by addition of waxes and thickeners which may become partially dispersed in the oil and partially dispersed in the water.

8.2 Experimental

8.2.1 Materials

Nine body butter formulations, both commercially available and prepared in our laboratories, were investigated. Of these, seven samples were O/W- and two were W/O-based body butters. A summary of the composition of the samples is given in Table 8.1 which only defines the oil phase, surfactant systems used, waxes and thickeners. The other components, e.g. preservatives and actives, are not given. For comparison, three food butter/margarine-type products were investigated only for the rheological characteristics.

8.2.2 Rheological Measurements

The rheological measurements were carried out using a Physica USD 200 universal dynamic spectrometer (Paar Physica, Germany). A cone-and-plate geometry with a cone angle of 2° and diameter 50 mm was used. Three rheological techniques were applied, namely flow–viscosity curve measurements, constant stress measurements (creep tests) and dynamic measurements (oscillatory tests).

8.2.2.1 Flow–Viscosity Curve Measurements

For these measurements the sample placed on the rheometer (29 °C) is sheared with a shear rate, $\dot{\gamma}$, from 0 to 500 s^{-1} in a linear ramp and then decreased from 500 to 0 s^{-1} at the same rate. Comparative tests were carried out at two different rates, namely 45 s and 4 min by period (which means 90 s and 8 min for the total test time). The shear stress obtained (calculated by the torque measured on the cone-plate) was measured as function of the shear rate applied.

Table 8.1 Description of the main components of the nine body butter formulations tested.

Product name	Emulsion type	Emulsifier system	Principal emollients used	Thickeners (waxes, fatty alcohol, ...)	Hydrocolloids
F388	O/W	Cetearyl Glucoside/ Methyl glucose sesquistearate	Octyldodecyl myristate/ Cyclomethicone/Myristyl Myristate/Isostearyl Neopentanoate/ Dimethicone/Phenyl trimethicone	Shea butter/ Cetearyl alcohol/ Stearic acid	Xanthan gum/ Acrylate cross polymer
F389	O/W	Sorbtian Stearate/ Sucrose Cocoate/ Glyceryl Stearate	Vegetable Oil/Sunflower seed oil/Squalane/ Zea Mays Oil/ Stearyl Dimethicone	Spent grain wax/ Candelilla wax	Carbomer/ Xanthan gum
F390	O/W	Glyeryl Stearate/ PEG-100 Stearate	Sesame Oil/ Cyclomethicone	Shea butter/ Cetearyl alcohol	Xanthan gum
F384	O/W	Glyeryl Stearate/ PEG-100 Stearaten	Isopropyl Isostearate/ PCA Dimethicone/ Squalane	Shea butter/ Cocoa butter/ Cetearyl alcohol/ Microcrystalline wax	Xanthan gum
F385	O/W	Glyeryl Stearate/ PEG-100 Stearate	Isopropyl Isostearate/ Squalane	Shea butter/ Cocoa butter/ Cetearyl alcohol/ Microcrystalline wax	Xanthan gum
F386	O/W	Sorbtian Stearate/ Sucrose Cocoate/ Glyceryl Stearate	Squalane/ Triethylhexanoin/ Caprylic/capric triglyceride/PCA Dimethicone/ Sweet Almond Oil	Shea butter/ Cocoa butter/ Behenyl alcohol	Xanthan gum
F284	O/W	SE GMS/Ceteth-20	Cycolmethicone/ Caprylic/capric triglyceride/Sunflower Oil/Cetyl palmitate/ Dimethicone	Cetyl alcohol/ Cetyl palmitate	Sodium Acrylate/Sodium Acryloyldimethyl Taurate Copolymer/ Carbomer
F391	W/O	PEG-30 Dipolyhydroxy-stearate	Caprylic/capric triglyceride/Dicetearyl dimer dilinoleate/ Dioctyldodecyl dodecanedioate/ Ethylhexyl Palmitate/ Cyclopentasiloxane	Beeswax/ Synthetic wax/ Shea butter	–

Table 8.1 *(continued)*

Product name	Emulsion type	Emulsifier system	Principal emollients used	Thickeners (waxes, fatty alcohol, ...)	Hydrocolloids
F463	W/O	PEG-30 Dipolyhydroxy-stearate	Caprylic/capric triglyceride/ Triehtylhexanoin/ Cyclopentasiloxane/PCA Dimethicone	Beeswax/ Microcrystalinne wax	Stearalkonium Hectorite

8.2.2.2 Dynamic (Oscillatory) Measurements

Amplitude Sweep For these measurements a sinusoidal strain (oscillating around the rotational axis of the geometry device) is applied with a fixed frequency (1 Hz) at a constant temperature (29 °C). The strain value is increased in a logarithmic ramp from 0.1 to 100% (50 measuring points every 10 s). For a viscoelastic system, the stress and the strain oscillate with the same frequency but out of phase. This is called the phase shift angle, δ, and describes the time delay between the oscillation which is preset and that which is determined as the test result (δ is simply given by the product of the time shift of the sine waves of stress and strain Δt and the frequency ω in rad s^{-1}, i.e. $\delta = \Delta t\omega$). A phase shift angle of 0° shows ideal elastic behavior (i.e. behavior of an extremely rigid solid) whereas a phase shift angle of 90° shows ideal viscous behavior (i.e. behavior of a pure liquid). Most disperse systems such as the body butters have $0° < \delta < 90°$, i.e. viscoelastic behavior between the above extremes. At $\delta = 45°$ the behavior is exactly in the middle of the two extremes. In this case, the viscous and elastic portions of the viscoelastic behavior are exactly the same size.

From the amplitudes of the stress and the strain amplitudes σ_0 and γ_0, respectively, and the phase angle shift one can obtain the complex modules G^*:

$$G^* = \sigma_0/\gamma_0 \tag{1}$$

From the phase angle shift δ, G^* can be resolved into an elastic or storage modulus (G') and the loss or viscous modulus (G''):

$$G' = G^* \cos\delta \tag{2}$$

$$G'' = G^* \sin\delta \tag{3}$$

Finally, the dynamic viscosity η' can be calculated from the loss modulus:

$$\eta' = G''/\omega \tag{4}$$

In this amplitude test, the rheological parameters (G^*, G', G'' and η' as a function of strain amplitude γ) are virtually constant up to a critical strain γ_{cr}. This constant region is referred to as the linear viscoelastic region (LVER). Above γ_{cr}, G^* and G' start to decrease whereas G'' and η' start to increase with further increase in strain amplitude. In the present study, η' was obtained at the value at the end of the LVER.

Frequency Sweep During frequency sweep measurements, the strain amplitude is kept constant at a value in the LVER, whereas the frequency is changed from 10 to 0.01 Hz (at 29 °C). The dependences of G' (elastic component) and G'' (viscous component) on frequency are recorded. In the low-frequency regime (long time) $G'' > G'$, whereas in the high-frequency regime (short time) $G' > G''$. At a characteristic frequency ω^*, $G' = G''$, referred to as the cross-over point of both moduli, and this is equal to the inverse of the characteristic relaxation time of structure present in the sample. For our purpose, em_1Hz designates the elastic modulus (G') obtained from the frequency sweep at a frequency of 1 Hz.

8.2.2.3 Constant Stress (Creep Test) Measurements

In this measurement, a constant stress is applied on the system (kept at 25 °C) and the deformation (strain) γ is followed as a function of time for 2 min. The compliance J calculated is simply the strain divided by the applied stress for each measured point. After this time, the stress was removed, keeping its value at zero, and the strain was followed for another 2 min to obtain the recovery curve of the sample. The total compliance could be resolved into an elastic component (J_e or el_comp) and a viscous component (J_v or vis_comp).

The critical stress, σ_{crit} (designated also Crit_strs), obtained from creep measurements, could be considered as the limit of the viscoelastic domain above which the viscosity of the system starts to decrease with further increase in the stress (flowing).

8.2.3 Schematic Representation of the Rheological Curves

A schematic representation of flow curves, oscillatory response and constant stress (creep) behavior is shown schematically in Figure 8.1.

8.2.4 Spectrum Descriptive Analysis

All body butters were evaluated using the Spectrum Descriptive Analysis™ method [3]. This sensory technique relies on obtaining accurate numbers by a well-trained sensory test panel. This panel consists of approximately 10 members and each panelist evaluates each product three times using well-defined attributes with a fixed meaning. The 21 attributes can be subdivided into several groups: appearance, pick-up, rub-out, immediate after-feel and after-feel after 20 min.

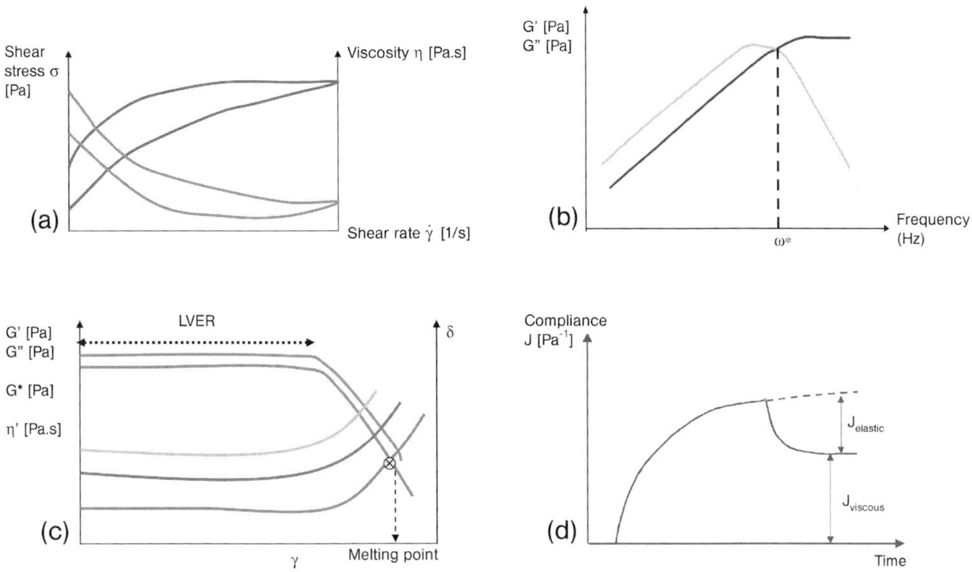

Figure 8.1 Illustration of typical (a) flow, (b) frequency sweep, (c) amplitude sweep and (d) creep curves obtained for viscoelastic materials.

8.2.5 Principal Component Analysis

Principal component analysis (PCA) was performed. This technique involves a mathematical procedure that transforms a number of (possibly) correlated variables into a (smaller) number of uncorrelated variables called *principal components*. The first principal component accounts for as much of the variability in the data as possible and each succeeding component accounts for as much of the remaining variability as possible. The powerful PCA tool provides easy-reference plots with the possibility of including additional information also.

8.3 Results and Discussion

Figure 8.2 shows a typical flow curve of the body butter F389 (O/W-based body butter). It can be seen that as the shear rate is increased above zero a maximum in the stress is observed, which is normally described as an elastic overshoot. As the shear rate is increased, the stress decreases from its maximum value and ultimately it shows a gradual increase in stress with increase in shear rate. On reducing the shear rate from 500 to 0 s^{-1} the stress decreases and it reaches a low value at nearly zero shear rate. This behavior is typical for a thixotropic system. On increasing the shear rate, the structure of the system is gradually destroyed.

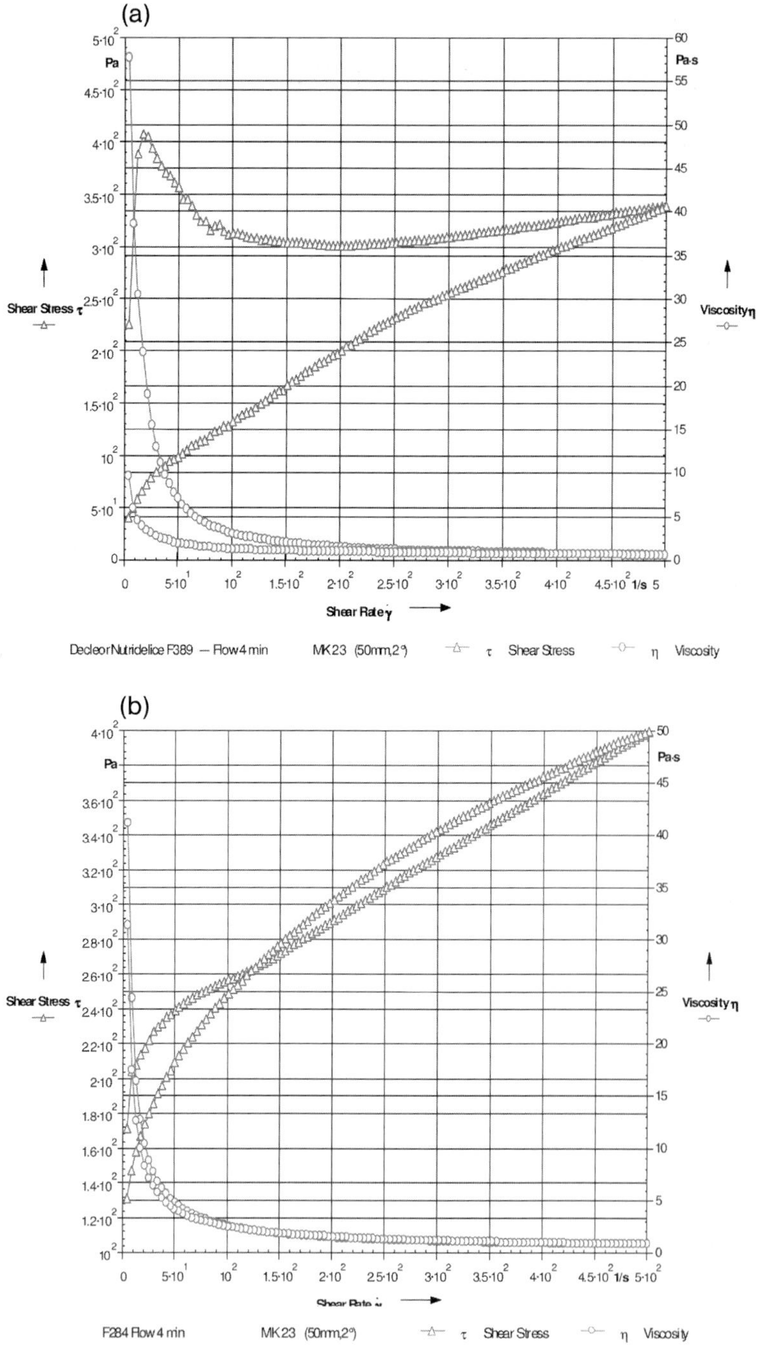

Figure 8.2 Flow curves 4 min up and 4 min down of body butters (a) F389 and (b) F284.

When the shear rate is reduced, this structure builds up only in part during this short time scale of the experiment (namely 4 min for the up curve and 4 min for the down curve). This behavior is also reflected in the plot of viscosity as a function of shear rate, which clearly shows that the up curve is much higher than the down curve. With the exception of sample F284, all other samples shows the same behavior. However, the magnitude of thixotropy was dependent on the nature of the sample and it is not possible to quantify this thixotropy in this case since the samples never reach a steady state showing this significant elastic overshoot. If the experiments are carried out within a shorter time scale (namely 45 s for the up curve and 45 s for the down curve), the difference between the up and down curves becomes even larger. However we cannot define a thixotropic index in this casc since, as mentioned above, all samples showed high elastic overshoot.

It is interesting that sample F284 (Figure 8.2) did not show any elastic overshoot and in addition the up and down curves are very close to each other. It is likely in this case that a steady state has been reached and the structure that is broken under shear recovers immediately when the shear is removed.

The above elastic overshoot can be explained in terms of the structure of the body butters similar to the texture of real food butter. The latter consist of water (or oil) droplets that are dispersed in oil (or water) and thickened by the addition of waxes and thickeners. These materials produce a three-dimensional gel network in the continuous phase which is similar to that produced in butter. A schematic illustration of the structure produced in butter is given in Figure 8.3 [4]. In this case, the disperse phase consists of water droplets containing salt and solids-not-fat (SNF) materials. The continuous oil phase (consisting of saturated and unsaturated glycerides of fatty acids) contains fat globules which may contain crystals of fat. As the system is agitated during preparation of butter, the crystals that already exist in the fat globules become disrupted and fragments take up

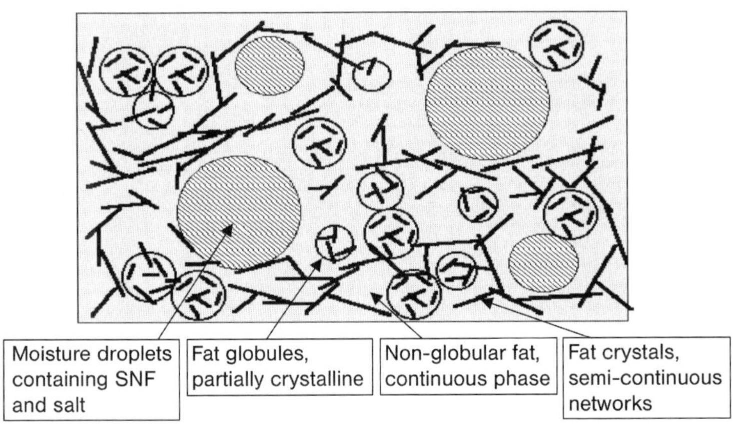

Figure 8.3 Diagrammatic sketch of the structure of butter (SNF means solids-not-fat, i.e. proteins, lactose, minerals, acids, enzymes, vitamins).

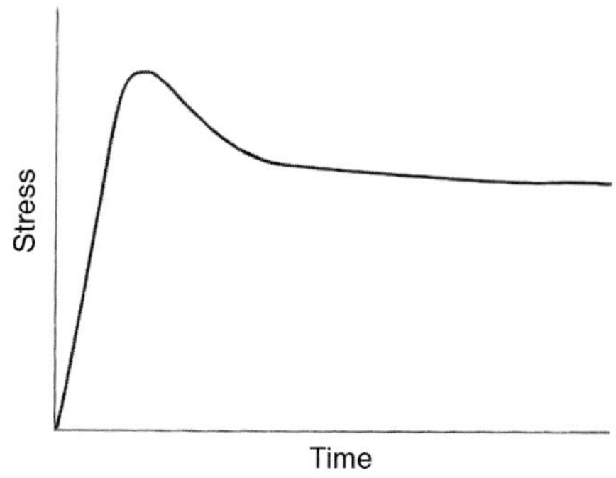

Figure 8.4 Build-up of stress at the commencement of shearing butter (overshoot phenomenon).

more or less random positions within the fat that eventually surround the newly formed water droplets. Such a network may have a considerably open structure and yet have considerable rigidity.

The structure of body butters used in personal care formulations is probably similar to that of food butter, but with relatively lower rigidity, which is essential to ensure rapid and good spreading on the skin. For an O/W body butter, the oil droplets are dispersed in an aqueous continuous phase that contains waxes to produce the required "three-dimensional" gel network structure. For W/O body butters, the disperse phase consists of water droplets that are surrounded by waxes to produce the gel structure, which is probably similar to that of food butter. If a body butter formulation is continuously sheared at a constant shear rate, the stress rises steeply at first, which could be attributed to an initial elastic rearrangement of the gel network. However, if the time scale of the experiment is increased (at constant shear rate), this stress will decrease and it reaches a steady state. This is shown schematically in Figure 8.4.

The above behavior is consistent with the response of body butters on application. At first, some pressure is required and the formulation resists spreading, but once this pressure has been overcome (on gentle pressing) the formulation begins to spread and does so readily. The formulation F284 (referred to as a body cream and without a body butter claim on the packaging) shows a different behavior, which as described above does not give any elastic overshoot and a steady state seems to be reached during the test time. This may be correlated with the structure of this formulation, which does not have any wax and the gel structure is only produced by cross-linked acrylates and fatty alcohol.

Figure 8.5 shows a typical creep curve for formulation F389 (O/W body butter). For comparison, Figure 8.6 shows a creep curve for a W/O body butter formulation (F391). It can be seen that for the O/W body butter formulation (F389) the elastic component is higher than the viscous component (82% versus 18%). By measuring creep curves at various stresses, a critical stress (σ_{crit} or Crit_strs) of 75 Pa was obtained. All other O/W body butter formulations gave a high elastic

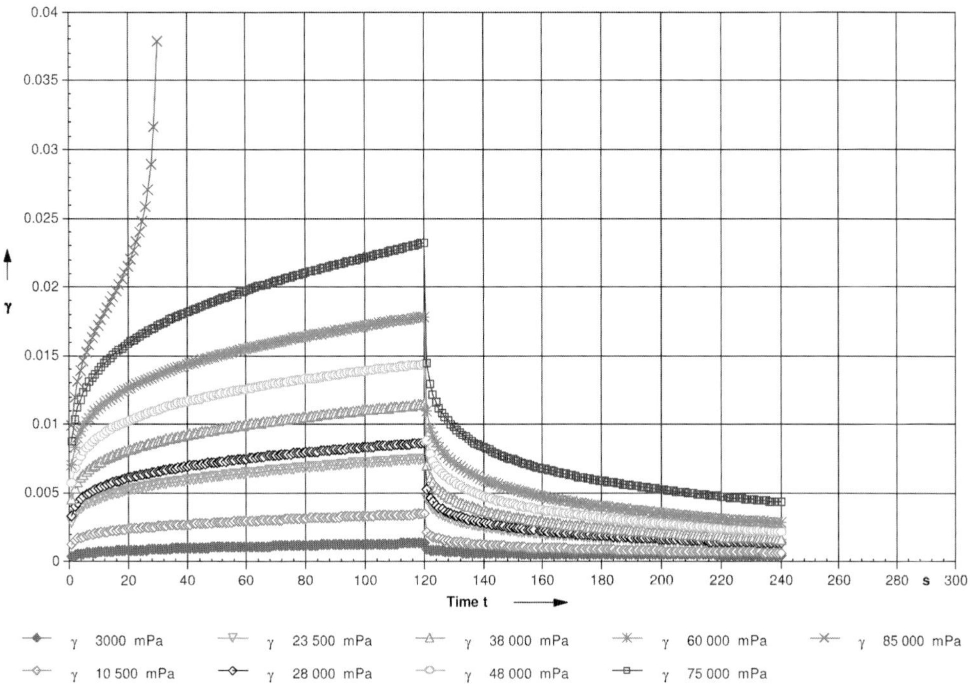

Figure 8.5 Creep curve for formulation F389 (O/W body butter).

component and also high critical stress. In contrast, the W/O body butter formulations gave a lower elastic component and a lower critical stress. A summary of the results is given in Table 8.2.

A summary of the elastic modulus at a frequency of 1 Hz (em_1Hz) and amplitude in the linear viscoelastic region is given in Table 8.3. With the exception

Table 8.2 Rheological parameters calculated from the creep curves.

Formulation	Emulsion type	Crit_strs (Pa)	el_comp (%)	vis_comp (%)
F388	O/W	105	76.97	23.03
F389	O/W	75	81.97	18.03
F390	O/W	58	75.05	24.95
F384	O/W	135	76.45	23.55
F385	O/W	120	74.93	25.07
F386	O/W	17.5	69.06	30.94
F284	O/W	70	91.89	8.11
F391	W/O	2.75	29.53	70.47
F463	W/O	3.5	16.63	83.37

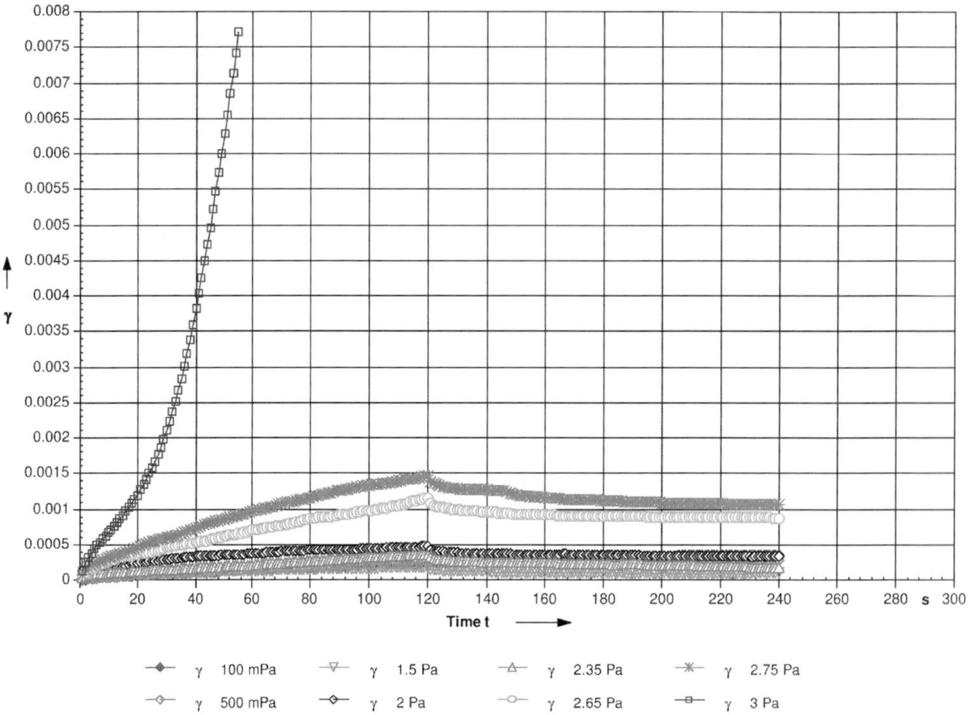

Figure 8.6 Creep curve for formulation F391 (W/O body butter).

of formulation F284 (which we have already identified as a "body cream"), the moduli values are generally higher for the O/W-based body butters than for the W/O-based body butters. This reflects the difference in structure between the two systems. The O/W body butters that contain waxes give a larger number of contact points and the bond strength between the points is also relatively high (more crystalline like in nature).

Figure 8.7 represents a PCA scores plot for the rheological parameters where several classical formulations (both O/W and W/O), body butter-based formulations and real food butter are plotted along two principal axes (PC 1 and PC 2). At first glance, it appears that almost all body butters and food butters are located on the right side of the PCA scores plot, indicating that they can be differently characterized from classical formulations by rheology. Only body butter F284 is separated from the rest of the body butter formulations, indicating that this formulation is behaving differently, as confirmed earlier in this chapter.

Looking at more details of body butter formulations in the scores plot, it can be clearly shown that the W/O emulsion (marked with a circle in Figure 8.7) is grouped apart from O/W-based body butters.

To find out how groups of body butters towards classical formulations and W/O towards O/W body butter formulations are characterized, relevant rheologi-

Table 8.3 em_1Hz calculated from frequency sweep.

Formulation	Emulsion type	em_1Hz (Pa)
F388	O/W	5511.7467
F389	O/W	6910.7222
F390	O/W	12315.193
F384	O/W	12376.517
F385	O/W	10732.305
F386	O/W	3974.8315
F284	O/W	1471.1714
F391	W/O	725.96868
F463	W/O	1786.722

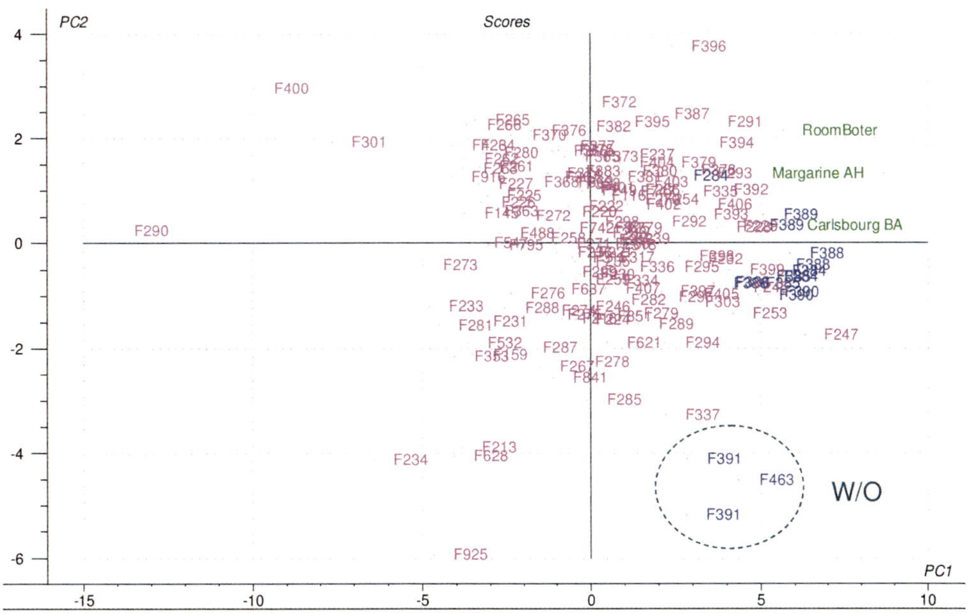

Figure 8.7 PCA scores plot based on rheological parameters for body butter formulations (in blue), classical emulsions (in pink) and real food butters (in green). Note: some body butters were measured twice on rheological properties to check for variance in the measurements which is acceptable.

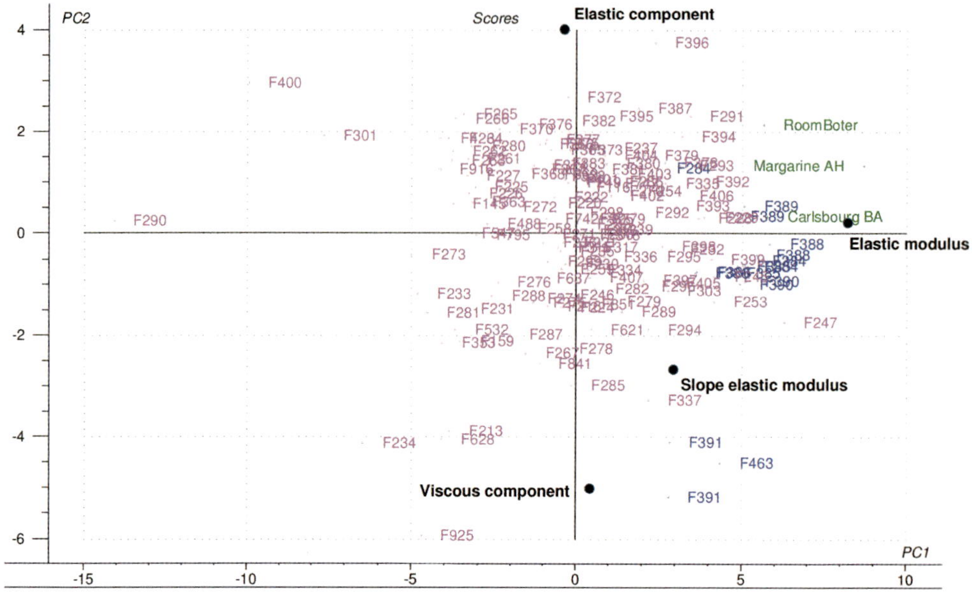

Figure 8.8 PCA scores and loadings plot. To keep the plot well organized, only the most relevant rheological variables are projected. Body butters, classical emulsions and food butters are characterized by the color codes blue, pink and green, respectively.

cal variables in the loadings plot are projected on the scores plot, as highlighted in Figure 8.8. Comparing the loadings and the scores, it can be seen that the elastic modulus (G') makes a large contribution to differentiating body butters from classical emulsions. In general, body butters have higher elastic modulus scores than classical formulations. This, as discussed above, is due to the stronger gel network with the body butter formulations.

The differentiation between O/W and W/O body butter formulations can be attributed to the viscous and elastic components (vis_comp and el_comp – measured during the creep curve) and to a lesser extent slope elastic modulus (extracted from the frequency sweep measurement). The W/O body butter formulations are characterized from O/W body butter formulations by high viscous component (or low elastic component) and higher slope elastic modulus scores.

Regarding the real food butters, they can be also characterized by high elastic modulus, like body butter formulations, with the remark that the elastic component is slightly higher.

Figure 8.9 represents a PCA scores plot for the sensorial parameters where several classical formulations (both O/W and W/O) and body butter-based formulations are plotted along two principal axes (PC 1 and PC 2). Again, body butter formulations can be reasonably grouped from classical formulations and characterized by sensory attributes.

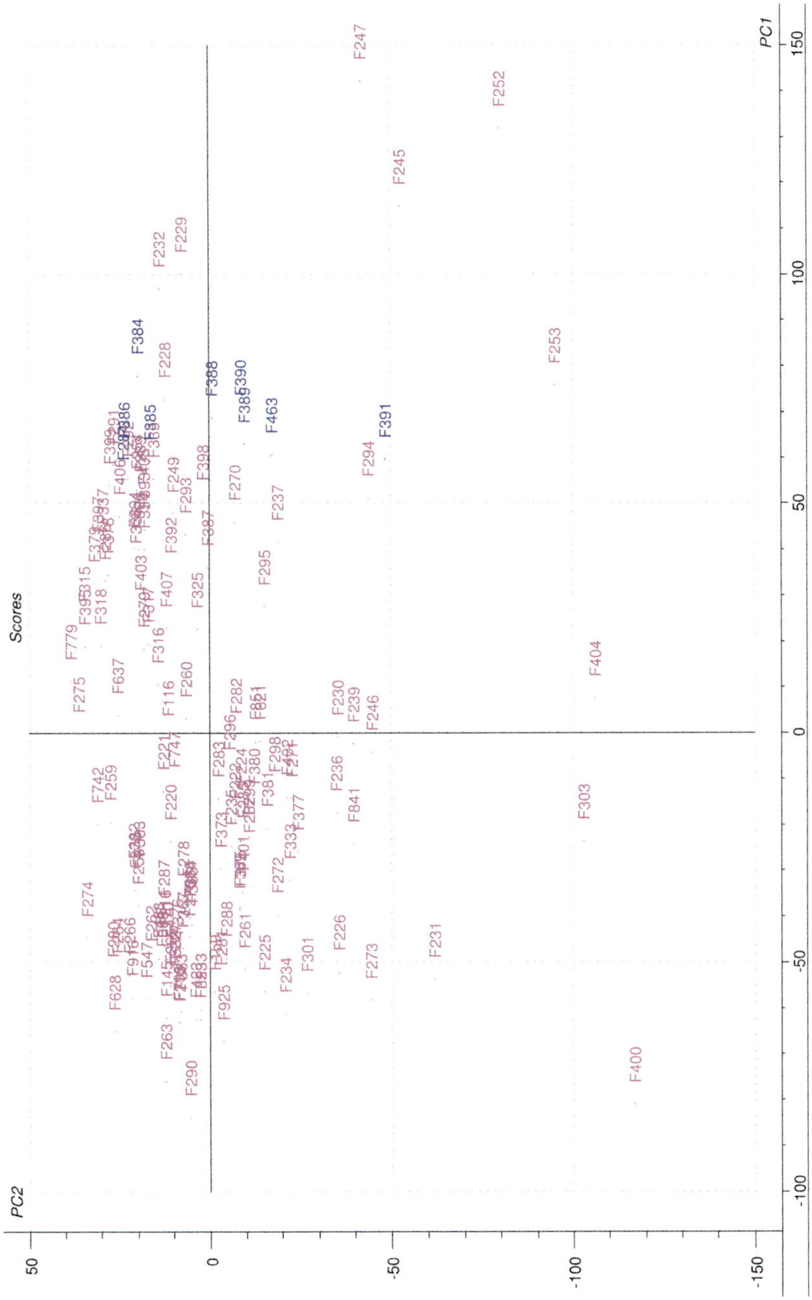

Figure 8.9 PCA scores plot of body butter formulations (in blue) and classical formulations (in pink) based on sensory attributes.

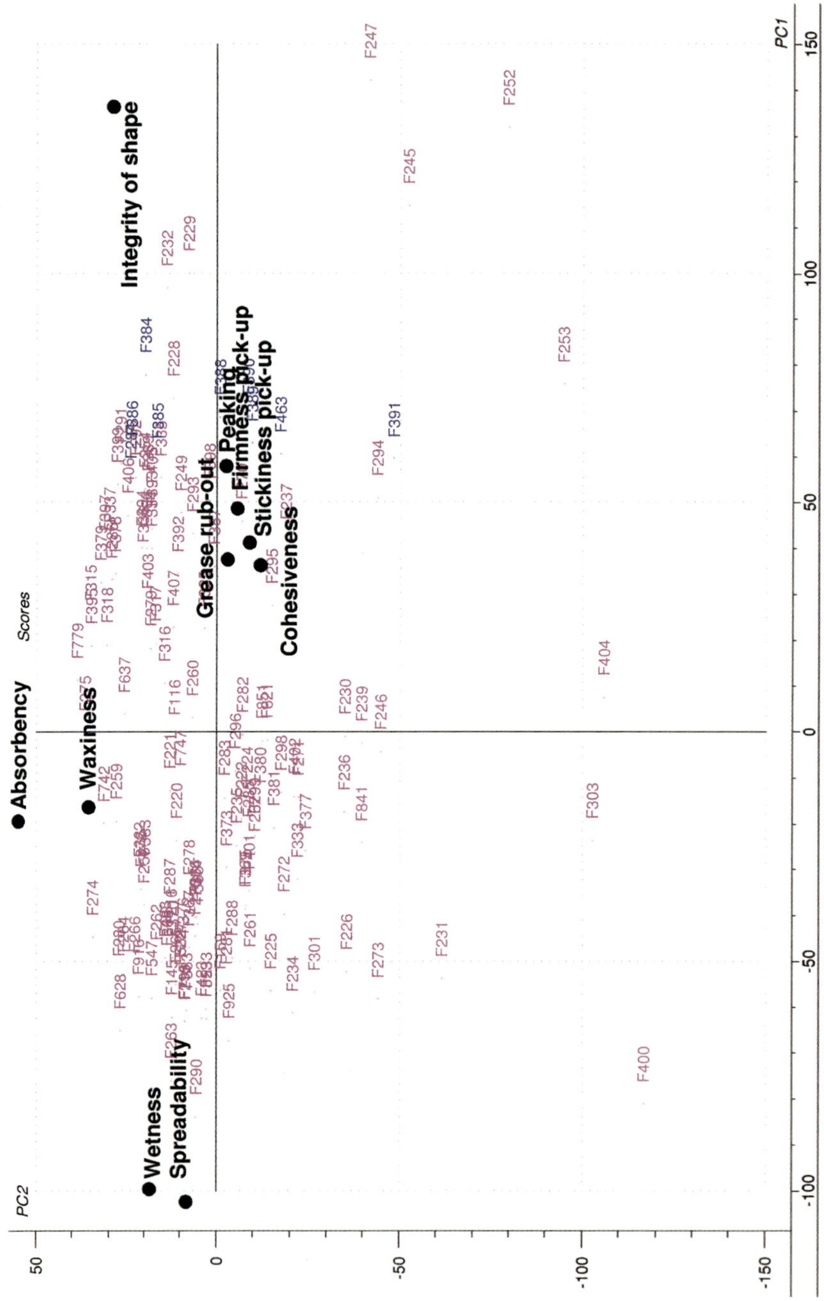

Figure 8.10 PCA scores and loadings plot. To keep the plot well organized, only the most relevant sensory attributes are projected. Body butter formulations and classical emulsions are colored blue and pink, respectively.

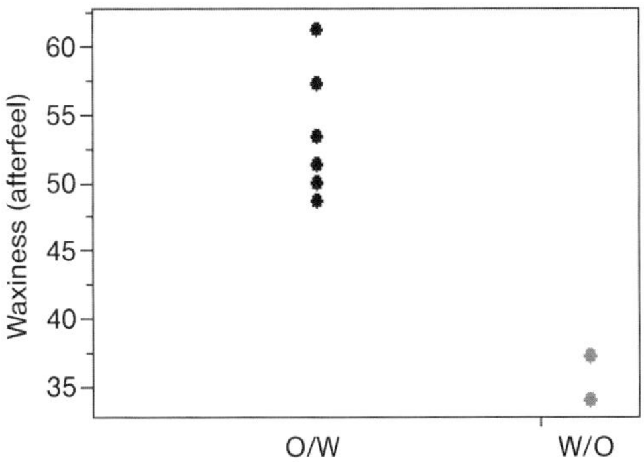

Figure 8.11 Waxiness (after-feel) sensory attribute can differentiate O/W from W/O body butter formulations.

To demonstrate the influence of sensory attributes on characterization, relevant sensory attributes in the loadings plot are projected on the scores plot, resulting in scores and loadings plot as highlighted in Figure 8.10.

The plot in Figure 8.10 reveals that body butter formulations have high firmness, stickiness, integrity of shape and grease as listed on the right side of the plot and low wetness and spreadability compared with the classical formulations.

In contrast to rheology, no grouping of W/O or O/W body butter formulations can be straightforwardly observed in the PCA scores plot in Figure 8.9. Application of hierarchical cluster analysis confirmed that only the waxiness attribute (after-feel) is able to differentiate W/O from O/W body butters, as plotted in Figure 8.11. O/W body butters show in general high wax residue after application compared with the W/O body butter emulsions.

8.4 Conclusion

Using our formulations data, we have confirmed in the sensorial and rheological investigations that body butter-based formulations can be characterized in two different ways, based on difference from classical formulations and the distinction between O/W and W/O body butter formulations.

Using rheological data, body butter formulations are characterized from classical formulations by high elastic modulus (the O/W systems, which contain waxes, give a larger number of contact points and the bond strengths between the points are relatively high; this gives a higher elastic modulus compared with the W/O systems). It was found that real food butters have rheological characteristics comparable to those of cosmetic body butter formulations, with the small difference of a slightly higher elastic component. The distinction between O/W and W/O body butter systems can be described by viscous component, elastic modulus and to a lesser extent slope elastic modulus. The O/W systems show high elastic component, low viscous component and to a lesser extent low slope elastic modulus compared with the W/O systems.

Using sensory attributes, body butter emulsions have in general lower wetness and spreadability and high grease, stickiness, firmness and integrity of shape in comparison with classical formulations. The O/W body butter systems show high waxiness after-feel in comparison with W/O body butter formulations.

References

1 Th.F. Tadros, *Adv. Colloid Interface Sci.*, **68**, 97 (1996).
2 Th.F. Tadros, *Adv. Colloid Interface Sci.*, **108–109**, 227 (2004).
3 M. Meilgaard, G.V. Civille, B.T. Carr, *Sensory Evaluation Techniques*, CRC Press, Boca Raton, FL, 1991.
4 J. H. Prentice, *Dairy Rheology – a Concise Guide*, 1992, pp. 67–84.

9
Interparticle Interactions in Color Cosmetics

Lorna M. Kessell and Tharwat F. Tadros

Abstract

Pigments are the primary ingredient of any color cosmetic and the way in which these particulate materials are distributed within the product will determine many aspects of product quality, including functional activity (color, opacity, UV protection) and also stability, rheology and skin feel.

The particulate distribution depends on many factors, such as particle size and shape, surface characteristics, processing and other formulation ingredients but ultimately is determined by the interparticle interactions. A thorough understanding of these interactions and how to modify them can help to speed up product design and solve formulation problems.

The fundamental principles of preparing pigment dispersions will be briefly described. These include wetting, dispersion (or wet milling including comminution) and stabilisation. This will be followed by a section on dispersion stability for both aqueous and non-aqueous media. The use of rheology in assessing particulate dispersions will be included, and finally the interaction with other formulation ingredients when this particulate is incorporated within a suspoemulsion will be discussed.

9.1
Introduction

The art of pigment dispersion has existed since prehistoric times. As long as 30 000 years ago Paleolithic humans probably used metal oxides as pigments and naturally occurring substances as binders and even high molecular weight dispersants such as egg white, animal fats and vegetable sugars for their cave paintings. Color *cosmetics* is also not new. Much could be achieved by trial and error and the ancient arts.

Pigments are the primary ingredient of any color cosmetic and the way in which these particulate materials are distributed within the product will deter-

Colloids and Interface Science Series, Vol. 4
Colloids in Cosmetics and Personal Care. Edited by Tharwat F. Tadros

ISBN: 978-3-527-31464-5

mine many aspects of product quality, including functional activity (color, opacity, UV protection) but also stability, rheology and skin feel.

Several color pigments are used in cosmetic formulations, ranging from inorganic pigments (such as red iron oxide) to organic pigments of various types. The formulation of these pigments in color cosmetics requires a great deal of skill since the pigment particles are dispersed in an emulsion (oil-in-water or water-in-oil). The pigment particles may be dispersed in the continuous medium, in which case one should avoid flocculation with the oil or water droplets. In some cases, the pigment may be dispersed in an oil, which is then emulsified in an aqueous medium. Several other ingredients are added, such as humectants, thickeners and preservatives, and the interaction between the various components can be very complex.

The particulate distribution depends on many factors, such as particle size and shape, surface characteristics, processing and other formulation ingredients, but ultimately is determined by the interparticle interactions. A thorough understanding of these interactions and how to modify them can help to speed up product design and solve formulation problems.

The fundamental principles of the preparation of pigment dispersions will be briefly described. These include wetting, dispersion (or wet milling, including comminution) and stabilization. This will be followed by a section on dispersion stability for both aqueous and nonaqueous media. The use of rheology in assessing dispersants will be included. The application of these fundamental principles for color cosmetics will be discussed. Finally, the interaction with other formulation ingredients when this particulate is incorporated within a suspoemulsion will be discussed.

9.2 Fundamental Principles of Preparation of Pigment Dispersions

9.2.1 Wetting of the Powder

The process of wetting involves the replacement of the solid/vapor interface with an interfacial tension γ_{SV} with the solid/liquid interface with interfacial tension γ_{SL}. Wetting can be described in equilibrium thermodynamics in terms of the contact angle θ by Young's equation [1] at the wetting line as illustrated in Figure 9.1. At the wetting line there is an equilibrium between solid, liquid and vapor and here the interfacial tensions can be balanced, resulting in the following expression:

$$\gamma_{SV} = \gamma_{SL} + \gamma_{LV} \cos\theta \tag{1}$$

$$\cos\theta = \frac{\gamma_{SV} - \gamma_{SL}}{\gamma_{LV}} \tag{2}$$

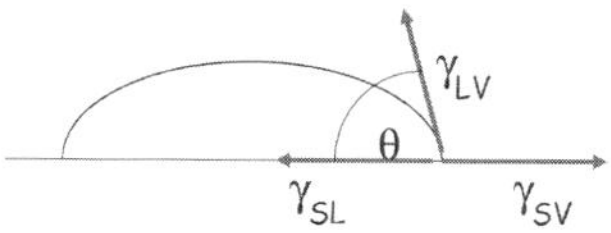

Figure 9.1 Schematic representation of the contact angle.

The energy required to achieve dispersion wetting, W_d, is given by the product of the external area of the powder, A, and the difference between γ_{SL} and γ_{SV}:

$$W_d = A(\gamma_{SL} - \gamma_{SV}) \tag{3}$$

Using Young's equation:

$$W_d = -A\gamma_{LV} \cos\theta \tag{4}$$

Thus wetting of the external surface of the powder depends on the liquid surface tension and contact angle. If $\theta < 90^\circ$, $\cos\theta$ is positive and the work of dispersion is negative, i.e. wetting is spontaneous.

9.2.2
Wetting of the Internal Surface

For agglomerates (represented in Figure 9.2), which are found in all powders, wetting of the internal surface between the particles in the structure requires liquid penetration through the pores. Assuming the pores to behave as simple capillaries of radius r, Δp is given by the following equation:

$$\Delta p = \frac{2\gamma_{LV}\cos\theta}{r} \tag{5}$$

For liquid penetration to occur, Δp must be positive and hence θ should be less than 90°.

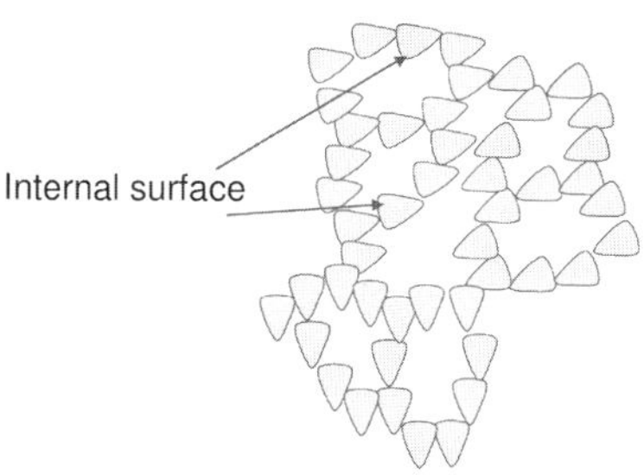

Figure 9.2 Schematic representation of an agglomerate.

The maximum capillary pressure is obtained when $\theta = 0$ and Δp is proportional to γ_{LV}, which means that a high γ_{LV} is required. Thus, to achieve wetting of the internal surface, a compromise is needed since contact angle only decreases as γ_{LV} decreases. We needs to make θ as close as possible to zero while not having a too low a liquid surface tension.

The rate of penetration of a liquid by a distance l through capillaries with radius r has been described by the Rideal–Washburn equation:

$$\frac{dl}{dt} = \frac{r\gamma_{LV}\cos\theta}{4\eta l} \tag{6}$$

where η is the viscosity of the liquid. Integration of this equation gives

$$l^2 = \left(\frac{r\gamma_{LV}\cos\theta}{2\eta}\right)t \tag{7}$$

Equation (7) shows that a plot of l^2 versus t gives a straight line and this forms the basis of measuring the contact angle on the surface of a powder, as will be discussed below.

For an agglomerate, the liquid pathway through the pores is complex and one cannot use a simple radius. In this case a tortuosity factor k must be introduced in Eq. (7):

$$l^2 = \frac{r\gamma_{LV}\cos\theta}{2\eta k^2}t \tag{8}$$

where $k = 1$ for cylindrical capillaries but for a more complex pathway it can reach values as high as 2.5.

9.3
Assessment of Wettability

9.3.1
Submersion Test – Sinking Time or Immersion Time

This is by far the simplest (but qualitative) method for the assessment of the wettability of a powder by a surfactant solution [2]. The time for which a powder floats on the surface of a liquid before sinking into the liquid is measured. A 100-mL volume of the surfactant solution is placed in a 250-mL beaker (of internal diameter 6.5 cm) and after standing for 30 min 0.30 g of loose powder (previously screened through a 200-mesh sieve) is distributed with a spoon on the surface of the solution. The time t for the 1–2-mm thin powder layer to disappear completely from the surface is measured using a stop-watch. Surfactant solutions with different concentrations are used and t is plotted versus surfactant concentration, as illustrated in Figure 9.3.

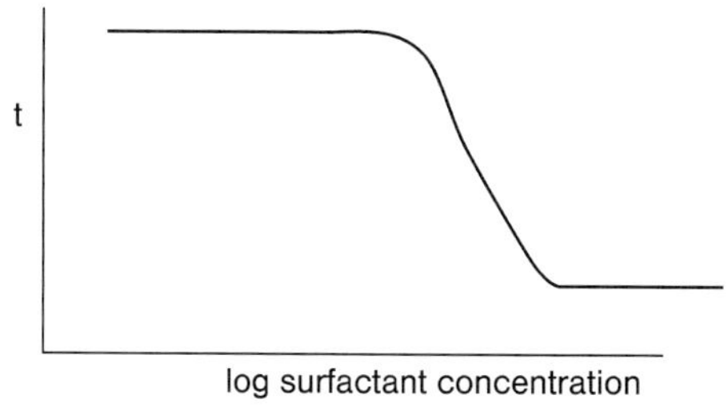

Figure 9.3 Sinking time as a function of surfactant concentration.

It can be seen from Figure 9.3 that the sinking time starts to decrease sharply above a critical surfactant concentration, reaching a minimum above this concentration. The above procedure can be used to select the most effective wetting agent. The lower the surfactant concentration above which a rapid decrease in sinking time occurs and the lower the minimum wetting time obtained above this concentration, the more effective the wetter is.

9.3.2
Contact Measurement for Assessment of Wettability

As discussed above, the contact angle θ can be used for quantitative assessment of a surfactant as wetting agent for a particular powder. The simplest procedure is to measure the contact angle on a flat surface of the powder. This requires preparation of a flat surface, for example by using a large crystal of the chemical or by compressing the powder into a thin plate (using high pressure as is commonly used, for example, for IR measurements). However, the above procedure is inaccurate since by compressing the powder its surface will change and the measured contact angle will not be representative of the powder in question. This procedure may be used to compare various wetting agents and the assumption is made that the lower the surfactant concentration required to reach a zero contact angle, the more effective the wetter is.

The contact angle determination for powders can also be carried out by measuring the rate of liquid penetration through a carefully packed bed of powder. By plotting l^2 (where l is the distance covered by the liquid flowing under capillary pressure) versus time t, a straight line is obtained [Eq. (8)] and its slope is equal to $r\gamma_{LV} \cos\theta/2\eta k^2$ (where r is the equivalent capillary radius, k is the tortuosity factor, γ_{LV} is the liquid surface tension and η is the liquid viscosity). From the slope, $\cos\theta$ is obtained provided that r/k^2 is known.

The tortuosity factor k and the ratio r/k^2 can be obtained by using a liquid that completely wets the powder giving a zero contact angle and $\cos\theta = 1$. The powder is carefully packed in a tube with sintered glass at the end using a specially designed cell fitted with a plunger for packing the powder (as supplied by Kruss). The suspended cell with its porous base is first placed in contact with liquid hexane, which gives a zero contact angle with most powders. The rate of penetration of hexane through the powder plug is measured by following the increase in weight ΔW of the cell with time. From the plot of ΔW versus t one can obtain r/k^2 from the slope of the straight line. The cell is then removed and the hexane is allowed to evaporate completely. The same cell with its powder pack is then suspended over surfactant solutions of various concentrations and this allows one to obtain the contact angle as a function of concentration. The most effective wetter will be the one that gives $\theta = 0°$ at the lowest concentration.

For $\theta = 0°$ or $\cos\theta = 1$, γ_{SL} and γ_{LV} have to be as low as possible. This requires quick reduction of γ_{SL} and γ_{LV} under dynamic conditions during powder dispersion (this reduction should normally be achieved in less than 20 s). This requires fast adsorption of the surfactant molecules both at the L/V and S/L interfaces. It should be mentioned that reduction of γ_{LV} is not always accompanied by simultaneous reduction of γ_{SL} and hence it is necessary to have information on both interfacial tensions, which means that measurement of the contact angle is essential in the selection of wetting agents. Measurement of γ_{SL} and γ_{LV} should be carried out under dynamic conditions (i.e. at very short times). In the absence of such measurements, the sinking time described above could be applied as a guide for wetting agent selection.

To achieve rapid adsorption, the wetting agent should be either a branched chain with a central hydrophilic group or a short hydrophobic chain with a hydrophilic end group – the most commonly used wetting agent for hydrophobic solids in aqueous media is Aerosol OT (diethylhexyl sulfosuccinate). This molecule has a low critical micelle concentration (cmc) of 0.7 g dm^{-3}, above which the water surface tension is reduced to ~25 mN m^{-1} in less than 15 s.

Several nonionic surfactants such as the alcohol ethoxylates can also be used as wetting agents. These molecules consist of a short hydrophobic chain (mostly C_{10}) which is also branched and a medium chain polyethylene oxide (PEO) mostly consisting of six EO units or fewer. These molecules also reduce the dynamic surface tension within a short time (<20 s) and they have reasonably low cmc.

In all cases one should use the minimum amount of wetting agent to avoid interference with the dispersant that needs to be added to maintain the colloid stability during dispersion and on storage.

9.4 Dispersing Agents

The above-mentioned wetting agents consisting of small molecules are seldom effective in stabilization of the suspension against flocculation. This is due to the small energy barrier produced by these molecules. For effective stabilization of the suspension against flocculation, one requires a dispersing agent, which will normally replace the wetting agent at the S/L interface and produce an effective repulsive barrier on close approach of the particles.

This repulsive barrier is particularly important for concentrated suspensions (that contain more than 50% by volume of solids). The following section looks more closely at the principles of stabilization of particulate suspensions.

9.5 Stabilization

All particles experience attractive forces on close approach, as illustrated schematically in Figure 9.4.

The strength of this van der Waals attraction, $V_A(h)$, depends upon the distance h between particles of radius R and is characterized by the Hamaker constant, A. The Hamaker constant expresses the attraction between particles (in a vacuum), and depends on the dielectric and physical properties of the material and, for some materials such as TiO_2, iron oxides and alumina, this is exceptionally high so that (in nonaqueous media at least), despite their small size, a dispersant is always needed to achieve colloidal stabilization.

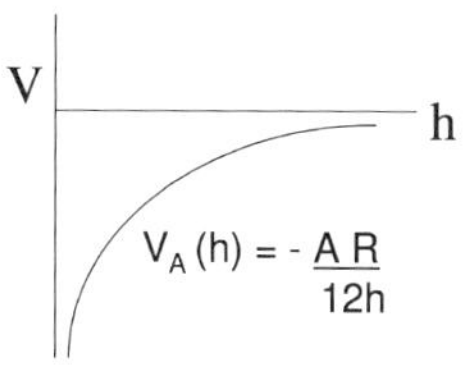

Figure 9.4 Schematic representation of van der Waals attraction.

The medium and/or a particle coating greatly affects the resultant force and an effective Hamaker constant can be expressed as

$$A = (A_p{}^{1/2} - A_m{}^{1/2})^2 \qquad (9)$$

It is not necessary to measure the actual attractive force (although this can be done by atomic force microscopy), but we must be aware of its existence and how it might be modified. Table 9.1 shows Hamaker constants for various particles and media [3].

In order to achieve stability, one must provide a balancing repulsive force to reduce interparticle attraction. This can be done in two main ways, by electrostatic or steric repulsion, as illustrated in Figure 9.5a and b (or a combination of the two, Figure 9.5c).

Table 9.1 Hamaker constants for various particles and media.

Material	A_p ($\times 10^{20}$ J)	Medium	A_m ($\times 10^{20}$ J)
PTFE	3.8	Water	3.7
Quartz	8.6	Pentane	3.8
TiO_2	19.5	Ethanol	4.2
Alumina	15.5	Hexadecane	5.1
Metals	20–40	Cyclohexane	5.2

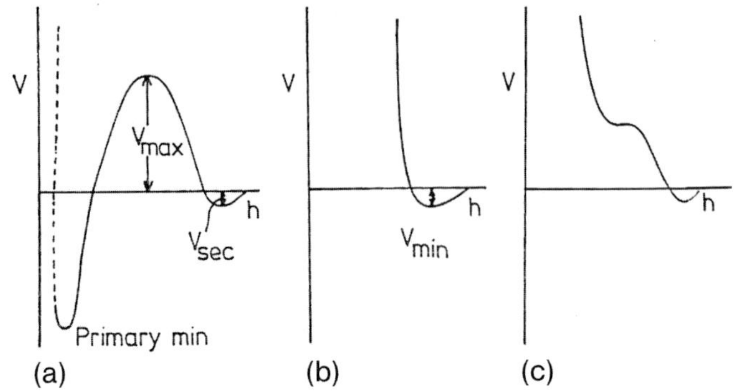

Figure 9.5 Energy–distance curves for three stabilization mechanisms: (a) electrostatic, (b) steric, (c) electrosteric.

9.5.1 Electrostatic Stabilization

Inorganic oxides have ionizable groups on their surface, which therefore means that in aqueous media they can develop a surface charge depending on pH, which affords an electrostatic stabilization to the dispersion. On close approach, the particles experience a repulsive potential overcoming the van der Waals attraction, which prevents aggregation. This stabilization is due to the interaction between the electric double layers surrounding the particles. This is illustrated in Figure 9.6. The double-layer repulsion depends on the pH and electrolyte concentration and can be predicted from zeta potential measurements (Figure 9.6). Surface charge can also be produced by adsorption of ionic surfactants.

This balance of electrostatic repulsion with van der Waals attraction is described in the well-known Deryaguin–Landau–Verwey–Overbeek theory of colloid stability (DLVO theory) [4, 5]. Figure 9.4a shows two attractive minima at long and short separation distances: V_{sec} that is shallow, a few kT units, and $V_{primary}$ that is deep and exceeds several hundred kT units. These two minima are separated by an energy maximum V_{max} that can be greater than $25kT$, thus preventing flocculation of the particles into the deep primary minimum.

When the pH of the dispersion is well above or below the isoelectric point or the electrolyte concentration is less than 10^{-2} mol dm^{-3} 1:1 electrolyte, the electrostatic repulsion is often sufficient to produce a dispersion without the need for added dispersant.

However, in practice, this condition often cannot be reached since at high solids content the ionic concentration from the counter- and co-ions of the double layer is high and the surface charge is not uniform. Therefore, a polyelectrolyte dispersant such as sodium polyacrylate is required to achieve this high solids content. This produces a more uniform charge on the surface and some steric repulsion due to the high molecular weight of the dispersant. Under these conditions, the dispersion becomes stable over a wider range of pH at moderate electrolyte

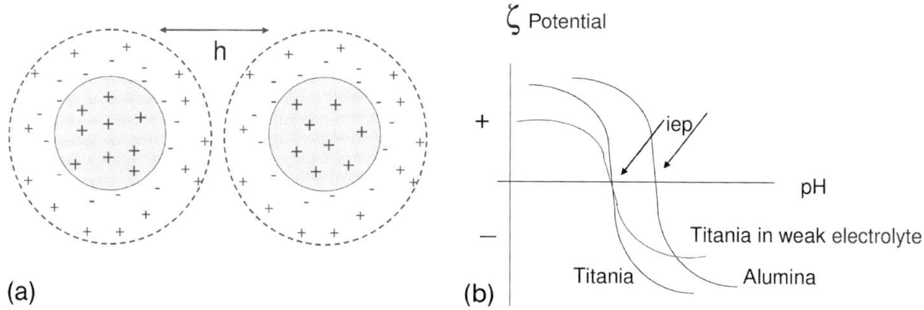

Figure 9.6 (a) Schematic representation of double-layer repulsion and (b) variation of zeta potential ζ with pH for titania and alumina.

concentration. This is electrosteric stabilization (Figure 9.5c shows a shallow minimum at long separation distances h, a maximum (of the DLVO type) at intermediate h and a sharp increase in repulsion at shorter h). This combination of electrostatic and steric repulsion can be very effective for stabilization of the suspension.

9.5.2 Steric Stabilization

This is usually obtained using adsorbed layers of polymers or surfactants. The most effective molecules are the A–B or A–B–A block or BA_n graft polymeric surfactants [5], where B refers to the anchor chain. This anchor should be strongly adsorbed on the particle surface. For a hydrophilic particle this may be a carboxylic acid, an amine or phosphate group or other larger hydrogen bonding-type block such as poly(ethylene oxide). The A chains are referred to as the stabilizing chains, which should be highly soluble in the medium and strongly solvated by its molecules. A schematic representation of the adsorbed layers is shown in Figure 9.7. When two particles with an adsorbed layer of hydrodynamic thickness δ approach a separation distance h that is smaller than 2δ, repulsion occurs (Figure 9.5b) as a result of two main effects: unfavorable mixing of the A chains

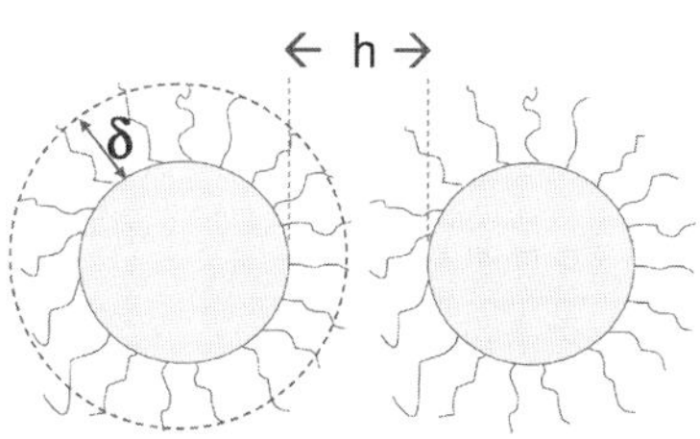

Figure 9.7 Schematic representation of steric layers.

when these are in good solvent conditions and reduction in configurational entropy on significant overlap.

9.5.3
Optimizing Electrosteric and Steric Stabilization

The efficiency of steric stabilization depends on both the architecture and the physical properties of the stabilizing molecule. Steric stabilizers should have an adsorbing anchor with a high affinity for the particles and/or insoluble in the medium. The stabilizer should be soluble in the medium and highly solvated by its molecules.

For aqueous or highly polar oil systems, the stabilizer block can be ionic or hydrophilic, such as poly(alkylene glycols) and for oils it should resemble the oil in character. For silicone oils silicone stabilizers are best; other oils could use a long-chain alkane, fatty ester or polymers such as poly(methyl methacrylate) (PMMA) or poly(propylene oxide).

9.6
Surface–Anchor Interactions

Various types of surface–anchor interactions are responsible for the adsorption of a dispersant on the particle surface: Ionic or acid–base interactions; sulfonic acid, carboxylic acid or phosphate with a basic surface, e.g. alumina; amine or quat with an acidic surface, e.g. silica; hydrogen bonding; surface esters, ketones, ethers, hydroxyls; multiple anchors – polyamines and polyols (H-bond donor or acceptor)or polyethers (H-bond acceptor). Polarizing groups, e.g. polyurethanes, can also provide sufficient adsorption energies and in nonspecific cases lyophobic bonding (van der Waals) driven by insolubility (e.g. PMMA). It is also possible to use chemical bonding, e.g. by reactive silanes.

For relatively reactive surfaces, specific ion pairs may interact, giving particularly good adsorption on a powder surface. An ion pair may even be formed *in situ*, particularly if in low dielectric media. Some surfaces are actually heterogeneous and can have both basic and acidic sites, especially near the iep isoelectric point. Hydrogen bonding is weak but is particularly important for polymerics which can have multiple anchoring.

The adsorption strength is measured in terms of the segment/surface energy of adsorption χ^s. The total adsorption energy is given by the product of the number of attachment points n and χ^s. For polymers the total value of $n\chi^s$ can be sufficiently high for strong and irreversible adsorption even though the value of χ^s may be small (less than $1kT$, where k is Boltzmann's constant and T is the absolute temperature. However, this situation may not be adequate, particularly in the presence of an appreciable concentration of wetter and/or other surfactants used as adjuvants. If the χ^s of the individual wetter and/or other surfactant molecules is higher than the χ^s of one segment of the B chain of the dispersant, these small molecules can displace the polymeric dispersant particularly at high wetter and/or other surfactant molecules, and this could result in flocculation of the suspension. It is therefore essential to make sure that the χ^s per segment of the B chain is higher than that of wetter and/or surfactant adsorption and that the wetter concentration is not excessive.

9.7 Optimizing Steric Potential

In order to optimize the steric repulsion, we consider the steric potential as expressed by Napper [6]:

$$V(h) = 2\pi kTR\Gamma^2 N_A \left(\frac{V_p{}^2}{V_s}\right)(0.5 - \chi)\left(1 - \frac{h}{2\delta}\right)^2 + V_{\text{elastic}} \tag{10}$$

where k is Boltzmann's constant, T is the absolute temperature, R is the particle radius, Γ is the amount adsorbed, N_A is Avogadro's number, V_p is the specific partial volume of the polymer, V_s is the molar volume of the solvent, χ is the Flory–Huggins parameter, δ is the maximum extent of the adsorbed layer and V_{elastic} takes account of the compression of polymer chains on close approach.

It is instructive to examine the terms in this relationship:

1. The adsorbed amount Γ; higher adsorbed amounts will result in more interaction/repulsion.
2. Solvent conditions as determined by χ, the Flory–Huggins chain–solvent interaction parameter; two very distinct cases emerge. We see maximum interaction on overlap of the stabilizing layers when the chain is in good solvent conditions ($\chi < 0.5$). Osmotic forces cause solvent to move into the highly concentrated overlap zone, forcing the particles apart. If $\chi = 0.5$, a theta solvent, the steric potential goes to zero and for poor solvent conditions ($\chi > 0.5$) the steric potential becomes negative and the chains will attract, enhancing flocculation. Note that a poorly solvated dispersant can enhance flocculation/aggregation.
3. Adsorbed layer thickness δ. The steric interaction starts at $h = 2\delta$ as the chains begin to overlap and increases as the square of the distance. Here the importance is not the size of the steric potential but the distance h at which it begins.
4. The final interaction potential is the superposition of the steric potential and the van der Waals attraction as shown in Figure 9.4b.

For sterically stabilized dispersions, the resulting energy–distance curve often shows a shallow minimum, V_{min}, at a particle–particle separation distance h comparable to twice the adsorbed layer thickness δ. For a given material, the depth of this minimum depends on the particle size R and adsorbed layer thickness δ.

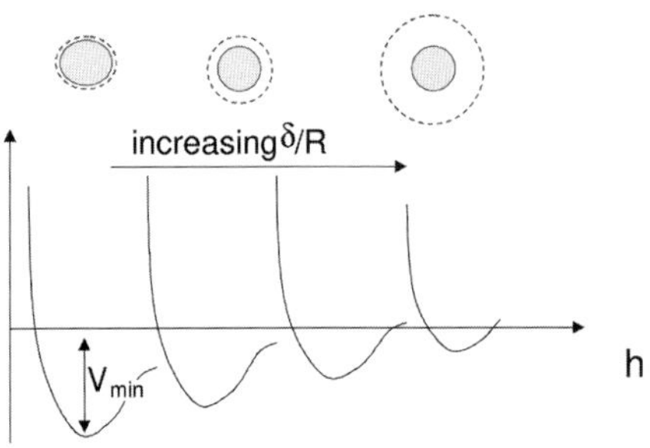

Figure 9.8 Variation of V_{min} with δ/R.

Hence V_{min} decreases with increase in δ/R, as illustrated in Figure 9.8. This is because as we increase the layer thickness the van der Waals attraction weakens so the superposition of attraction and repulsion will have a smaller minimum. For very small steric layers, V_{min} may become deep enough to cause weak flocculation, resulting in a weak attractive gel. Hence we can see how the interaction energies can also determine the dispersion rheology.

On the other hand, if the layer thickness is too large, the viscosity is also increased due to repulsion. This is due to the much higher effective volume fraction ϕ_{eff} of the dispersion compared with the core volume fraction. We can calculate the effective volume fraction of particles plus dispersant layer by geometry and we see that it depends on the thickness of that adsorbed layer, as illustrated in Figure 9.9. The effective volume fraction increases with relative increase in the

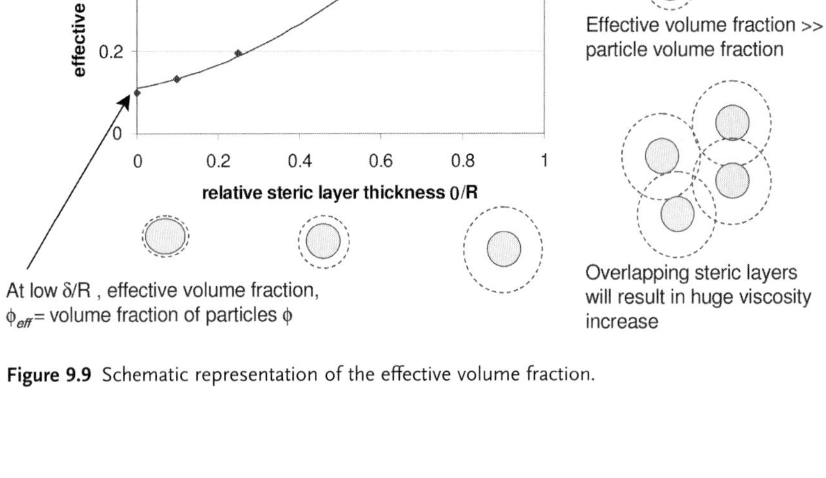

Figure 9.9 Schematic representation of the effective volume fraction.

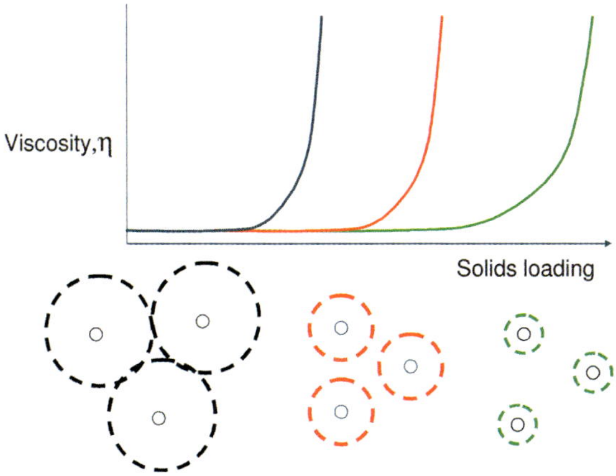

Figure 9.10 Dependence of solids loading on adsorbed layer thickness.

dispersant layer thickness. Even at 10% volume fraction we can soon reach maximum packing ($\phi = 0.67$) with an adsorbed layer comparable to the particle radius. In this case, overlap of the steric layers will result in significant viscosity increases. Such considerations help to explain why the solids loading can be severely limited, especially with small particles. In practice, solids loading curves can be used to characterize the system and will take the form of those illustrated in Figure 9.10.

A higher solids loading might be achieved with thinner adsorbed layers but may also lead to interparticle attraction, resulting in particle aggregation. Clearly a compromise is needed: choosing an appropriate steric stabilizer for the particle size of the pigment.

9.8 Classes of Dispersing Agents

One of the most commonly used type of dispersants for aqueous media is nonionic surfactants.

The most common nonionic surfactants are the alcohol ethoxylates, $R{-}O{-}(CH_2{-}CH_2{-}O)_n{-}H$, e.g. $C_{13/15}(EO)_n$, with n being 7, 9, 11 or 20. These nonionic surfactants are not the most effective dispersants since the adsorption by the $C_{13/15}$ chain is not very strong. To enhance the adsorption on hydrophobic surfaces, a poly(propylene oxide) (PPO) chain is introduced in the molecule, giving $R{-}O{-}(PPO)_m{-}(PEO)_n{-}H$.

The above nonionic surfactants can also be used for stabilization of polar solids in nonaqueous media. In this case the PEO chain adsorbs on the particle surface, leaving the alkyl chains in the nonaqueous solvent. Provided that these alkyl chains are sufficiently long and strongly solvated by the molecules of the medium, they can provide sufficient steric repulsion to prevent flocculation.

A better dispersant for polar solids in nonaqueous media is poly(hydroxystearic acid) (PHS) with molecular weight in the range 1000–2000 Da. The carboxylic group adsorbs strongly on the particle surface, leaving the extended chain in the nonaqueous solvent. With most hydrocarbon solvents the PHS chain is strongly solvated by its molecules and an adsorbed layer thickness in the range 5–10 nm can be produced. This layer thickness prevents any flocculation and the suspension can remain fluid up to high solids content.

The most effective dispersants are those of the A–B, A–B–A block and BA_n types. A schematic representation of the architecture of block and graft copolymers is shown in Figure 9.11.

B, the "anchor chain" (red), is chosen to be highly insoluble in the medium and has a strong affinity to the surface. Examples of B chains for hydrophobic solids are polystyrene (PS), poly(methyl methacrylate) (PMMA), poly(propylene oxide) (PPO) and alkyl chains provided that they have several attachments to the surface. The A stabilizing (blue) chain has to be soluble in the medium and strongly solvated by its molecules. The A chain–solvent interaction should be strong, giving a Flory–Huggins χ-parameter < 0.5 under all conditions. Examples of A chains for aqueous media are poly(ethylene oxide) (PEO), poly(vinyl alcohol) (PVA) and polysaccharides (e.g. polyfructose). For nonaqueous media, the A chains can be poly(hydroxystearic acid) (PHS).

One of the most commonly used types of A–B–A block copolymers for aqueous dispersions are those based on PEO (A) and PPO (B). Several molecules of PE–PP–PEO are available with various proportions of PEO and PPO. The commercial name is followed by a letter L (Liquid), P (Paste) or F (Flake). This is followed by two numbers that represent the composition – the first digit represents the PPO molar mass and the last digit represents the percentage of PEO: F68 (PPO molecular mass 1508–1800 + 80% or 140 mol EO); L62 (PPO molecular

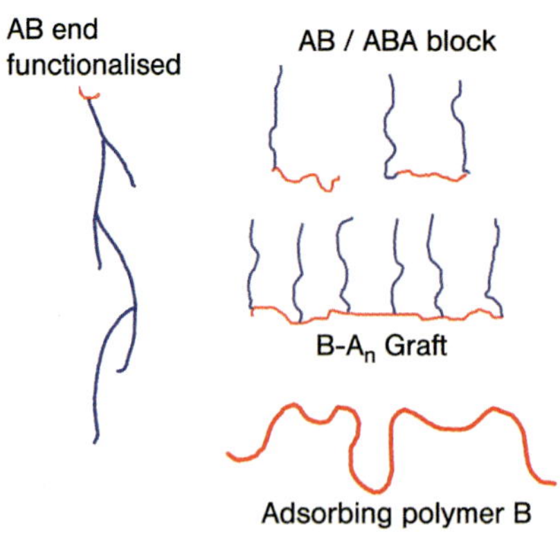

Figure 9.11 Schematic representation of the architecture of block and graft copolymers.

mass 1508–1800 + 20% or 15 mol EO). In many cases two molecules with high and low EO content are used together to enhance the dispersing power.

An example of a BA_n graft copolymer is based on a PMMA backbone [with some poly(methacrylic acid)] on which several PEO chains (with an average molecular weight of 750) are grafted. It is a very effective dispersant, particularly for high solids content suspensions. The graft copolymer is strongly adsorbed on hydrophobic surfaces with several attachment points along the PMMA backbone and a strong steric barrier is obtained by the highly hydrated PEO chains in aqueous solutions.

Another effective graft copolymer is hydrophobically modified inulin, a linear polyfructose chain A (with degree of polymerization > 23) on which several alkyl chains have been grafted. The polymeric surfactant adsorbs with multi-point attachment with several alkyl chains.

9.9 Assessment of Dispersants

9.9.1 Adsorption Isotherms

This is by far the most quantitative method for the assessment and selection of a dispersant. A good dispersant should give a high affinity isotherm, as illustrated in Figure 9.12. The amount adsorbed Γ is recorded as a function of the equilibrium solution concentration, i.e. left in solution after adsorption.

In general, the value of Γ_∞ is reached at lower C_2 for polymeric surfactant adsorption compared with small molecules. The high-affinity isotherm obtained with polymeric surfactants implies that the first added molecules are virtually completely adsorbed and such a process is irreversible. The irreversibility of adsorption is checked by carrying out a desorption experiment. The suspension at the plateau value is centrifuged and the supernatant liquid is replaced by pure carrier medium. After redispersion, the suspension is centrifuged again and the concentration of the polymeric surfactant in the supernatant liquid is determined analytically. For lack of desorption, this concentration will be very small, indicating that the polymer remains on the particle surface.

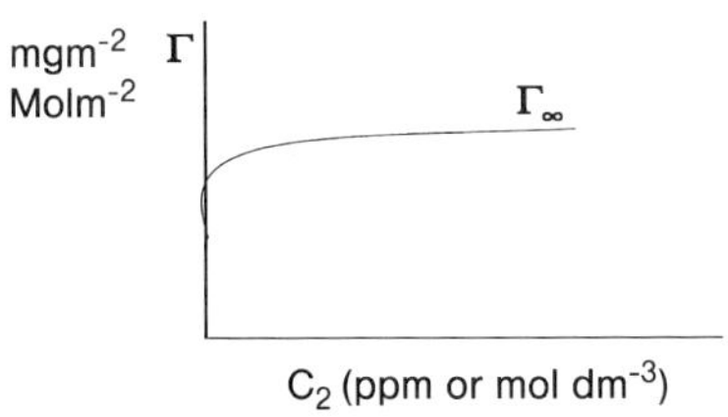

Figure 9.12 High-affinity isotherm.

9.9.2
Measurement of Dispersion and Particle Size Distribution

An effective dispersant should result in complete dispersion of the powder into single particles. In addition, on wet milling (comminution) a smaller particle distribution should be obtained (this could be assessed using a Malvern Master Sizer). The efficiency of dispersion and reduction in particle size can be understood from the behavior of the dispersant. Strong adsorption and an effective repulsive barrier prevent any aggregation during the dispersion process. It is necessary in this case to include the wetter (which should be kept at the optimum concentration). Adsorption of the dispersant at the solid/liquid interface results in lowering of γ_{SL} and this reduces the energy required for breaking the particles into smaller units. In addition, by adsorption in crystal defects, crack propagation occurs (the Rehbinder effect) and this results in the production of smaller particles.

9.9.3
Rheological Measurements

Although Brookfield viscometers are still widely used in industry, they should be used with caution in the assessment of dispersion stability; high shear viscosity (as measured by Brookfield viscometer) can be misleading as a predictor of sedimentation velocity. Low or zero shear viscosity measured on a rheometer is the best indicator. Figure 9.13 demonstrates how at high shear, dispersion B has the highest viscosity and might be expected to give the best resistance to sedimentation. However, dispersion A has the highest low shear viscosity and will hence have the best sedimentation stability.

Concentrated dispersions are viscoelastic, that is, they have both viscous and elastic characteristics. Oscillatory rheometry can therefore give us much more information about the interparticle interactions than viscometry. For example, an elastic modulus which dominates the shear sweep can confer significant stability to a formulation with dispersed solids.

Rheological techniques are often the most informative techniques for assessment and selection of a dispersant. The best procedure is to follow the variation of relative viscosity η_r with the volume fraction ϕ of the dispersion. For this pur-

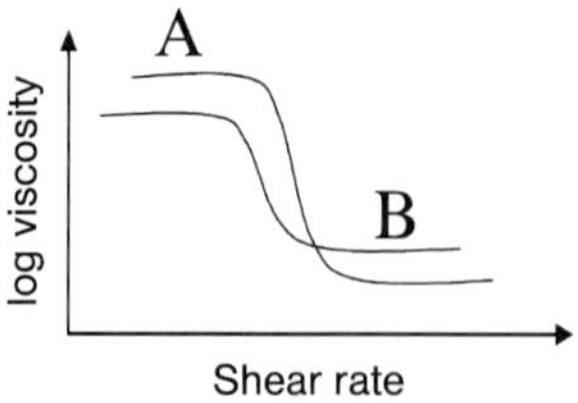

Figure 9.13 Schematic flow curve for particulate dispersions.

pose a concentrated suspension (say 50% w/w) is prepared by milling using the optimum dispersant concentration. This suspension is further concentrated by centrifugation and the sedimented suspension is diluted with the supernatant liquid to obtain volume fractions ϕ in the range 0.1–0.7. The relative viscosity η_r is measured for each suspension using the flow curves. η_r is then plotted as a function of ϕ and the results are compared with the theoretical values calculated using the Dougherty–Krieger equation, as discussed below.

Dougherty and Krieger [7] derived an equation for the variation of the relative viscosity η_r with the volume fraction ϕ of suspensions assumed to behave like hard spheres:

$$\eta_r = \left(1 - \frac{\phi}{\phi_p}\right)^{-[\eta]\varphi_p} \tag{11}$$

where $[\eta]$ is the intrinsic viscosity, which is equal to 2.5 for hard spheres, and ϕ_p is the maximum packing fraction, which is ~0.6–0.7. The maximum packing fraction ϕ_p is obtained by plotting $1/\eta_r^{1/2}$ versus ϕ and in most cases a straight line is obtained, which is then extrapolated to $1/\eta_r^{1/2} = 0$ and this gives ϕ_p.

$\eta_r - \phi$ curves are established from the experimental data using the flow curves. The theoretical $\eta_r - \phi$ curves obtained from the Dougherty–Krieger equation are also established using a value of 2.5 for the intrinsic viscosity $[\eta]$ and ϕ_p calculated using the above extrapolation procedure. As an illustration, Figure 9.14 shows a schematic representation for results for an aqueous suspension of hydrophobic particles that are dispersed using a graft copolymer of PMMA backbone on which several PEO chains have been grafted [8]. Both the experimental and theoretical $\eta_r - \phi$ curves show an initial slow increase in η_r with increase in ϕ, but at a critical ϕ value η_r shows a rapid increase with further increase in ϕ.

It can be seen from Figure 9.14 that the experimental η_r values show a rapid increase above a high ϕ value (>0.6). The theoretical $\eta_r - \phi$ curve [using Eq. (11)] shows an increase in η_r at a ϕ value close to the experimental results. This shows a highly deflocculated (sterically stabilized) suspension. Any flocculation will cause a shift in the $\eta_r - \phi$ curve to lower values of ϕ. These $\eta_r - \phi$ curves can be used for the assessment and selection of dispersants. The higher the value of ϕ at

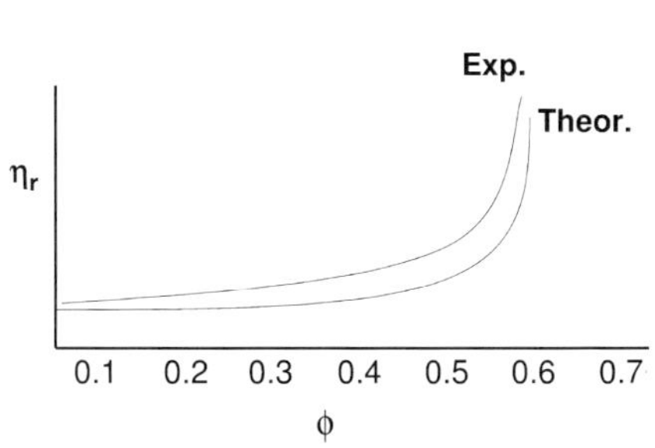

Figure 9.14 Variation of η_r with ϕ for suspensions stabilized with a graft copolymer.

which the viscosity shows a rapid increase, the more effective the dispersant is. Strong adsorption of the graft polymeric surfactant and the high hydration of the PEO chains ensure such high stability. In addition, such polymeric surfactant is not likely to be displaced by the wetter surfactant molecules provided that these are not added at high concentrations. It is essential to use the minimum wetter concentration that is sufficient for complete wetting of the powder.

9.10
Application of the Above Fundamental Principles to Color Cosmetics

Pigments are in fact the primary ingredient of any modern color cosmetic. Pigments need to be incorporated first into slurries and for most color chemists the primary objective is to reduce the viscosity and improve the ease of use of these slurries. It is important to remember that both attractive and repulsive interactions result in a viscosity increase. The aim is therefore to reduce particle–particle interactions.

It is not just in the processing where optimization is required; the particle distribution in the *final* cosmetic will determine its functional activity (color, opacity, UV protection), stability, rheology and skin feel. The particle distribution depends on a number of characteristics such as particle size and shape, surface characteristics, processing and compatibilities but is ultimately also determined by interparticle interactions.

Let us consider some of the potential benefits of controlling particle–particle interactions.

Concerning dispersion stability, there are two main consequences of *instability* in particulate dispersions: flocculation or agglomeration and sedimentation. For color cosmetics, insufficient deagglomeration (all pigments are agglomerated as supplied) can manifest itself as poor color consistency or streaking, with color being liable to change on application. Sedimentation effects can appear as color flotation or plate-out. Sedimentation is determined by gravity and is not necessarily a sign of colloidal instability. It simply needs to be controlled. The sedimentation velocity tends to increase with particle size (hence aggregation is bad), but is reduced by increased fluid viscosity. Dispersion stability may manifest itself in different ways and for the formulator one can expect:

- *Lower viscosity in manufacture.* Figure 9.15 demonstrates the potential benefits (for viscosity dependence on pigment concentration) when a suitable dispersant is added. This can be liberating in removing formulation restrictions and more practically in reducing processing times and cost.
- *Higher pigment concentrations.* These can be achieved, giving increased functionality.
- *Improved color strength.* Color often improves with milling time but again can be stepped up by the incorporation of suitable dispersants.
- *Improved product quality.* One can expect improvements in stability, consistency and function.

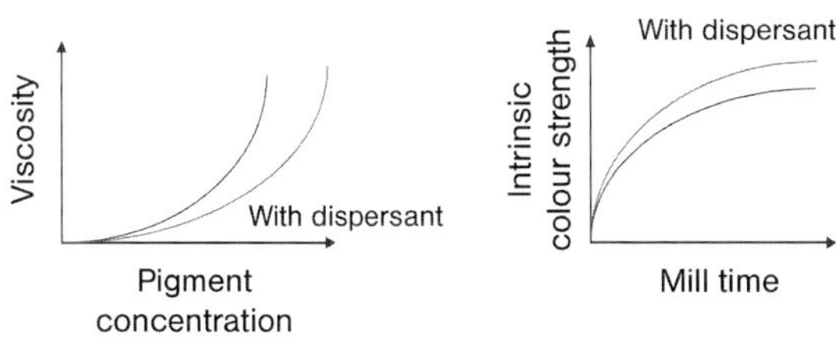

Figure 9.15 Effect of dispersant on viscosity and intrinsic color strength.

Product quality is the key to product differentiation in the market and it is highly desirable, therefore, to reduce flooding and floating caused by flocculation of differing pigments. The control and reproducibility of gloss/shine and brightness and the ability to control rheology and skin feel, particularly at high solids loadings, are all within reach here.

Finally, the optimization of functionality can often depend strongly on the state of dispersion. Opacity and UV attenuation of TiO_2, for example, is strongly dependent on particle size [9] (Figure 9.16). A titanium dioxide pigment, designed to provide opacity in a formulation, will not realize its maximum hiding power unless it is dispersed and remains dispersed in its constituent particles of 200–300 nm. A UV-attenuating grade of TiO_2, on the other hand, must be dispersed down to its primary particle size of 50–100 nm in order to be optimally functional as a sunscreen agent. Both powders as supplied (in order to be handleable), however, have similar agglomerate sizes of several microns.

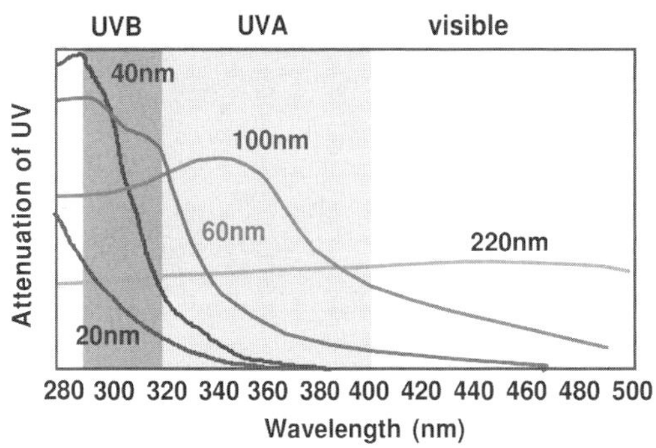

Figure 9.16 UV attenuation versus wavelength for TiO_2 dispersion.

9.11
Principles of Preparation of Color Cosmetics

As mentioned above, the first task is to obtain complete wetting of the powder. Both external and internal surfaces of the agglomerates must be adequately wetted by using a suitable surfactant. For aqueous dispersions, the above-mentioned

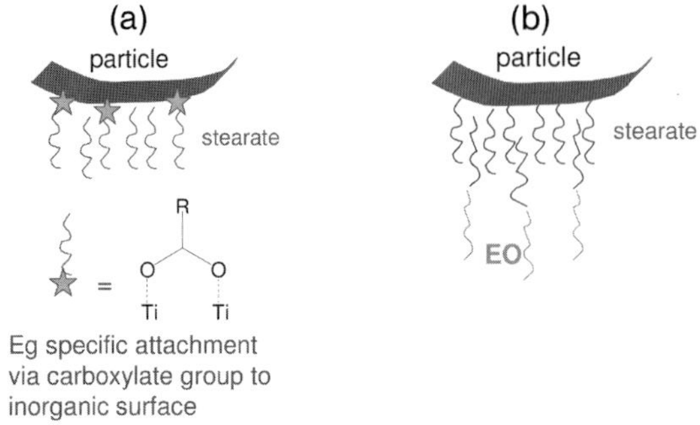

Figure 9.17 (a) Schematic representation of specific interaction of stearate to TiO_2 and (b) effect of addition of alcohol ethoxylate.

wetting agents such as Aerosol OT and alcohol ethoxylates are generally efficient. For hydrophilic pigments in oil, one can use coated particles (with a hydrophobic coating) or sodium stearate, which strongly binds to the hydroxyl surface. A schematic representation for binding of stearate to hydrophilic TiO_2 is shown in Figure 9.17, thus rendering it easily wetted and dispersed in oils. This figure also shows the effect of addition of an alcohol ethoxylate to this coated TiO_2, which can then be dispersed in an aqueous medium.

This process is followed by complete dispersion and/or comminution and adequate stabilization of the resulting single particles, as illustrated in Figure 9.18.

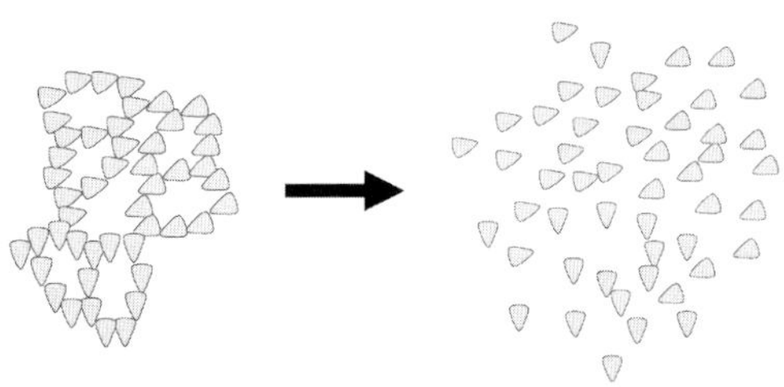

Figure 9.18 Schematic representation of the dispersion process.

9.11.1 Dispersion/Comminution

Simple mixing of inorganic powders can produce a fluid dispersion even at high solids. However, this is not necessarily an indication of a "well-dispersed" material and indeed a particle size analysis (and, for UV attenuators, spectral analysis) demonstrates that particle dispersion is not optimized. Particulate powders are supplied in an aggregated state. However, they must be milled down to their individual units in order to provide their designed function. This process must allow transport of the dispersant to the particle surface and adsorption there. Finally, the dispersion must remain stable to dilution or addition of further formulation components. The presence of a suitable dispersant/stabilizer at the right level can be critical in achieving a usable and stable dispersion and preventing re-aggregation on standing.

9.11.2 Optimizing Dispersion in Practice

In practice, the dispersion chemist may use some simple laboratory tools to assess dispersion quality and arrive at an appropriate dispersion recipe. Having assessed wetting as previously described, one will often plot a dispersant demand curve in order to establish the optimum dispersant loading. The pigment is processed (milling or grinding) in the presence of the carrier oil and wetting agent with varying levels of dispersant. The state of dispersion can be effectively monitored by rheology and/or some functional measurement (e.g. color strength, UV attenuation)

Figure 9.19 shows the results for some fine particle TiO_2 in isopropyl isostearate as dispersing fluid and poly(hydroxystearic acid) as dispersant [10].

Dispersions were produced at 30% w/w solids so that they could be prepared on a bead mill at all dispersant loadings and their UV attenuation properties compared. Zero shear viscosities give an indication of interparticle interactions and were found to be at a minimum at around 5% dispersant. UV attenuation was used as an indicator of particle size.

The unmilled dispersions (1) appeared very fluid, but UV measurement revealed poor attenuation properties, implying that the particles are still aggregated. The solid particles quickly settled out of suspension to form a sediment in the bottom of the beaker. An improvement of UV attenuation properties, along with an increase in viscosity, was observed upon milling. The aggregates were broken down into their constituent particles in the mill (2), but in the absence of dispersant they quickly reaggregate by van der Waals attraction in a more open structure. This caused the mill to block. Further improvements in UV properties were observed when the dispersion was milled in the presence of the dispersant (2), but viscosity was still high.

Addition of sufficient dispersant allows the particles to disperse to single particulates (3) which are well stabilized and the viscosity drops. This is an optimized dispersion. UV properties are well developed.

On addition of further dispersant, the particles gain an extended stabilization layer (4), causing potential overlap of stabilization layers which is sufficient to produce a weak repulsive gel. The viscosity again rises and the dispersion has a measurable yield value. UV properties are still well developed but the solids loading becomes very limited.

These dispersant demand curves, particle size monitoring in addition to solids loading curves (Figure 9.15) are very useful tools in optimizing a pigment dispersion in practice. Further examples are given in Chapter 4.

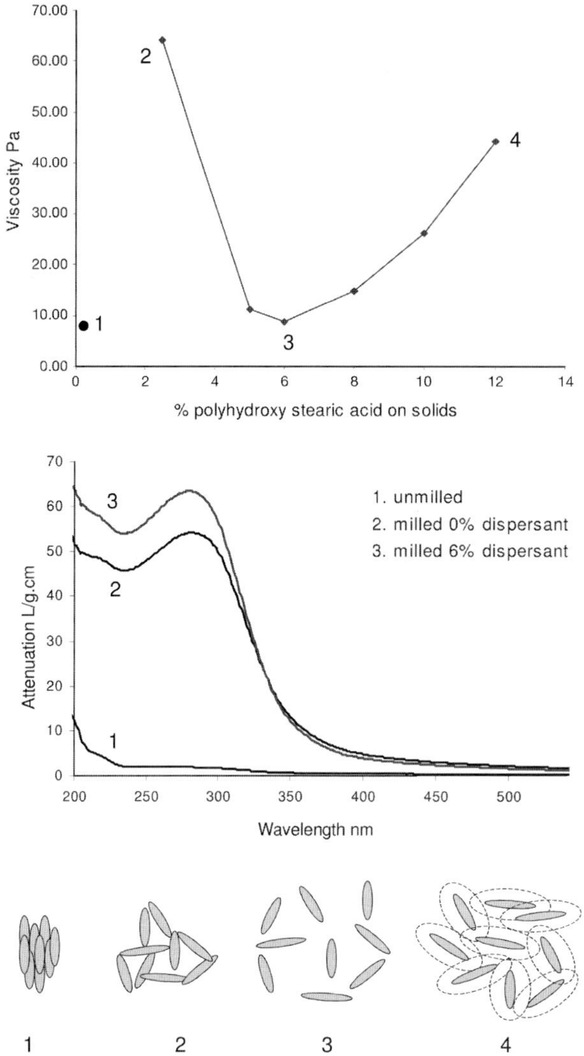

Figure 9.19 Zero shear viscosity (dispersant demand curve), UV attenuation curves and a schematic of the milling process.

9.11.3 Suspoemulsions

Color cosmetic pigments are added to oil-in-water (O/W) or water-in-oil (W/O) emulsions. The resulting system is referred to as a suspoemulsion. The particles can be in the internal or external phases or both, as illustrated in Figure 9.20. An understanding of competitive interactions are also important in optimizing formulation stability and performance.

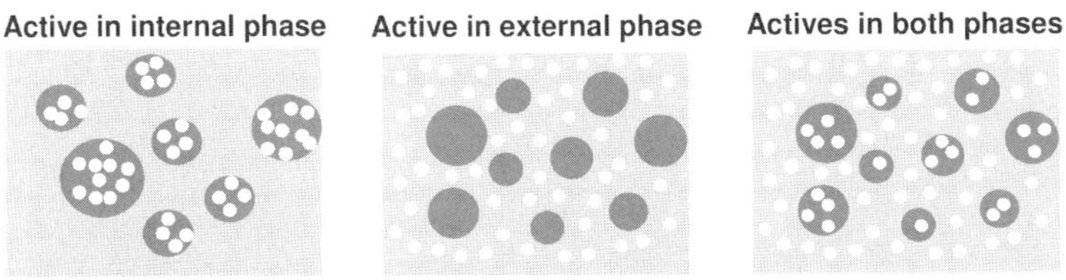

Figure 9.20 Schematic representation of suspoemulsions.

Possible instabilities which might arise in final formulations are as follows:

- heteroflocculation from particles of differing charge;
- electrolyte intolerance of electrostatically stabilized pigments;
- competitive adsorption/desorption of a weakly anchored stabilizer (homoflocculation and emulsion coalescence);
- interaction between thickeners and charge-stabilized pigments;

Several steps can be taken to improve the stability of suspoemulsions, which are in fact very similar to those for optimal steric stabilization: (1) use of a strongly adsorbed ("anchored") dispersant, e.g. by multi-point attachment of a block or graft copolymer; (2) use of a polymeric stabilizer for the emulsion (also with multi-point attachment); (3) preparation of the suspension and emulsion separately and allowing enough time for complete adsorption (equilibrium); (4) using low shear when mixing the suspension and emulsion; (5) use of rheology modifiers; (6) increasing dispersant and emulsifier concentrations to ensure that the lifetime of any bare patches produced during collision is very short; (7) use of the same molecule for emulsifier and dispersant; (8) reducing the emulsion droplet size.

9.12 Conclusions

In this chapter, what we have tried to demonstrate is that optimization of color cosmetics can be achieved through a fundamental understanding of colloid and interface science.

The dispersion stability and rheology of particulate formulations depend on interparticle interactions, which in turn depend on the adsorption and conformation of the dispersant at the solid/liquid interface. Dispersants offer the possibility of being able to control the interactions between particles such that consistency is improved. We have also shown that it is not possible to design a universal dispersant due to specificity of anchor groups and solvent–steric interactions. As color chemists we should be encouraged to take a step back and look at what is stabilizing the particles and how to improve that. Finally, in order to optimize performance in the final formulation, we must consider the interactions between particles, dispersant, emulsifiers and thickeners and strive to reduce the competitive interactions through proper choice of the modified surface and also the dispersant to optimize adsorption strength.

References

1 T. Blake, in *Surfactants*, Th.F. Tadros (ed.), Academic Press, London, 1984.

2 Th.F. Tadros, *Applied Surfactants*, Wiley-VCH, Weinheim, 2005.

3 J. Visser, *Adv. Colloid Interface Sci.*, **3**, 331 (1972).

4 B.V. Deryaguin, L. Landau, *Acta Physicochem. USSR*, **14**, 633 (1941).

5 E.J.W. Verwey, J.Th.G. Overbeek, *Theory of Stability of Lyophobic Colloids*, Elsevier, Amsterdam, 1948.

6 D.H. Napper, *Polymeric Stabilization of Dispersions*, Academic Press, London, 1983.

7 I.M. Krieger, *Adv. Colloid Interface Sci.*, **3**, 111 (1972).

8 Th.F. Tadros, *Adv. Colloid Interface Sci.*, **104**, 191 (2003).

9 J.L. Robb, L.A. Simpson, D.F. Tunstall, *Drug Cosmet. Ind.*, (1994).

10 L.M. Kessell, B.J. Naden, Th.F. Tadros, *Attractive and repulsive gels*, presented at the IFSCC Congress, Orlando, FL, 2004.

10
Starch-Based Dispersions*

Ignác Capek

Abstract

Starch is the most promising raw material for the production of biodegradable plastics, and is a natural renewable polysaccharide obtained from a great variety crops. The native starch granule is heterogeneous both chemically (e.g. amylose and amylopectin) and physically (e.g. crystalline and noncrystalline regions). The presence or absence of crystalline order is often a basic factor underlying starch properties. When starch is heated in excess water, the crystalline structure is disrupted (due to breakage of hydrogen bonds) and water molecules become linked by hydrogen bonding to the exposed hydroxyl groups of amylose and amylopectin. Starch is not a true thermoplastic but in the presence of plasticizers (water, glycerin, sorbitol, etc.) at high temperatures and under shear, it readily melts and flows, allowing its use as an injection, extrusion or blow molding material, similar to most conventional synthetic thermoplastic polymers. The possible use of starch as a thermoplastic, biodegradable, non-food material depends on its attainable properties. To fulfill the various demands for the functionality in different starch products, industrially processed starch is modified enzymatically, physically or chemically. This allows other potential uses in many different industries. Chemical substitution and chemical crosslinking are the main types of modifications that are carried out. Chemically modified starches with improved properties are gaining increasing importance in industry, not only because they are inexpensive, but also mainly because the polysaccharide portion of the product is biodegradable. Their applications relate to agriculture, industry, medical treatment and sanitation, etc., which make them important polymeric materials and dispersions in the fields of dehumidification, dehydration, water preservation and water absorption. For the last few decades, chemical modification of starch by graft copolymerization of vinyl monomers on to it has been a subject of both academic and industrial interest. Depending on the extent of crosslinking,

* A list of symbols and abbreviations is given at the end of this chapter.

Colloids and Interface Science Series, Vol. 4
Colloids in Cosmetics and Personal Care. Tharwat F. Tadros

ISBN: 978-3-527-31464-5

granule swelling will be more or less important, leading to a wide range of rheological properties. The properties of crosslinked starch suspensions are very similar to those of closed-packed microgel dispersions and can be described on a similar basis. Grafting has also been used as an important technique for modifying the physical and chemical properties of polymers and colloidal properties of dispersions. The grafting efficiency of starch by copolymerization varies with the type of monomer (the water-soluble or oil-soluble unsaturated monomers) and starch. The grafting can proceed on the surface granules, which can lead to core–shell structures. When a water-soluble starch (carboxymethyl starch) is used, the grafting is more effective because starch molecules (their radicals) dissolved in the reaction medium interact with monomer. Much effort has been made in recent years to develop biodegradable materials, particularly compostable plastics, i.e. plastics that degrade easily under well-defined environmental conditions. These materials may be synthetic, natural or a combination of both.

10.1 Introduction

After cellulose, starch is the most abundant carbohydrate in the world. The raw material is available in sufficient amounts and in high purity. Its total annual world production is estimated to be between 25 and 45 million tons. This is reason why starch has become one of the most studied and promising raw materials for the production of biodegradable plastics – it is a natural renewable polysaccharide obtained from a great variety of crops. The applications of these material are varied, ranging from packaging to agricultural fields. Starch-based composites, reinforced with some additives, can be injection-molded and exhibit properties that might be useful in biomedical applications [1]. In recent decades, studies concerning the total or partial replacement of synthetic plastics by biodegradable materials have been increasing steadily [2, 3]. To fulfill the various demands for the functionality in different starch products, industrially processed starch is modified enzymatically, physically or chemically. In most chemical modifications of starch, usually referred to as chemical derivatization, the granule form is maintained and the hydroxy groups are partially substituted, yielding starch ethers and esters, in addition to anionic and cationic starches. Other types of chemical derivatization are oxidation, crosslinking and grafting of starches [4]. Grafting and copolymerization have been used as important techniques for modifying the physical and chemical properties of polymers. Starch is the most promising raw materials for the production of biodegradable plastics, which is a natural renewable polysaccharide obtained from a great variety of crops.

Starch, the main energy reserve of higher plants, consists of amylose and amylopectin. Amylose is considered as an essentially $(1 \rightarrow 4)$-linked α-D-glucan, whereas amylopectin contains up to 5% of branched α-D-glucose residues. The short amylopectin side-chains are linked to longer chains [5] and arranged in double helices, building up the organized crystalline framework of the starch granule [6, 7]. Amylose is essentially a linear polymer in which the anhydroglucose units are predominantly linked through glucosidic bonds. It may contain about 200–2000 anhydroglucose units. Amylopectin is a branched polymer containing, in addition to anhydroglucose units linked together as in amylose through glucosidic bonds, periodic branches at the carbon–6 position. Each branch contains about 20–30 anhydroglucose units. The molecular mass distribution, amylose and amylopectin content and the degree of branching of these molecules depend on the botanical source of the starch granules [8]. Light-scattering measurements indicate molecular weights in the millions. Depending on the botanical source of the starch, the amylopectin/amylose ratio can vary from 1 to near 0.3. Potato starch, for example, contains about 79% amylopectin and 21% amylose. Chemical modifications of starches provide starch products that fulfill various demands [9]. Detailed information on the distribution of substituents can contribute to the understanding of relations between molecular structure and functional properties, thus opening up ways to more rational derivatization processes. It has been reported that in methylated starches, crystalline linear amylopectin side-chains, which play an important role in the retrogradation of gelatinized starches, contain fewer substituents than amorphous branched parts [10, 11].

Starch is an inexpensive material in comparison with most synthetic plastics and is readily available. It is produced by nearly all green plants to store energy. Starch occurs as granules in grains, roots and tubercules and is composed of a mixture of amylose and amylopectin, both of high molar mass and consisting of $(1 \rightarrow 4)$-α-D-glucopyranose repeating units (Figure 10.1). Whereas amylose is essentially linear, amylopectin is highly branched via additional $(1 \rightarrow 6)$-α linkages, with a cluster-type structure [12]. However, a slight degree of branching (9–20 branch [α-$(1 \rightarrow 6)$] points per molecule) has been reported in amylose (the major component) from various starch sources [13]. Furthermore, the extent of branching increases with the molecular size of amylose. The amylose/amylopectin ratio varies with the origin of the starch and is responsible, to a large extent, for its functional characteristics. Amylopectin is the major component with a weight-average molecular weight (M_w) of the order of 10^7–10^9. It is composed of linear chains of $(1 \rightarrow 4)$-α-D-glucose residues connected through $(1 \rightarrow 6)$-α linkages (5–6%). The average size of the unit chains of amylopectin is 20–25. The amylopectin molecule contains several distributions of chains (*a*, *b* and *c*) which differ in their chain length. The *a* chains are linked to *b* chains and do not carry any other chains; the *b* chains carry one or more *a* chains and/or *b* chains; and the *c*-chain has the reducing end group of the molecule. The branch points in the amylopectin molecules are not randomly distributed, but are clustered and the inter-adjacent linear segments form thin crystalline lamellar domains having a width of 5–7 nm [14]. When observed under polarized light, the starch granules show birefringence, which indicates a certain degree of molecular organization. X-ray scattering has shown that granular starch has an overall crystallinity of about 20–45%. The amorphous regions are formed by amylose and the residues around the branch points of amylopectin. The linear segments

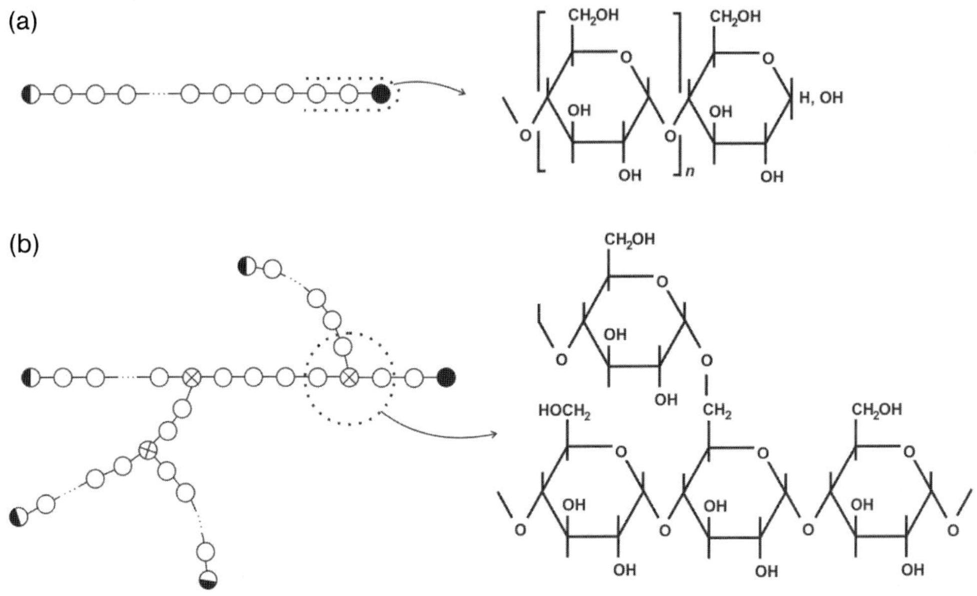

Figure 10.1 Amylose (a) and amylopectin (b). An empty circle denotes D-glucoside units and their connection via an α-glucoside bond, a full circle the reducing ending unit and a half-empty circle the non-reducing unit. A circle with a cross denotes the point of grafting.

of amylopectin are present in the form of double helices crystallized into thin lamellar domains [15].

The native starch granule is heterogeneous both chemically (e.g. amylose and amylopectin) and physically (e.g. crystalline and noncrystalline regions). The presence or absence of crystalline order is often a basic factor underlying starch properties [16]. It is very well known that amylose exists in three different helical conformations in aqueous solutions. These polymorphs are named A-, B- and V-forms. The A- and B-forms comprise parallel-packed, left-handed double helices. The A- and B-forms can be considered as extended helices with, unlike the V-form, no hydrogen bonding between consecutive turns of the helices. The V-form occurs only in the presence of some complex-forming ligand. It has been known for a long time that amylose forms helical inclusion complexes with a variety of organic compounds such as lipids [17], carboxylic acids [18] and different surfactants [19]. The helical coil formed by the amylose–lipid inclusion complex in aqueous solution has been demonstrated to comprise 6–8 repeating glucose units per helical turn. Amylose takes the so-called V-form conformation when forming the inclusion complexes. The V amylose is a generic name for crystalline amyloses, obtained as single helices co-crystallized with compounds such as water, iodine, dimethyl sulfoxide (DMSO), alcohols or fatty acids. In the

V-form, a single chain of amylose forms a helix with a relatively large cavity. The central cavities of the V-forms have a pitch of about 0.8 nm per turn. A detailed X-ray conformational analysis of the hydrated helix with a guest molecule inside has been reported [20]. Yamamoto et al. [19] found that the binding of sodium alkyl sulfates to amylose is cooperative when the degree of polymerization (DP) of the polymer is high enough and that its cooperativeness depends on the carbon chain length of the surfactant. Also, the outer branches of the amylopectin molecule are assumed to form the same kind of inclusion complexes with suitable lipids [21]. Depending on their origins, various types of native starches present specific morphologies giving distinctive X-ray powder patterns [6, 16, 22]. The sharpness of the X-ray diffraction pattern of starch granules depends on their water content, the B type being more sensitive to hydration than the A type starch [6, 16, 22–24]. When the crosslinking degree (cld) changes, the morphology of powder, tablet or film forms also changes.

The overall water suspension properties of starches are strongly influenced by their botanical origin and by the shearing and/or heating conditions under which they are prepared [25]. In its native form, a starch granule is insoluble in cold water and most uses involve a heating treatment in the presence of an excess of water. Below a critical temperature (about 60 °C), that is, the so-called gelatinization temperature, starch granules absorb water and undergo swelling to many time their original size. This process is often attributed to the diffusion of linear amylose chains outside the swollen granule. Beyond this critical temperature, the swollen starch granules can undergo a disruption into smaller aggregates or particles. Complete solubilization of all the starch granules or macromolecules may occur when starch is heated at temperatures beyond 100 °C. The heating treatment is therefore associated with large changes in the viscosity characteristics. When starch is heated in excess water, the crystalline structure is disrupted (due to breakage of hydrogen bonds) and water molecules become linked by hydrogen bonding to the exposed hydroxyl groups of amylose and amylopectin. This causes an increase in granule swelling and solubility. The swelling power and solubility provide evidence of the magnitude of interaction between starch chains within the amorphous and crystalline domains. The extent of this interaction is influenced by the amylose/amylopectin ratio and by the characteristics of amylose and amylopectin in terms of molecular weight/distribution, degree of length of branching and conformation. Amylose–lipid complexes have been shown to restrict swelling and solubilization. The swelling power and solubility of different starches (at 95 °C) ranging from 14.6 to 51% and from 7.8 to 26.7%, respectively [14].

Starch, when heated in the presence of excess water, undergoes an order–disorder phase transition called gelatinization over a temperature range characteristic of the starch source. The above phase transition is associated with the diffusion of water into the granule, water uptake by the amorphous background region, hydration and radial swelling of the starch granules, loss of optical birefringence, uptake of heat, loss of crystalline order, uncoiling and dissociation of double helices and amylose leaching [26, 27]. Gelatinization in excess water is primarily a swelling-driven process. This swelling acts to destabilize the amylopectin crystalline lamellae, which are ripped apart. This process occurs rapidly for an individual crystallite, but over a wide range for the whole granule. The same mechanism occurs in conditions of limiting water. However, there is insufficient water for gelatinization to proceed to completion. Many methods are available for the determination of starch gelatinization, such as Kofler hot-stage microscopy, differential scanning calorimetry (DSC), pulsed nuclear magnetic resonance, enzymatic digestibility, small-angle X-ray scattering (SAXS) and small-angle scattering. For example, DSC measures the gelatinization transition temperatures onset (T_o), midpoint (T_p), conclusion (T_c) and the enthalpy (ΔH) of gelatinization [14, 28]. Almost any application of starch dispersion involves processes that lead to the disruption of the molecular order within the granules. This may be achieved by heating a starch suspension in water above a temperature denoted the gelatinization temperature. Gelatinization, which is influenced by the presence of the other solutes, is preceded by swelling and provides irreversible changes in properties [29]. When disruption of the crystalline regions occurs in the presence of a low content of water, the process is denoted melting. Industrially, thermal and mechanical energy input associated with the addition of plasticizers, such as water and glycerol, are necessary to transform granular starch into a homogeneous matrix. Starch is not a true thermoplastic but in the presence of plasticizers (water, glycerin, sorbitol, etc.), at high temperatures (90–180 °C) and under shear, it readily melts and flows, allowing its use as an injection, extrusion or blow molding material, similar to most conventional synthetic thermoplastic polymers. In this respect, it must be remembered that starch is a natural material with varying properties and that generally is a mixture of branched molecules (amylopectin) and linear molecules (amylose). To assess starch as a technical polymer, comprehensive knowledge of the processing behavior, structure and properties of the separated amylose and amylopectin components and mixtures of both starch fractions and different additives in relation to their molecular weight is required.

Chemically modified starches with improved properties are gaining increasing importance in industry, not only because they are inexpensive, but also mainly because the polysaccharide portion of the product is biodegradable. Their applications relate to agriculture, industry, medical treatment, sanitation, etc., which make them important polymeric materials in the fields of dehumidification, dehydration, water preservation and water absorption. Chemical modification of starch by graft copolymerization of vinyl monomers on to it has given a series of composites materials [30–34]. Also much work has been reported on the grafting of acrylonitrile [35, 36], acrylic acid [37–39], methacrylates [40–42] and acrylamide (AAm) on to starch [43].

Although mostly used as foodstuffs, the versatility of starch allows its use in a multitude of non-food uses. These include its use in the formulation of products such as adhesives, detergents, ceramics, paper coatings, aids in textile industry and in oil recovery operations. To improve its versatility further, chemical modification of polysaccharide chains is the focus of current research in order to derive products with "tailor-made" characteristics for a variety of applications. Homogeneous processes using organic solvents can be used to obtain derivatives from native starch with uniform and selective substitution. DMSO is a solvent for both starch and many synthetic polymers and is used to solubilize a number of graft copolymers based on starch. The water–DMSO mixed solvent system has been used as a solubilizing agent for starch and its components [44]. Amylose forms a stable solution in DMSO. Although starch dispersions in water are unstable, with the phenomena of retrogradation being observed, dispersions in aqueous DMSO have been reported to be relatively stable [45].

The possible use of starch as a thermoplastic, biodegradable, non-food material depends on its attainable properties. The desirable biodegradability of extruded starch films has to be weighed against inadequate mechanical properties and stability problems as compared with synthetic polymer films. As compared with synthetic polymers, little is yet known about the structure–property relationships in extruded starch. For example, van Soest et al. [46] showed that the structure of several native starch extrudates was dependent on the starting compound and the conditions of extrusion. The influence of ambient humidity on the crystallinity and stress–strain behavior of thermoplastic potato starch materials has been reported [47], and Fritz et al. [48] studied the use of starch as a component in a compound with synthetic polymers or renewable raw materials.

Traditional plastic materials, produced from synthetic polymers, are known for their inertness to immediate attack by microorganisms. Although this characteristic is associated with their multipurpose applications, increasing environmental concerns have arisen from their disposal in nature. As a response, the development of biodegradable materials that could replace synthetic polymers, at least in some of their applications, has been encouraged. Much effort has been made in recent years to develop biodegradable materials, particularly compostable plastics, i.e. plastics that degrade easily under well-defined environmental conditions. These materials may be synthetic, natural or a combination of both. Polysaccharide-based polymer is one of the most promising materials to achieve this object. It is produced worldwide from several crops and is considered truly biodegradable [49]. Much of the research efforts on the subject are focused on the development of thermoplastic materials composed essentially of starch [50]. The use of starch to produce biodegradable plastics began in the 1970s. In the granular state, it was used as a filler for polyolefins [51] and as a component in synthetic polymer blends with a totally disrupted starch granule structure [52]. Thermoplastic starch has two main disadvantages compared with most plastics currently in use, i.e. it is mostly water soluble and has poor mechanical properties. These can be improved by mixing starch with certain synthetic polymers and adding crosslinking agents such as Ca and Zr salts [53].

Despite the efforts to recycle used plastics, recycling is neither practical nor economical for certain applications, such as waste bags, agricultural mulch films or food packaging. For this kind of application, plastics are expected to degrade into safe by-products after their usage under normal composting conditions. Extended studies have been undertaken to replace partially or totally the synthetic nonbiodegradable plastics. Replacement of petroleum-based plastics with materials from agroresources, especially starch, is attractive from the standpoint of providing biodegradation properties to the end product. This replacement will permit us to conserve our petrochemical resources and to find out new nonfood uses of starch. Indeed, starch is inexpensive (about US$ 0.2 per pound), is totally biodegradable and is available in large quantities from certain crops produced in abundance beyond available markets [54].

Most of the unique properties of starch and its dispersions arise from the fact that it is composed of two distinct polymer fractions, amylose and amylopectin, which are packed as spherocrystalline granules. The granules are built up by alternating semicrystalline and crystalline shells and amylopectin is mainly responsible for the crystallinity of starch [55]. On heating starch under excess water conditions, the granules swell, the starch polymers are partially solubilized and leached from the granules and finally the starch granules disintegrate. The irreversible changes taking place on heating are commonly denoted gelatinization. Depending on the gelatinization conditions, namely on the type of starch, starch concentration, time–temperature conditions and the amount of shear, large differences in the extent of granule swelling and solubilization may be obtained [56]. This is important for the preparation of starch model systems, which are used to investigate different aspects of starch in food, such as the textural properties, the interactions between starch and other food components and the enzymatic degradation of starch.

Different physicochemical techniques, such as differential scanning calorimetry [57], wide-angle X-ray diffraction [58] and nuclear magnetic resonance [59] may be applied to study the changes taking place at the molecular level on heating starch. On the other hand, microscopic and rheological methods provide structural information over a long distance scale, i.e. at the supramolecular level. The characterization of starch granules and their changes on gelatinization by light microscopy, mostly using phase contrast, is a classical method in starch physicochemistry [60]. More detailed knowledge on the microstructure of starch systems and the distribution of amylose and amylopectin was obtained by microscopic investigation of cryo-sectioned systems stained with iodine [61]. Likewise, rheological methods, empirical and fundamental, are sensitive to differences in the swelling state of starch granules. They were successfully applied to identify the differences in the swelling behavior of starches of different botanical origin [62] and the influence of the preparation procedure [63].

The classical method of Leach et al. [64], which comprises a centrifugation step to separate the starch granules from the solubilized material, allows the quantification of the swelling capacity and solubility of starches. This method and modifications of it were successfully applied to study the differences between starches, the influence of complex-forming lipids [65] and the preparation procedure [66].

10.2
Starch-Based Nanomaterials

10.2.1
Modification Approaches

Starch modification is an attractive way to produce thermoplastic materials and has been well known since the early 1940s [67, 68]. Modified and unmodified starch products are extensively used for a variety of applications such as sizing agents for textiles and paper, as adhesives for corrugated and laminated paper boards and wall papers, flocculants, binders, fabric printing aids, thickeners and many other non-food industrial applications. For the production of surface coatings and polymeric molding compounds, industry relies heavily on synthetic polymeric resins produced from petrochemical resources. Petrochemicals are depletable resource and the ever-increasing demand for petroleum-based products has adversely affected their cost and availability in recent times. Starches are readily available and renewable. The potential of starch-based products can be substantially increased if it is possible to alter or correct the inherent defects which limit applications in coatings and shaped articles applications. Hence the use of starch as a partial substitute for petroleum-derived polymers is currently an active area of research. A further advantage of such a substitution is the low cost and biodegradability of synthetic polymers incorporating the plant-derived materials [69]. The starchy portion of the polymer can be easily attacked by microorganisms, leading to environmental breakdown of the material, thereby losing its integrity resulting in particles small enough to cause minimal damage to the environment [41, 70].

To fulfill the various demands for the functionality in different starch products, industrially processed starch can be modified chemically. Chemical substitution and chemical crosslinking are the main types of modifications that are carried out. In most chemical modifications of starch, usually referred to as chemical derivatization, the granule form is maintained and the hydroxy groups are partially substituted, yielding starch ethers and esters, in addition to anionic and cationic starches. Other types of chemical derivatization are oxidation, crosslinking and grafting of starches [4]. For example, chemical crosslinking is particularly used to inhibit granule disruption. Depending on the extent of crosslinking, granule swelling will be more or less important, leading to a wide range of rheological properties. The properties of crosslinked starch suspensions are very similar to those of close-packed microgel dispersions and can be described on a similar basis [71]. Grafting has also been used as an important technique for modifying the physical and chemical properties of polymers. Graft polymerization originates from the formation of an active site at a point on a polymer chain other than its end and exposure of this site to a monomer. Most graft copolymers are formed by radical polymerization. In many instances, chain transfer reactions are involved by the abstraction of hydrogen atoms [72]. Starch is modified by grafting with vinyl monomers (e.g. methyl acrylate) on to the starch backbone, yielding thermoplastic materials that can be injection molded or extruded into films with properties similar to those of low-density polyethylene [73]. Thermoplastic starch is a relatively new material for application as a biodegradable plastic and is one of the main polymers studied today in this field. It is used alone or compounded, usually with polar synthetic polymers, in contents that usually exceed 50%. Starch graft copolymers are becoming increasingly important because of their potential application in industry. The wide range of available vinyl and other monomers suggests that the grafting is a powerful method for effecting substantial modification to starch properties, thereby widening its range of applications [72]. Several synthetic polymers have been used to improve the mechanical properties of thermoplastic starch, such as ethylene–acrylic acid copolymer and ethylene–vinyl alcohol copolymer [74].

Various neutral starch derivatives have been prepared by reacting maize starch with mono- and dimethylol resins based on urea, thiourea and melamine [75]. On reacting starch (StOH) with mono-(RCH_2OH) or dimethylol-containing resins ($HOCH_2RCH_2OH$) in the presence of magnesium chloride, the following reactions occur:

$$\mathrm{StOH + RCH_2OH \rightarrow StOCH_2R + H_2O} \tag{1}$$

$$\mathrm{2StOH + HOCH_2RCH_2OH \rightarrow StOCH_2RCH_2OSt + 2H_2O} \tag{2}$$

These reactions occur in addition to self-polymerization of the resins used in mono- and dimethylolurea, thiourea and melamine, respectively. Starch was reacted with various resins using different magnesium chloride concentrations. The data show that (1) the nitrogen content of prepared samples [e.g. monomethylolurea (MMU)] increases with increasing catalyst concentration to reach a maximum value and decreases on using higher concentrations:

$$\%\mathrm{N}/\mathrm{MgCl_2}\ (\mathrm{g}):\ 4.6/0.5,\ 5/1\ (\text{maximum}),\ 4.7/1.5,\ 4.5/2,\ 4.2/2.5,\ 4/3 \tag{3}$$

and (2) the maximum occurs at different concentrations depending on the resin type.

Starch was reacted with various resins using different resin:starch molar ratios (0.05–1.0) in the presence of magnesium chloride at 150 °C. The data [the nitrogen content and reaction efficiency (%) of prepared samples] show the following:

1. The nitrogen content of prepared samples increases with increasing resin: starch molar ratio (R_{rst}), e.g. for MMU $R_{rst}/\%N$ varies as follows:

$$R_{rst}/\%\mathrm{N} = 0.05/0.85,\ 0.07/1.13,\ 0.1/1.48,\ 0.2/2.1,\ 0.25/2.5,\ 0.33/3,\ 0.5/3.9,\ 1/4.8 \tag{4}$$

2. The nitrogen content of the starch–dimethylol resin reaction product is higher than that of the starch–monomethylol resin reaction product. This can be explained as follows. On reacting starch with resin two reactions occur:
   a) a reaction between starch and resin and
   b) self-polymerization of the resin. Self-polymerized monomethylol resin acquires a lower molecular weight and higher solubility than that of the self-polymerized dimethylol resin. On the other hand, self-polymerized dimethylol resin acquires higher molecular weight and lower solubility, which is reflected in the high nitrogen content of the extracted samples.

3. The nitrogen content of starch derivatives prepared by using thiourea resins is lower than that prepared by urea resins. This is valid in the case of mono- and dimethylol resins. This is attributed to
   a) higher solubility values of self-polymerized thiourea resins than that of urea resins or
   b) lower reactivity of thiourea resins than that of urea resins towards reaction with starch.

4. The reaction efficiency ($R_{rst}.E$ in %) decreases with increasing R_{rst}, for example, for MMU $R_{rst}/\%R_{rst}.E$ varies as follows:

$$R_{rst}/\%R_{rst}.E = 0.05/100,\ 0.07/96,\ 0.1/89,\ 0.2/65,\ 0.25/64,\ 0.33/60,\ 0.5/55,\ 1/40 \quad (5)$$

   This behavior was discussed in terms of
   a) the limited available hydroxyl groups on the starch macromolecule;
   b) the crosslinking reactions, which decrease the available surface area and the hydroxyl groups on using higher resin concentrations; and
   c) the effect of steric hindrance.

5. The reaction efficiency (%) on using dimethylol resins is higher than that of monomethylol resin. This is due to the crosslinking effect of the dimethylol resins that serves to lower the solubility values of self-polymerized resins.

6. The reaction efficiency (%) of different starch derivatives follows the order

$$\text{urea resins} > \text{thiourea resins} > \text{melamine resins} \quad (6)$$

   This was attributed to steric hindrance and differences in the solubility of self-polymerized resins, which is higher in the case of melamine resins and lower in the case of urea resins.

Several starch derivatives were prepared by reacting starch with mixtures of dimethylol resin and the resin base (1:1) using different mixture:starch molar ratios. On reacting starch with dimethylolurea–urea, dimethylolthiourea–thiourea or dimethylolmelamine–melamine mixtures in the presence of magnesium chloride, the following reaction products are formed:

1. $StOCH_2HN{-}C{=}ONHCH_2HN{-}C{=}ONH_2$ (7)
2. $StOCH_2HN{-}C{-}ONHCH_2OSt$
3. the reaction product of dimethylolurea with urea
4. self-polymerization product of dimethylolurea

The same reactions occur in the case of thiourea and melamine derivatives. All of the starch derivatives (MMU, monomethylolurea; DMU, diethylolurea; MMTU, monomethylolthiourea; DMTU, dimethylolthiourea; MMM, monomethylolmelamine; DMM, dimethylolmelamine) thus prepared were water insoluble and were shown to have zero swellability [75].

T. Yamada et al. [76] have shown that treatment of potato starch with HCl solution, followed by treatment with either KOH or saturated $Ca(OH)_2$ solution, reduced the Brabender viscosity curves. The extent of this reduction followed the order

$$Ca^{2+} > H^+ > K^+ \tag{8}$$

The higher reduction in viscosity observed with Ca^{2+} was attributed to a crosslinking effect between Ca^{2+} and phosphate groups on amylopectin. In the authors' opinion, the influence of ions on starch viscosity is probably due to the interplay of two factors:

1. structuring of water molecules around the ions (this would reduce granular swelling, resulting in a lower viscosity); and
2. interactions of ions with the phosphate groups on amylopectin.

It has been reported [77–79] that the introduction of reactive functional groups into the backbone of starches brings about products (anionic and cationic starches) that are capable of removing heavy metal ions from industrial waste water.

Thus, anionic starches could be prepared in any of several ways:

1. via reacting crosslinked starch with monochloroacetic acid [80];
2. via alkali treatment of polyacrylonitrile–starch graft copolymer [81];
3. via alkali treatment of polyacrylamide–starch graft copolymer [82];
4. by treating a poly(glycidyl methacrylate)–starch graft copolymer with phosphoric acid [2]; and
5. via treatment of a methylolated polyacrylamide–starch graft copolymer with acidic salts [83].

On the other hand, cationic starch can be prepared as follows:

1. via treatment of starch with dialkylaminoalkyl chloride [84];
2. or a chlorohydrin or a compound with an epoxy group containing amines in the presence of alkali [85];
3. or treatment of a poly(glycidyl methacrylate)–starch graft copolymer [86];
4. or methylolated polyacrylamide–starch graft copolymer with different amines [87].

Table 10.1 Variation of properties of esterified starch and amylose products with the degree of substitution (ds) and the reaction conditions for starch and amylose esterification [88][a].

Ester/alkyl chain code name ds	Contact angle (°)	%Water absorption[b]
Octanoated starch/C_8 OCSt1.8	87	3.6
Octanoated starch/C_8 OCSt2.7	92	3.8
Dodecanoated starch/C_{12} DODSt2.7	95	0.25
Octadecanoated starch/C_{18} OCDSt1.8	93	3.4
Octadecanoated starch/C_{18} OCDSt2.7	95	0.2
Octanoated amylose/C_8 OCAm0.57	Dissolves	–
Octanoated amylose/C_8 OCAm2.7	83	2.5
Dodecanoated amylose/C_{12} DODAM2.7	90	0.6
Octadecanoated amylose/C_{18} OCDAM2.7	132	0.4

a) Reaction time: 3 h at 105 °C, 0.12 mol (potato native) starch or amylose, 0.14–0.5 mol chloride.
b) Swelling time: 35 days.

Aburto et al. prepared and characterized starch and amylose esters with higher acid chlorides (C_8, C_{12}, C_{18}), thus having a longer side-chain [88]. The appearance of the esters produced depends on their degree of substitution (ds). Those with high ds have the form of a fluffy yellowish mass and behave like thermoplastic materials, whereas those with a low ds have the appearance of a white powder (Table 10.1). The degree of substitution for a starch or amylose derivative is defined as the moles of substituents of hydroxyl groups per D-glucopyranosyl structural unit of the polymer. Since each repeating unit contains three hydroxyl groups, the theoretical maximum ds is three. Table 10.1 shows the calculated ds of the synthesized starch esters as determined by elemental analysis and ^{1}H NMR spectroscopy.

The ^{1}H NMR spectrum of the esterified starch shows the three protons of the terminal methyl group of the acyl chain as a triplet at 0.86 ppm (Figure 10.2). The peaks between 1.23 and 1.67 ppm correspond to the 10 protons of the methylene groups in the acyl chain; whereas at a chemical shift of 2.2 ppm, the signal of the two protons of the α-methylene group is observed. Additionally, the NMR spectrum of the esterified starch reveals the presence of the seven protons of the glycoside structure between 3.5 and 5.5 ppm that are also found in the NMR spectrum of the native starch.

The FTIR spectra of native and esterified starch or amylose confirm the extent of esterification, as shown in Figure 10.3, depicting some representative spectra of amylose esters. In the spectrum of pure amylose, a strong, broad band between 970 and 1200 cm^{-1} with three peaks is the most characteristic band for a polysaccharide and is attributed to COO stretching. This band is also observed in amylose esters where the three peaks are better resolved. Another characteristic

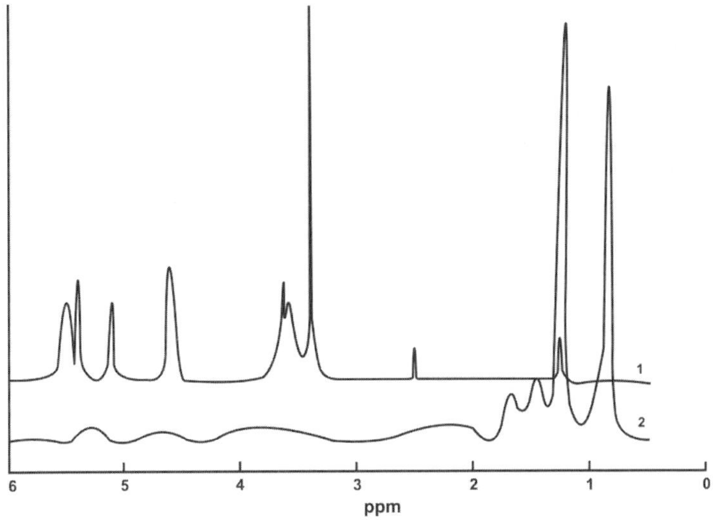

Figure 10.2 ^{1}H NMR spectra of (1) pure starch and (2) octanoated starch with ds 1.8 [88].

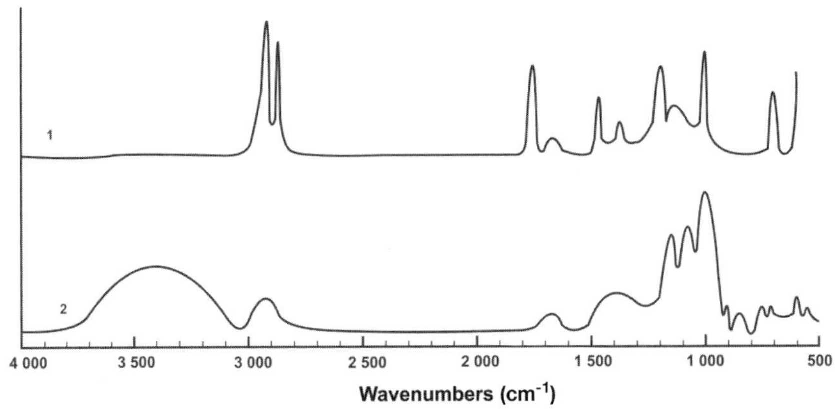

Figure 10.3 FTIR spectra of pure amylose and its octanoated (C_8) and octadecanoated (C_{18}) esters with ds 2.7 [88].

band is that between 3000– and 3700 cm^{-1}, due to hydroxyl bond stretching. The intensity of this peak decreases in the esterified derivatives. This peak's maximum is shifted towards higher wavenumbers, from 3391 cm^{-1} for pure amylose to 3459 cm^{-1}. This happens because there is a decrease in the concentration of hydrogen-bonded hydroxyls, as they are converted into ester groups during the reaction. An intense ester carbonyl band appears at 1746 cm^{-1} in the final products.

The absence of a shoulder in lower wavenumbers in this area verifies that hydrogen bonds between the remaining hydroxyl groups and the carbonyl groups of the esters are absent or very sparse. This band, and also the band at 2800–2950 cm^{-1}, corresponding to methylene group deformation, increase with the degree of substitution. The new peak that appears at 3022 cm^{-1} is due to the methyl groups of the ester.

All esters produced are soluble in common solvents, such as chloroform, in contrast to starch, which is soluble only in warm DMSO. On the other hand, the most prominent feature of the esterified starches is their reduced hydrophilicity as determined by contact angle measurements (Table 10.1). These values are comparable to that of poly(methyl methacrylate) (PMMA), which is a hydrophobic synthetic polymer and has a contact angle of 85°. The increased hydrophobicity of esterified esters is attributed to the replacement of hydrophilic hydroxyls by the relatively hydrophobic ester groups. Hydrophobicity increases with the degree of substitution, as can be seen in the case of octanoated and octadodecanoated starch. The contact angle is about 2–5° higher than for the corresponding esters with a lower degree of substitution. It increases also with side alkyl chain length. This is clearer in the case of amylose esters.

The loss of hydrophilic character is also reflected in water absorption measurements (Table 10.1). The main factor affecting water uptake seems to be the ds. Thus, starch esters, like the octanoated and the octadecanoated esters with ds 1.8, show significantly higher water absorption than the corresponding esters with ds 2.7. Even so, the maximum increase does not exceed 4 wt.%, indicating essentially hydrophobic materials. When it is immersed in water, native starch has the ability to absorb water to about 4–5 times its own weight. It seems that even the replacement of about half of the hydroxyl groups in esters with ds 1.8 is sufficient to changing drastically the hydrophilic character of the starch. It must be noted, however, that octanoated amylose with ds 0.54 shows a weight loss (about 3.3% of the initial weight) in the same period. Obviously, with such low ds, the product still remains hydrophilic. As a result, it may swell and part of it can be partially extracted by cold water. The side-chain length also has a minor effect on the water uptake properties since water absorption seems to decrease with increasing side-chain length. Dodecanoated and octadecanoated esters with ds 2.7 show a similar behavior, which is different from that of octanoated esters. The latter shows a very small water absorption, which seems to grow steadily, without reaching a plateau as in the case with esters having ds 1.8. This behavior could be attributed to the shorter chain length of the ester, which cannot hinder effectively the unreacted hydroxyl groups, and, as a result, they can absorb water. Finally, amylose and native starch esters show similar water absorption characteristics. The final conclusion is that the esters have low water absorption characteristics, making them appropriate for applications where water absorption must be minimal.

One negative point in all those efforts was the molecular weight reduction of starch during modification, due to the high susceptibility of starch to solvents and acid chlorides or anhydrides used for the acylation. Lately, there has been renewed interest in the preparation of modified starches with acetate [89], hydroxypropyl [90], alkyl siliconate [51] and fatty-acid ester (C_4–C_6) [91] groups. The main aim is to produce a fully biodegradable thermoplastic material, which will have the appropriate properties (especially mechanical), for replacing, whenever possible, the nonbiodegradable plastics used in the plastics industry.

The ds of carboxymethyl starch (CMS) can be calculated using the following equation:

$$\mathrm{ds} = 162 W_{\mathrm{Na}}/(23 - 81 W_{\mathrm{Na}}) \tag{9}$$

where W_{Na} (%) is the content of sodium and 162, 23 and 81 are the relative molecular masses of dehydrated glucose, sodium atom and substituent group, respectively.

10.2.2
Crosslinking/Gelatinization

The crosslinking can be performed with different modified starches originating from native potato starch at a temperature above the gelatinization temperature due to which the degree of crosslinking (cld) varied between 0.05 and 0.75 wt.% [92]. The samples were characterized by the swelling power Q and the average particle size D and observed by light microscopy and low-temperature scanning electron microscopy (SEM) (Table 10.2).

For samples with higher degrees of crosslinking, cld 0.5 and 0.75, individual particles were clearly displayed. In the case of cld 0.5, some of them are apparently more swollen than in the case of cld 0.75. However, some particles of lower size are also observed. The large part of starch particles (cld 0.5 and 0.75) apparently keep their initial structure, whereas the lowest crosslinked samples (cld 0.05 and 0.1) are broken down. Thus, a higher degree of crosslinking probably makes the starch granules more resistant to rupture than does a lower degree of crosslinking. The particle size distributions are rather broad and all samples exhibit a similar distribution. The average particle size increases with the extent of crosslinking, whereas the swelling power decreases. Low-temperature SEM observation shows the presence of both phases: particles and a continuous suspending

Table 10.2 Granule size (diameter, D), swelling power (Q) and critical concentration (C^*) of crosslinked starch derivatives [92].

Parameter	cld 0.05	cld 0.1	cld 0.25	cld 0.5	cld 0.75
D (μm)	20	22	45	60	60
Q ($\mathrm{cm^3\,g}$)	41	30	24	17	15
$C^* = 1/Q$ ($\mathrm{g\,dm^{-1}}$)	24	33	42	58	66

medium. The particles are not spherical and their size increases with the degree of crosslinking. The average particle size was about 20 μm for cld 0.1 and 60 μm for cld 0.5 (Table 10.2). Moreover, it could be assumed from the molecular weight determination that the continuous suspending medium (between 10 and 30 wt.%) is composed of water and polymer chains ($M_w = 3.5 \times 10^4$ g mol^{-1}), which are more solubilized than are the particles.

The role of crystallinity in release control, a series of powders, tablets and films, was analyzed for high-amylose starches (HASs) with different crosslinking degrees (cld, defined here as the amount of epichlorohydrin (g) used to crosslink 100 g of polymer under specific conditions, i.e. CLHAS-6 is obtained with an initial ratio of 6:100 crosslinking agent:high-amylose starch) [93, 94]). The diffraction spectra of crosslinked high-amylose starches (CLHASs) in powder, tablet and film forms showed differences with varying cld. For native high-amylose starch Hylon VII powder, a predominant B-type with elements of V-type diffraction pattern was found. The diffraction maxima at 0.57, 0.52, 0.39 and 0.37 nm are typical for the B-type diffraction pattern [95], whereas the shoulder at 0.68 nm and the peak at 0.45 nm are related to the V-type structure [23]. For CLHAS polymers with increasing cld, the intensity of 0.57, 0.52, 0.39 and 0.37 nm peaks diminishes, whereas the 0.45 nm peak becomes more important. At the same time, the shoulder at 0.68 nm becomes more and more separated and for the CLHAS-6 powder becomes a fairly well-defined peak. The small peak appearing at 1.18 nm is also characteristic of V-type single-helix structure. The general feature for CLHAS powder diffractograms is the loss of crystallinity with increasing cld. CLHAS-20 shows no discrete diffraction between 18 and 30° and the broader profile suggests a more amorphous structure.

For CLHAS powders with low and moderate cld (CLHAS-3, CLHAS-6), both B- and V-type patterns are still present, but the proportion between them changed in comparison with native high-amylose starch and gelatinized (but not cross-linked) CLHAS-0, for which B-type is predominant. For higher cld (CLHAS-20), the broader diffractogram can be ascribed to a low-ordered structure that may contain some single V-helices. The crosslinking procedure involves a gelatinization step that leads to partial or complete disruption of the predominant B-type order existing in native high-amylose starch. As a general feature, gelatinization induces changes in double-helix conformation, even an unraveling or unwinding of the double helices [96]. It was reported that amylose also exists as an interrupted helix in aqueous solution [97]. With increase in pH, the helix–coil transformation occurs and the molecule is regarded as a flexible coil [96]. By crosslinking, neutralization and drying, a new type of order and a new structure become possible. In solution, amylose is supposed to change first in single helices (pseudo-V-type diffraction pattern). However, under particular conditions, insoluble amylose can keep the initial B-type pattern [96]. For CLHAS powders, the transition from a predominant double-helix B-type diffraction pattern in native Hylon VII and CLHAS-0 to a predominant pseudo-V form (single-helix conformation) of CLHAS can be followed by X-ray diffraction analysis. The CLHAS-0 diffractogram is similar to that of native high-amylose starch. Since only gelatinization and no crosslinking was carried out (CLHAS-0), the native arrangements can be almost restored. Only when the chemical structure was modified by crosslinking did the X-ray profiles gradually change. Therefore, crosslinking appears to be the main treatment that induces structural modifications.

For higher crosslinked starches (CLHAS-6–CLHAS-8), the amorphous part becomes more extended and the less-ordered chains have more flexibility. When compressed there are probably rearrangements, which can generate a structure favorable to inducing stable network formation at swelling. When the cld is too high, the high density of transversal crosslinking between polysaccharide chains can hinder achieving a favorable conformation during swelling and the structure remains almost unordered. Hydroxyl groups of chains are not involved in network stabilization by interchain hydrogen bonding; they are only available for fast hydration [94]. It is well known that starches have good film-forming properties [98, 99]. Interpretation of the X-ray diffraction patterns of films cast from the different CLHASs indicates a morphological change, suggesting a different structural order (Figure 10.4). When CLHAS powders are suspended in water and boiled, the amorphous part can swell and on cooling can adopt new conformations. By slow water evaporation, the resulting structures, spontaneously achieved, become stable. In the X-ray patterns obtained from films, the double-helical order is present in CLHAS-0 and CLHAS-3 films and becomes less evident in CLHAS-6.

With increase in crosslinking density, the bands at 1022 and 1047 cm^{-1} increased, whereas a decrease in the 1000 cm^{-1} band was observed. On the basis of the results of the correlation of X-ray data on crystallinity and morphological changes with the variation of bands in the 1200–900 cm^{-1} region, it is possible to

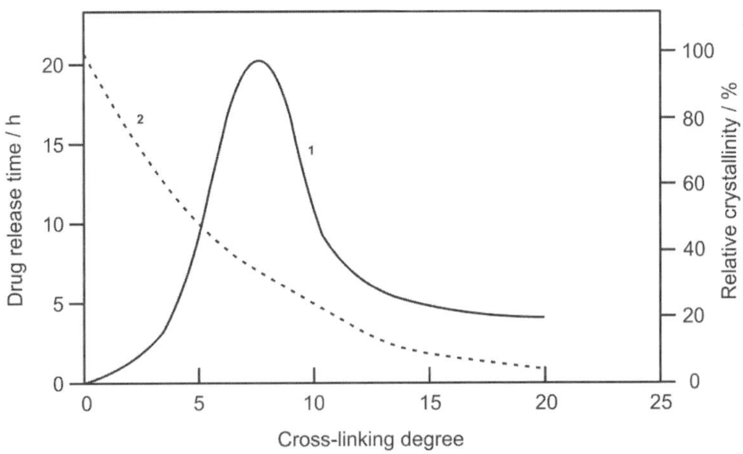

Figure 10.4 Relationship between drug-release time (curve 1), relative crystallinity (curve 2, B peak at 0.52 nm) and cld of CLHAS. The relative crystallinity was evaluated from X-ray analysis of CLHAS powders with various clds [93].

give an FTIR band assignment for the polymers analyzed. As observed by X-ray data, increasing the degree of crosslinking induced a decrease in the B-type double-helix morphology, whereas the presence of the pseudo-V-type and amorphous structures was increased. The decreasing band at 1000 cm^{-1} was associated with the crystalline order (B-type morphology), becoming less important in starches with higher clds. The increase of the 1022 and 1047 cm^{-1} bands could be related to the amorphous phase and to a pseudo-V-type structure. Both bands have almost the same evolution and because the helix conformation of polysaccharidic chains in noncrystalline and V-type structure is the same in solid starch [100]. Gidley and Bociek [101] suggested, via ^{13}C NMR spectroscopy of solids, that double- and single-helical chains can be associated with ordered and non-ordered conformational states, respectively.

The FTIR analysis shows that the major tendency is the loss in crystallinity until a certain cld, with no important changes at higher clds. It clearly appears that for moderate clds, a stable structure with a moderate crystallinity is responsible for the best swelling and release properties. At the same time, another region at 1500–1350 cm^{-1} of the FTIR spectra was deconvoluted for powders and films with different clds. The same method was used and from the nine bands obtained by deconvolution, the band around 1254 cm^{-1} exhibited a shift with increase in cld. Similar data were found for powders and films. It was shown [102] that the band at 1265–1254 cm^{-1}, assigned to a $-CH_2OH$-related mode, shifts for various polymorphic forms of amylose, from 1263 cm^{-1} in V-amylose to 1254 cm^{-1} in B-amylose. It was reported that when the cld increases, the tendency is for a conformational change from a B-type helix (characteristic of zero or low clds) to a V-type helix conformation (CLHAS-11). If the best-organized and compact structures (B-type) have smaller wavenumbers and those less-organized (V-type) higher wavenumbers, the shift to higher values (1267 cm^{-1} for CLHAS-20) could be interpreted as a B- to V-type transition.

CLHAS was introduced as an excipient (Contramid) for controlled drug release [103, 104]. It swells in water to form an elastic gel and the ability to regulate the swelling and thus regulate drug release in aqueous media as a function of crosslinking density makes this hydrogel particularly suitable as a pharmaceutical excipient. The permeability of solutes across the hydrogel barrier depends on the texture of the hydrogel. Thus, polymer hydrophilicity and crystallinity play an important role in drug release [105]. Appropriate resistance of swollen tablets and good control of the drug release (over 15–20 h) were only obtained for CLHASs with moderate cld: CLHAS-6 and CLHAS-8 (Figure 10.4). Higher clds (CLHAS-20 and above) generate a sharp decrease in the release time (1–3 h) and under certain conditions can even afford disintegrant properties for CLHAS [106]. The nonmonotonic variation of the drug release time with cld is a particular characteristic of the CLHAS matrix that differs from those of other classical polymeric matrices for which increasing clds lead to longer release times [107, 108]. This behavior of CLHAS was ascribed to the particular structure of the matrix where, in the case of low cld, covalent linkages, interchain hydrogen bonds and water-promoted hydrogen associations stabilize the network [94, 108, 109], thus controlling the access of water into the matrix. Aspects of the water uptake as a function of cld [106] and its role in the release behavior of CLHAS matrix [94, 105, 109] were reported.

The gelatinization of native starch can be carried out by heating either under atmospheric conditions or at higher pressures [110]. Depending on the type of starch, the dispersions attained different temperatures after the heat treatment, owing to different rheological properties. On heating of potato starch to 80 °C, the starch granules become strongly swollen and distorted, but the individual starch granules are still distinguishable. Blue regions, which indicate accumulations of amylose, are inside the starch granules and in the intergranular space. The outer zone of the starch granules is stained brownish-violet, as it is mainly composed of amylopectin. By further heating to 95 °C, the starch granules appear partially melted together, but amylose-rich and amylopectin-rich regions are still observed. After a holding time of 30 min at 95 °C, the starch dispersion had a more homogeneous appearance. However, remnants of swollen starch granules are present. A similar starch structure is observed for potato starch heated at 112 °C. On the other hand, on solubilizing commercial pregelatinized starch at 95 °C, only small starch aggregates and no granular structure are recognizable. The latter dispersion can therefore be considered a molecularly dispersed starch system. In contrast to potato starch, the swelling capacity of pea starch is restricted. On heating pea starch to 80 °C, the starch granules are swollen but less distorted than potato starch. After a heat treatment at 95 °C, the starch granules are swollen, but still no disintegration is observed. Even with a high-temperature treatment at 121 °C, the starch granules are only partially disintegrated and appear melted together. All pea starch samples show a strong blue staining due to the high amylose content. On heating regular wheat starch to 80 °C, most granules are swollen and lentil shaped, but not disintegrated. The starch granules are not stained homogeneously and the center of several starch granules appears dark. After heat treatment at 95 °C, the individual starch granules are still recognizable and solubilized starch material is located in the intergranular space. A significant disintegration of the starch granules occurred only after treatment at 1 bar overpressure for 30 min. In contrast to potato starch, the solubilization of commercial pregelatinized wheat starch did not yield a dispersion free of supramolecular structure, since fragments of starch granules are visible. The small granules (B-type) of wheat starch did not show a significantly different microstructure on heating at 80 and 121 °C compared with regular wheat starch. The tendency for granules to disintegrate decreases in the following order:

$$\text{potato} > \text{wheat, large granules} > \text{wheat, small granules} > \text{pea starch} \tag{10}$$

According to Tester and Morrison [111], starch granule swelling is essentially a property of amylopectin while amylose acts as a diluent. As shown by Prentice et al. [112], the remaining structures after partial dissolution of starch are mainly composed of amylopectin and do not show signs of either birefringence or crystallinity. Since the undissolved structures of cereal starches resemble the original starch granules, they are also termed "ghosts", whereas remaining potato starch structures are described as "gel", due to the little ghost-like integrity [112]. The swelling of starch granules is known to be affected by different factors, such as the crystalline structure, the length of amylopectin side-chains, the presence of endogenous amylose–lipid complexes, amylose/amylopectin ratio, granule size and the organization of the starch polymers in the granule. Consequently, it is not surprising to find that depending on the botanical origin of starch, similar time–temperature and shear conditions lead to large variations in the microstructure, ranging from extensive disintegration to limited swelling of the starch granules. The generation of molecularly dispersed starch systems by heat treatment is more easily accomplished for starches with a high swelling capacity, such as potato starch. On the other hand, the complete disintegration of the granular structure for starches with low swelling capacity, such as cereal and leguminous starches, requires prolonged heating at temperatures higher than 100 °C. Furthermore, shearing contributes to the disintegration of starch granules and potato and waxy starches are especially shear sensitive, owing to their high swelling capacity [113]. As shown by Babock et al. [114] and Dintzis and Bagley [63], the complete breakdown of the granular structure of maize and amylomaize starch, which have a low swelling capacity, requires heat treatment at temperatures higher than 100 °C in combination with mechanical energy input. According to Bechtel [115], fragments of starch granules are still visible after prolonged alkali dispersion of defatted corn starch under static conditions. This confirms that even drastic treatments, such as alkali solubilization, require mechanical energy input, if the starch is to be molecularly dispersed.

Amylose enrichment is found in the intergranular space and in the center of starch granules. The accumulation of amylose inside the starch granules is especially pronounced for potato starch dispersions. The preferential leaching of amylose and their accumulation in the intergranular and intragranular space has been shown by characterizing the starch solubles by iodine titration [116] and by microscopy [111]. The demixing of amylose and amylopectin is not surprising considering their thermodynamic incompatibility [117]. Furthermore, cooling and aging were shown to enhance the separation into amylose-rich and amylopectin-rich domains, as assessed by light microscopy [118]. The separation of the two starch polymers is favored by the fact that amylose is located in the amorphous regions of starch and that it is a small polymer compared with amylopectin. The leaching of amylose is reported to occur preferentially at the equatorial groove [111]. Evidence exists for the presence of radially arranged amorphous channels in the starch granules with pores at the surface, which allow the exit of the amylose [55].

In the case of potato starch, the variation in the shear viscosity was in the range of three orders of magnitude. From the results of the microstructural characterization and the respective rheological data, it is evident that the flow properties are closely related to the microstructure of the system, which in turn are determined by the preparation conditions and the botanical origin of the starch. It

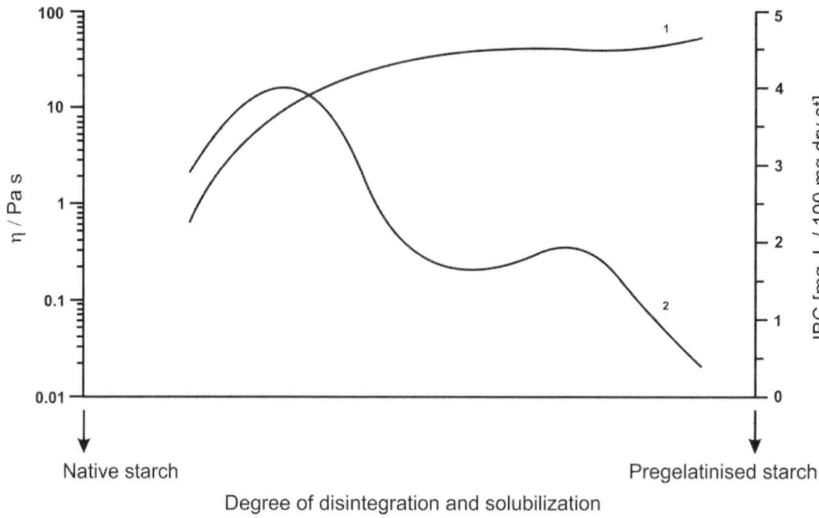

Figure 10.5 Schematic representation of the microstructure of potato starch dispersions in relation to viscosity and IBC [110].

was concluded that the rheological properties are primarily determined by the volume fraction occupied by the particles, their shape and deformability [119]. Furthermore, the solubilized material was also shown to contribute to the observed viscosity. A schematic representation of the relation between granule swelling, iodine binding capacity (IBC) and viscosity is presented in Figures 10.5 and 10.6. The IBC of the starch dispersions was determined after sample preparation by amperometric iodine titration. In selected cases the titration rate was varied. The curves were evaluated graphically and the IBC was calculated as follows [110]:

$$\text{IBC} = (I_b/St_{tot}) \times 100 \text{ (mg iodine/100 mg dry starch)}$$

where I_b = mg bound iodine and St_{tot} = mg dry starch in the titration vessel.

As a general trend, it was found that increased starch granule swelling leads to higher viscosity. Interestingly, the disintegration of the supramolecular structure

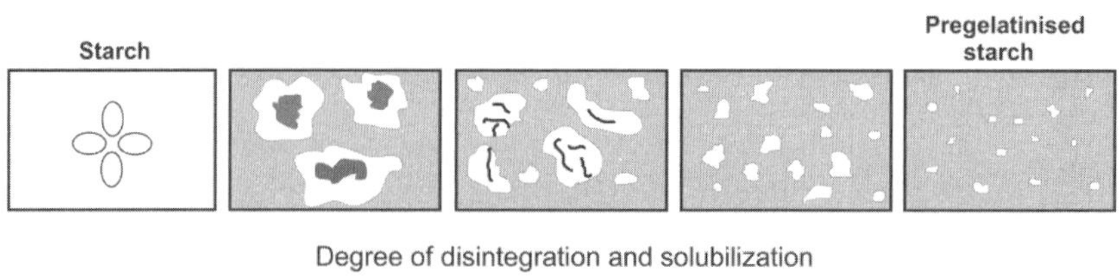

Figure 10.6 Schematic representation of starch disintegration and solubilization [110].

had the opposite effect on the viscosity of potato and wheat starch: the complete disintegration of potato starch granules was accompanied by a viscosity decrease of three orders of magnitude, while the viscosity of wheat starch increased as the starch granules disintegrated, which is most probably the result of the formation of a weak amylose network. This is conceivable, since low-concentration wheat starch dispersions have a stronger tendency to form gels on aging than potato starch, which is thought to be related to differences in the gelling fraction of starch, the amylose [120]. The viscosities of wheat starch systems heated to 95 °C and at 1 bar overpressure, both for 30 min, suggest that processing may have contributed to further release of amylose, promoting the build-up of a weak intergranular network. Amylopectin may also play an important role, since rheological properties indicative of flow-induced structures were also found for waxy maize starch dissolved in DMSO [63]. In connection with the importance of leached amylose, it has to be emphasized that the presented flow curves correspond to starch dispersions at 25 °C shortly after preparation. At this early stage of aging, the rheological properties of low-concentration starch dispersions are not as much influenced by the extent of amylose solubilization as after an aging period of several days [120]. Amylose has a tendency to aggregate on aging, leading to the formation of an intergranular network, and induces the gelation of starch, provided that a significant amount of amylose is leached from the starch granules [121]. On the other hand, Conde-Petit et al. [122] showed that the solubilization of amylose inside the starch granules does not contribute significantly to starch gelation. It is possible, however, to release the amylose from the swollen starch granules by degrading the supramolecular starch structure with mechanical energy input [122].

10.2.3 Grafting

The grafting of monomer on to starch is assumed to follow the same reaction scheme as has been elucidated for the graft copolymerization of vinyl monomers on to macromolecules, as follows (Scheme 10.1) [123, 124]:

1. Free radicals are formed on the C_2 of the anhydroglucose ring when a ceric ion is used to initiate grafting. The ceric ions are attached to sago starch to produce a sago starch–ceric complex.
2. Ceric ions are reduced to Ce^{3+} ion with the release of a proton.
3. As a result, the bond between C_2 and C_3 is broken and free radicals of sago starch are formed.
4. Free radicals so formed then react with monomer to produce the graft copolymer:

$$\text{starch}^* + \text{monomer (e.g. AAm)} \rightarrow \text{starch-g-AAm}^* \quad (11)$$

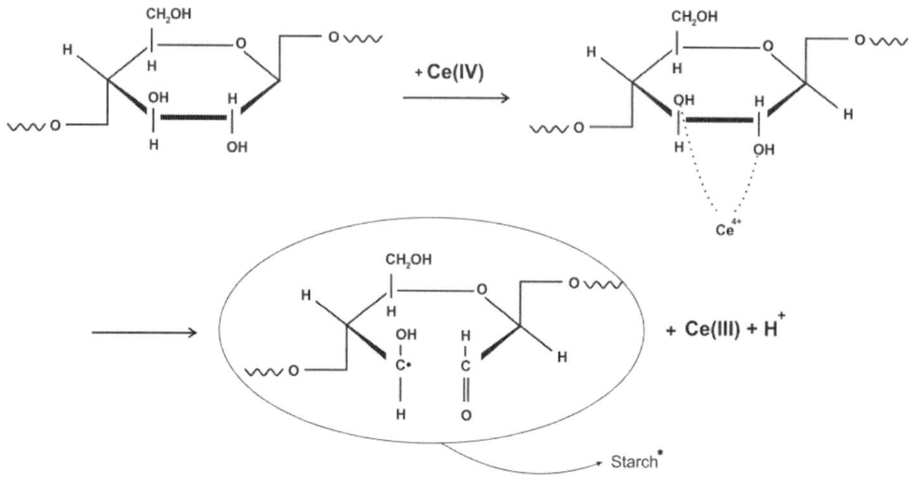

Scheme 10.1 Grafting of monomer on to starch [123, 124].

5. Bimolecular termination of active propagating radicals leads to the formation of inactive graft polymer:

starch-g-AAm* + starch-g-AAm* → inactive graft polymer (12)

starch* + starch-g-AAm* → inactive graft polymer (13)

starch* + PAAm* → inactive graft polymer (14)

Bimolecular termination given by the following reactions decreases the grafting efficiency:

starch* + starch* → inactive starch (15)

PAAm* + PAAm* → inactive polymer (16)

The grafting efficiency of starch by copolymerization varies with the type of monomer (the water-soluble or oil-soluble unsaturated monomers) and starch. In its native form, a starch granule is insoluble in cold water and most uses involve a heating treatment in the presence of an excess of water. Below a critical temperature (about 60 °C), that is, the so-called gelatinization temperature, starch granules absorb water and undergo swelling to many times their original size. It can turn into adhesive gel during the reaction, which is difficult to stir uniformly and the concentration of used starch is very low for grafting. In this case, the grafting can proceed on the surface granules and core–shell structures can appear. When a water-soluble starch is used (carboxymethyl starch) the grafting is more effective because starch molecules (their radicals) dissolved in the reaction medium interact with monomer. The grafting of a water-soluble starch with a water-soluble monomer leads to the water-soluble graft copolymer or its gel.

An example of one type of reaction is the grafting of water-soluble monomers [such as acrylamide (AAm)] on to carboxymethyl starch (CMS, water-soluble) [43]. The water-soluble monomer (AAm) can be selected as the copolymerization monomer since it easily forms hydrogen bonds with water, and it is expected to obtain the copolymer rich in AAm units with greater water absorption capacity. Dispersions (resins) of carboxymethyl starch graft acrylamide (CMS-g-AAm) were synthesized by copolymerization based on a free radical reaction. AAm was grafted on to CMS by using ceric ammonium nitrate (CAN) as an initiator. The results showed that the degree of substitution (ds) of starch first increased substantially and then decreased gradually with increased addition of sodium hydroxide and the water absorption capacity of CMS-g-AAm depended greatly on the ds of CMS, and its maximum with a ds of 0.75 of starch was $350\ \mathrm{g\,g^{-1}}$. The relationships between ds of CMS and synthesis conditions are given in Table 10.3. It can be seen that as the ratio of starch to chloroacetic acid (ClAA) was maintained at 1:1 mol, the ds of CMS first increased and then decreased gradually with increased addition of sodium hydroxide and depended on the reaction time. The maximum ds of CMS was 0.93 and the corresponding ratio of reactant was 2 mol NaOH to 1 mol ClAA to 1 mol AGU (anhydroglucose units of starch). Because the hydroxyl group on the starch chain was used in graft copolymerization with AAm, a proportion of the ds of starch was not too great. The concentration of initiator not only affected the graft ratio, but also the degree of crosslinking of the hydrogel of the product.

The IR spectra of pure starch and CMS-g-AAm as shown in Figure 10.7 indicate that both have a broad absorption band characteristic of the glucosidic ring of starch between 3700 and 3200 and between 1160 and $1030\ \mathrm{cm^{-1}}$. Moreover, there is an increment in the intensity of this band in the case of a CMS-g-AAm

Table 10.3 Reaction conditions versus degree of substitution (ds) [43].

No.	Ratio of materials[a)]	ds under different conditions			
1	1 mol AGU:1 mol ClAA, reaction time 4 h	$n(\mathrm{NaOH})/n(\mathrm{ClCH_2COOH})$	1.0	2.0	3
		ds	0.2	0.93	0.90
2	1 mol AGU:1 mol ClAA:1.5 mol NaOH	Reaction time (h)	2	4	6
		ds	0.50	0.75	0.76
3	1 mol AGU:1 mol ClAA:2 mol NaOH	Reaction time (h)	2	4	6
		ds	0.70	0.93	0.95

a) AGU, anhydroglucose units of starch; ClAA, chloroacetic acid. When the total volume of mixture was 200 mL, the concentration of initiator was controlled at $0.008\ \mathrm{mol\,L^{-1}}$.

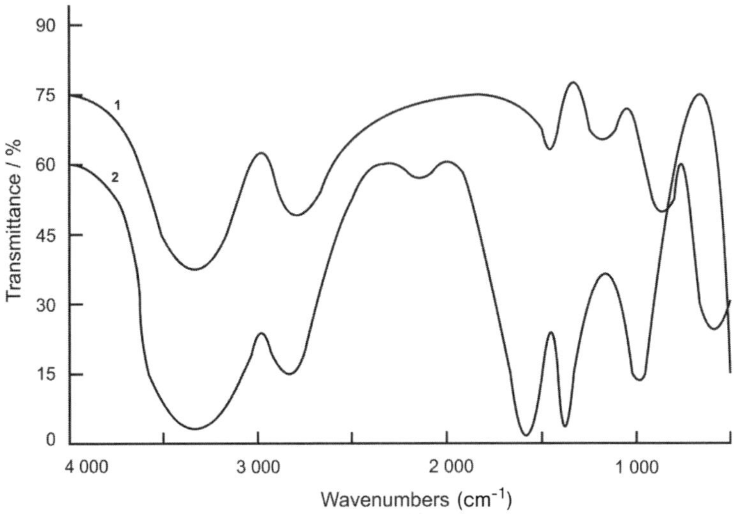

Figure 10.7 IR spectra of pure starch (1) and CMS-g-AAm (2) [43].

sample, owing to the utilized hydroxyl groups of the side-chain carboxyl. There was a remarkable difference in the wavenumber range 1500–1700 cm^{-1}, where strong absorption peaks at 1675 and 1575 cm^{-1} are present, for $CONH_2$ and COOH groups, respectively. The strong absorption peaks at 1411 and 1318 cm^{-1} indicated the symmetrical flexing vibration of carboxyl and carboxylate groups, respectively.

The mole ratio of AGU of starch to sodium hydroxide was maintained at 1:1 and only the relative molar quantity of ClAA was changed to prepare CMS with a different ds. The relationships between the molar addition of ClAA, the ds of CMS and the water absorption capacity of CMS-g-AAm are shown in Figure 10.8. It can be seen that, with increased addition of ClAA, the ds of CMS first increased slowly and then declined considerably and the water absorption capacity of CMS-g-AAm first increased substantially and then decreased gradually. The maximum water absorption capacity of CMS-g-AAm occurred with the addition of 0.75 mol ClAA. The ds of CMS mainly affected the graft density of CMS-g-AAm. The optimum addition of ClAA was 0.75 mol, which resulted in a ds of 0.75 for CMS and a water absorption of 350 g g^{-1} for CMS-g-AAm.

The mole ratio of starch, sodium hydroxide and chloroacetic acid was maintained at 1:1:0.75. The relationship between the water absorption capacity of CMS-g-AAm [C_{wt} (g g^{-1})] and the addition of AAm (g) is as follows:

$$C_{wt}/\text{AAm: } 160/5,\ 200/7.5,\ 240/10,\ 330/12.5 \text{ and } 300/15 \qquad (17)$$

When the addition of AAm was less than 12.5 g, the water absorption capacity of CMS-g-AAm increased drastically with increase in addition of AAm and the max-

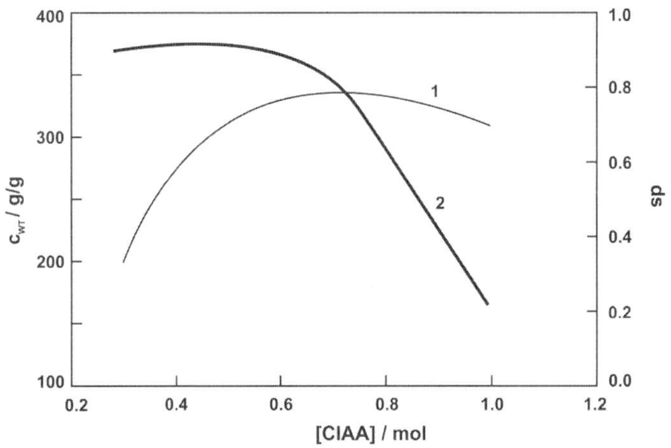

Figure 10.8 Relationships between water absorption capacity (C_{wt}), degree of substitution (ds) and addition of chloroacetic acid (ClAA).

imum water absorption capacity was obtained at an addition of 12.5 g of AAm. With further increase in addition of AAm, the water absorption capacity of CMS-g-AAm began to decline.

The results can be summarized as follows:

1. The ds of CMS depends greatly on the ratio of starch, sodium hydroxide and ClAA. For the preparation conditions of CMS as 1 mol St:1 mol NaOH:0.75 mol ClAA, the resulting best water absorption capacity of CMS-g-AAm is 350 g g^{-1} and the corresponding ds of CMS is 0.75.
2. The characteristic absorbing peaks in the IR spectra have proven that CMS participates in graft copolymerization with AAm.

The aqueous phase polymerization of vinyl acetate and butyl acrylate monomers produces the surface-active grafted starch radicals which agglomerate between themselves and form the primary polymer particles [125–127]. The radical polymerization of alkyl acrylates [43, 128] in the aqueous phase led to the formation of grafted polysaccharides. The core–shell polymer particles are stabilized by the graft copolymer and the adsorbed emulsifier molecules [Tween 20 (Tw 20, nonionic emulsifier, polyoxyethylene sorbitan monolaurate)] (Figure 10.9). The core is supposed to be formed by hydrophobic polymer [poly(butyl acrylate), (PBA)], the inner shell by the graft copolymer and the outer shell by the hydrophilic poly(ethylene oxide) chains of emulsifier Tw 20 and the starch segments. Both hydrophilic fragments project to the aqueous phase. The chain transfer to the emulsifier (Tw 20) initiates the accumulation of the covalently bound hydrophilic poly(ethylene oxide) chains of emulsifier Tw 20 at the particle surface (which project to the aqueous phase, Figure 10.10).

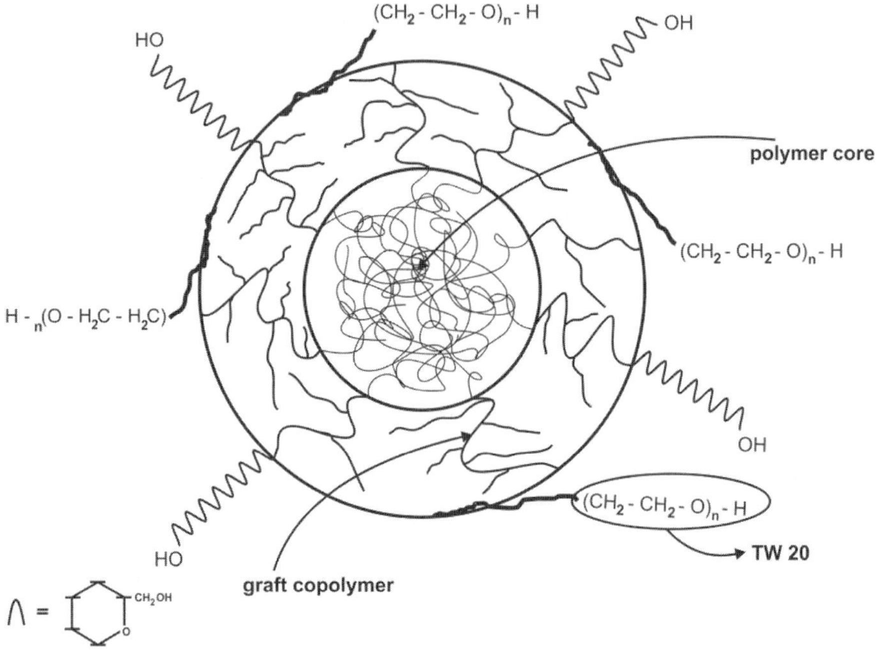

Figure 10.9 Polymer–starch composite particles [125–127].

The surface of starch granules has been modified by the graft polymerization of the hydrophobic glycidyl methacrylate monomer (GMA) [128]. This approach led to the preparation of a novel starch product (the core–shell composite particles) containing reactive, pendant glycidyl groups with the ability to copolymer-

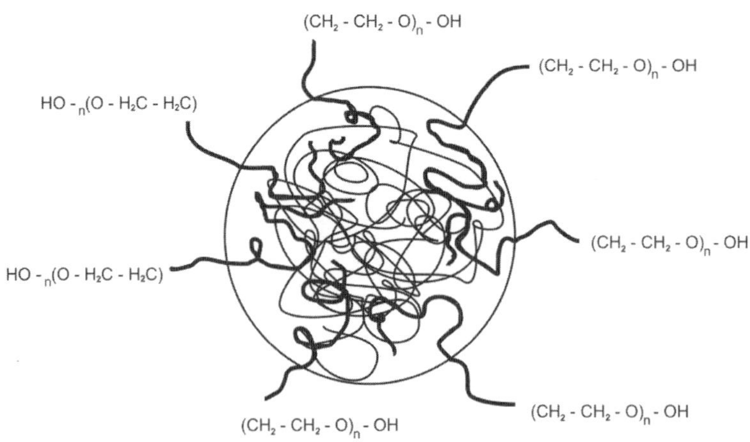

Figure 10.10 Poly(butyl acrylate) particles stabilized by Tw 20 [125–127].

ize with epoxy resins. GMA was chosen because of its dual functionality – an acrylic group and an epoxy group in the same molecule. A free radical initiator was used to graft the acrylic group on to the starch, leaving the pendant glycidyl group for effecting photopolymerization on exposure to UV radiation during a subsequent step.

For the purpose of preparing the UV-curable compositions, the grafted starch was mixed with cycloaliphatic diepoxide (CAE) and exposed to UV radiation. Photocuring was induced using a cationic photoinitiator. The modified starch due to the presence of residual free hydroxyl groups can form an interpenetrating polymer network by a chain transfer mechanism and this can impart flexibility to the otherwise brittle cycloaliphatic epoxide system. The study consisted of two parts, with the following objectives:

1. A photoreactive starch-g-GMA was produced.
2. The cycloaliphatic diepoxide was partly replaced with starch-g-GMA in UV-curable formulations and followed the cure behavior of the system induced by a cationic photointiator.

The FTIR spectra of the raw sago starch and starch-g-GMA are shown in Figure 10.11. The FTIR spectra of starch-g-GMA indicated the appearance of new peaks at 1731–1735 and 907–909 cm^{-1}, confirming the presence of ester carbonyl group (C=O) [124] and the epoxy groups [129]. These two peaks are absent in the raw starch. FTIR spectra of the samples treated with GMA contained the ester carbonyl groups but the epoxy absorption at 907–909 cm^{-1} were not signif-

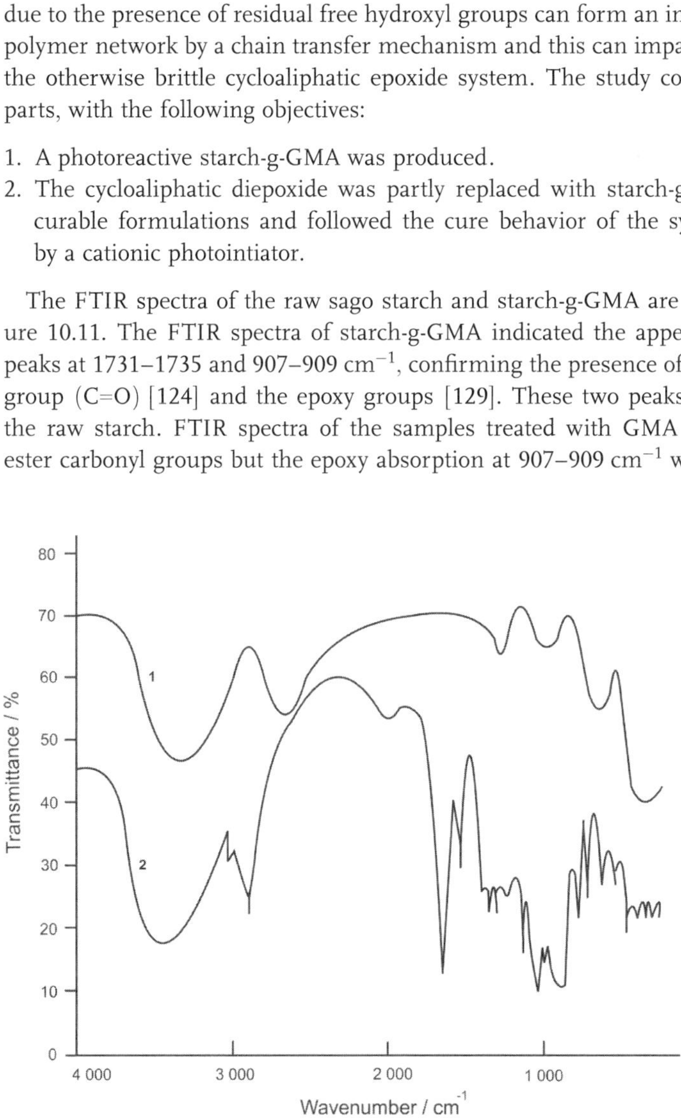

Figure 10.11 FTIR spectra of starch (1) and starch-g-GMA (2) (15 g GMA/25 g starch) [128].

icant. Either these peaks had been masked by the starch bands in the region or more probably the epoxy groups had been consumed in a parallel ring opening reaction under the acidic conditions caused by the release of protons in the redox reaction. The FTIR spectra of sago starch-g-GMA therefore gave supporting evidence that the GMA had been successfully grafted on to sago starch.

The percentage of grafting increases from as low as 2.0% at 1 g GMA/25 g starch to as high as 48.8% at 15 g GMA/25 g starch:

$$\%/\text{GMA(g)}: 2/1, 8/2, 11/3, 15/4, 20/5, 35/10, 49/15 \qquad (18)$$

The same trend was reported by Shukla and Athalye for the grafting of GMA on to cellulose using CAN as the initiator [130]. The increase in percentage of grafting could be associated with the greater availability of GMA molecules in the proximity of starch macro-radicals (sites for grafting).

Since the grafting reaction is carried out at room temperature, the overall reaction is essentially heterogeneous for the following reasons:

1. During the period of reaction no dissolution of starch is possible and the particles remain discrete and retain individuality throughout the reaction.
2. Further, as the grafting proceeds, the hydrophobic GMA layer gradually and continuously formed on the starch particles is expected not only to prevent any dissolution of starch but also to impart steric stability to the particles.

Under these conditions, it is highly likely that the grafting reaction of starch and the GMA results in a core–shell configuration consisting of a hydrophobic shell around the starch core. In addition to the epoxy groups which constitute the shell, there are also free OH groups of starch. Both of these groups are important in the UV-curable formulations induced by cationic photoinitiators. While the epoxy groups on the starch particles can copolymerize with the CAE, the OH groups can act as chain transfer groups and are potential sites for the formation of an interpenetrating polymer network. Formulations were therefore made by incorporating GMA-g-starch with cycloaliphatic diepoxide in UV-curable coatings induced by cationic photoinitiator and the following properties of the cured films were determined.

The results in Table 10.4 show that the pendulum hardness decreased from 142 to 104 on addition of raw starch to the coatings formulation. However, by using the starch-g-GMA, a slightly higher pendulum hardness of 112 is obtained. The decrease in hardness in both cases is due to the increased flexibility imparted to the rigid cycloaliphatic diepoxide. While achieving increased flexibility, it is generally inevitable that there is a trade-off in film hardness. Starch-g-GMA, while conferring adequate flexibility, can impart better hardness to the cured films than the addition of raw starch, thereby achieving a good compromise between these properties.

Both the raw and modified starch increased the flexibility of the cured film from 7.51 to 31.33 and 30.54 by as much as 318 and 307%, respectively. This

Table 10.4 Comparison of properties of UV-cured films based on modified and unmodified starch [128][a].

Sample	Water absorption (%)	Pendulum hardness	Mandrel flexibility
A	0.5875	104	31.36
B	0.4025	112	30.54
C	0.1696	142	7.51

a) Recipe (mixing conditions): 10 g cycloaliphatic diepoxide (CAE, ERL 4221), 0.3 g photoinitiator (UVI-6990), 1.5 g raw starch (A), 1.5 g starch-g-GMA (B).

improvement of the flexibility is due to the fact that starch, and also the starch-g-GMA, being basically polyols, function as a chain transfer agent, and become integrated into the polymeric network, thereby serving as internal plasticizers. Grafting of glycidyl methacrylate on to starch provides the potential for the chemical bonding between the starch and the cycloaliphatic epoxide through the epoxide functionality of the GMA. The epoxide can react with the epoxy group of cycloaliphatic epoxide and result in covalent bonding between the cycloaliphatic epoxide matrix and the starch filler. This imparts some rigidity to the film. Hence the flexibility of film containing starch-g-GMA is somewhat less than that of pure starch. A compromise between the hardness and flexibility can thus be achieved, and the mechanical properties of the coatings are generally subsequently improved. Furthermore, the core–shell configuration can contribute to better water resistance than pure starch. The gel content refers to the percentage of crosslinking that occurs in the coatings film. Almost all of the formulations show an excellent gel content result of more than 99%. The high gel content even at relatively high concentrations of starch and starch-g-GMA is again due to the formation of an interpenetrating polymer network as a result of chain transfer caused by the OH groups of starch and the starch-g-GMA.

van der Burgt et al. examined the substitution pattern in branched regions of methylated starches in more detail to determine whether preferences exist for substitution sites at branched glucose residues [131]. Methylated potato starches (MP) and methylated amylopectin potato starches (MAP) were prepared by methylation of starch granules in an alkaline aqueous suspension using dimethyl sulfate [11]. The molar substitution (MS) values of the starch derivatives studied are given in Table 10.5.

Gelatinized methylated starches MP10–MP30 and MAP10–MAP30 were extensively digested with α-amylase from *Bacillus subtilis* [132], yielding mixtures of (1 → 4)-α-D-glucans of different sizes with varying degrees of (1 → 6) branching (DB). The so-called α-limit dextrins, which are highly branched and have a degree of polymerization (DP) of >8 [10], were separated from the linear oligomers by precipitation with methanol. The molar substitution (MS) values of the α-limit

Table 10.5 MS values of α-limit dextrins and α,β-limit dextrins from MP10 and MP30 and from MAP10 and MAP30 [131].

Samples/code	MS[a)] (granule)	MS (α-limit dextrins)	MS (α,β-limit dextrins)	ΔMS[b)] (%)	MS_{calc}[c)] (α,β-limit dextrins)	DB[d)] (α-dextrin fractions)	DB[d)] (α,β-limit dextrins)	ΔDB[e)] (%)
Methylated potato starch/MP10	0.103	0.142	0.179	26.1	0.20–0.24	11.9	14.8	24.4
Methylated potato starch/MP30	0.296	0.380	0.415	9.2	0.35–0.38	9.7	11.5	18.6
Methylated amylopectin potato starch/MAP10	0.097	0.172	0.198	15.1	0.22–0.25	11.8	14.2	20.3
Methylated amylopectin potato starch/MAP30	0.293	0.421	0.449	6.7	0.33–0.37	8.7	10.9	25.3

a) Molar substitution (MS) is defined as moles of substituents/mole of glucose residues.
b) $\Delta MS = [MS(\alpha,\beta\text{-limit dextrins}) - MS(\alpha\text{-limit dextrins})/MS(\alpha\text{-limit dextrins})] \times 100\%$. (19)
c) Calulated MS_{calc} of virtual α,β-limit dextrins using the following equation:

$$MS(\alpha,\beta\text{-limit dextrins}) = DB \times MS_{branch} + DB \times MS_{terminal} + [100 - (2 \times DB)] \times MS_{chain} \quad \text{with } 0 < MS_{terminal} < 0.15 \qquad (20)$$

d) The degree of branching (DB) is defined as the percentage of glucose residues that are branched (calculated by dividing the amount of residues linked at O-1, O-4 and O-6 by the total amount of methyl glucosides in dextrin multiplied by 100%).
e) $\Delta DB = [DB(\alpha,\beta\text{-limit dextrins}) - DB(\alpha\text{-dextrin fractions})/DB(\alpha\text{-dextrin fractions})] \times 100\%$. (21)

dextrins were determined after quantifying the amount of cross-contamination during methanol precipitation (Table 10.5) [10]. The methanol precipitates thus obtained (further referred to as α-dextrin fractions) of MP10–MP30 and MAP10–MAP30, mainly containing α-limit dextrins (84–99%) [10], were extensively digested with β-amylase from *Bacillus cereus*, yielding α,β-limit dextrins and small oligomers [132, 133]. For each sample, the DP decreased during β-amylolysis, as could be demonstrated by ^{1}H NMR spectroscopy [134, 135]. Small (1→4)-linked oligomers with or without single (1→6) branching, originating from cross-contamination during the methanol precipitation [10], and with DP up to 4 were obtained. Determination of the substitution level by using monosaccharide analysis shows that the MS values of the α,β-limit dextrins are higher than those of the corresponding α-limit dextrins (Table 10.5). This can be rationalized from the mode of action of β-amylase, because the binding of the maltosyl groups that are subsequently cleaved will be sterically hindered by the presence of methyl substituents. Not only branching points stop the digestion, but also substituted glucose residues. Since predominantly unsubstituted maltose is released from the α-limit dextrins, the α,β-limit dextrins will have higher MS values. As can be seen from Table 10.5, the relative increases in MS values

(ΔMS) are larger for the lower substituted α-limit dextrins. This observation is in agreement with the steric hindrance of β-amylase, which is expected to increase with the MS. β-Amylolysis of α-limit dextrins from MP10–MP30 and MAP10–MAP30 results in a relative enrichment of partially methylated glucose residues in the generated α,β-limit dextrins as compared with the α-limit dextrins.

The degree of branching (DB) values of all α-dextrin fractions increase after β-amylolysis. However, this increase shows no correlation with the MS. Using MS_{branch} and MS_{chain} and DB values, the MS values of "α,β-limit dextrins" (reconstructed virtual α,β-limit dextrins) were calculated according to Eq. (20) (see footnote c to Table 10.5). The substitution level of terminal glucose residues ($MS_{terminal}$) is estimated to be between 0 and 0.15. As can be seen in Table 10.5, the measured MS value of an α,β limit dextrin differs significantly from that of a reconstructed virtual α,β-limit dextrin. This is mainly due to the inaccurate value of $MS_{terminal}$.

10.3 Dispersions

The micelles solubilize oil or water in the micelle volume, thereby introducing a heterogeneity in the local concentration of the reactants. Likewise, polymer (starch) micelle-like aggregates start to form along the polymer chain at a critical aggregation concentration (CAC). The CAC is thus an analogue of the critical micellar concentration (CMC), but in the solution with an added polymeric compound. A characteristic feature of this parameter is that it is always lower than the CMC of the corresponding emulsifier (surfactant) [136, 137]. The lower CAC is particularly pronounced in solutions of polyelectrolytes with an opposite charge to the emulsifier. The emulsifier often interacts cooperatively with polymers at the CAC, forming micelle-like aggregates within the polymer. Non-cooperative association between emulsifier and polymer is characterized by the simple partitioning of emulsifier between polymer and the aqueous phase. The addition of emulsifiers to aqueous solutions of amphiphilic polymers can either induce or break up interpolymer aggregation. The emulsifier can interact cooperatively with polymers at the CAC, forming micelle-like aggregates within the polymer. Emulsifiers with a relatively long tail bind to the amphiphilic copolymers by simple partitioning between the aqueous phase and the polymer (non-cooperative association) [138].

Polymer–emulsifier systems are commercially important in a number of applications. At higher concentration, the micelles are bridged by the polymer chains, forming a network that exhibits interesting rheological behavior. The interaction between emulsifiers and polymers can the fore be described by two critical aggregation concentrations. The first concentration, CAC, corresponds to the emulsifier concentration when binding interaction between emulsifiers and polymer molecules first occurs, which represents the onset of the formation of a polymer–emulsifier aggregation complex [136, 137]. The second critical concentration, C_2, is more obscure. It is commonly used to represent the emulsifier concentration when the polymer becomes saturated with emulsifier aggregates in polymer–emulsifier aggregation complexes. In addition, another critical concentration, C_m, representing the formation of free emulsifier micelles in the polymer solution, was previously reported in the literature [139]. For some of polymer–emulsifier systems, free emulsifier micelles start to form after the saturation concentration, C_2. Under this condition, C_m is analogous to C_2. However, for other polymer–emulsifier systems, the formation of free emulsifier micelles proceeds and C_2 or C_m is less than C_2. In this case, there is a competition between the formation of free emulsifier micelles and that of polymer–emulsifier aggregation complexes at emulsifier concentrations between C_m and C_2 [139].

Zhen and Tung studied extensively interactions of NaCMA (sodium carboxymethylamylose) with SDS [140]. The critical association concentration (CAC) was found to be 3×10^{-3} M for sodium dodecyl sulfate (SDS). Dynamic fluorescence quenching measurements indicated that the aggregation number of associated micelles is smaller than for free micelles ($N_{agg} \approx 47$; NaCMA = 1.6 wt.%, [SDS] = 5×10^{-3} M). For hexadecyltrimethylammonium bromide (C_{16}TAB) N_{agg} ranges from 36 to 55, compared with 147 for the free micelles. There have been many studies aimed at the determination of surfactant aggregation numbers in polyelectrolyte–surfactant systems.

The CMC of a mixed emulsifier [Slovasol 2430, nonionic emulsifier, alkyl polyoxyethylene ether-type emulsifier: $C_{24}H_{49}O(CH_2CH_2O)_{29}CH_2CH_3$)–starch] system was reported to increase with increase in the mass fraction of starch up to a certain critical concentration of starch (Figure 10.12) [126]:

$$\text{CMC} \times 10^4/(\text{mol dm}^{-3})/\text{wt.\% starch: } 2.36/0,\ 3.12/0.1,\ 4.4/0.25,\ 1.3/0.5,\ -/0.75 \qquad (22)$$

At 0.5 wt.% of starch the CMC decreases and at 0.75 wt.% of starch the micellar aggregation does not appear. The increase in the starch concentration increases the viscosity of the reaction systems. The deviation of the CMC from linearity at 0.5 and 0.75 wt.% of starch can be attributed to the very high viscosity of the reaction system. The increased immobilization of emulsifier molecules in the viscous starch gel depresses the aggregation of emulsifier molecules.

The addition of 0.5 wt.% of starch slightly increased the CMC of Tween 20 (Tw 20, nonionic emulsifier, polyoxyethylene sorbitan monolaurate) (Figure 10.13). The self-aggregation of emulsifier molecules is therefore much stronger in the Tw 20 solution than in the Slovasol 2430 solution or the interaction between emulsifier and starch is stronger in the latter. This indicates that Slovasol 2430 is a more hydrophilic than Tw 20. The additional hydrophobic methyl group (as oxypropylene) in the emulsifier (PEO) shifts the formation of polymer (PEO)–emulsifier (SDS) aggregates to lower emulsifier concentration [141]. In both Slovasol and Tw 20 systems the values of the surface tension at very low emulsifier concentrations (much below the CMC) do not retain the classical

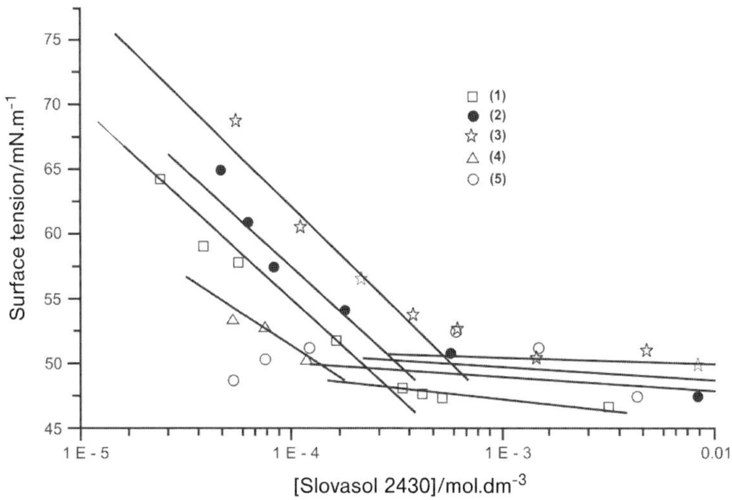

Figure 10.12 Variation of surface tension of aqueous emulsifier Slovasol 2430 solution with the emulsifier and starch contents at 25 °C. (1) without starch; (2) 0.1 wt.% starch; (3) 0.25 wt.% starch; (4) 0.5 wt.% starch; (5) 0.75 wt.% starch) [126].

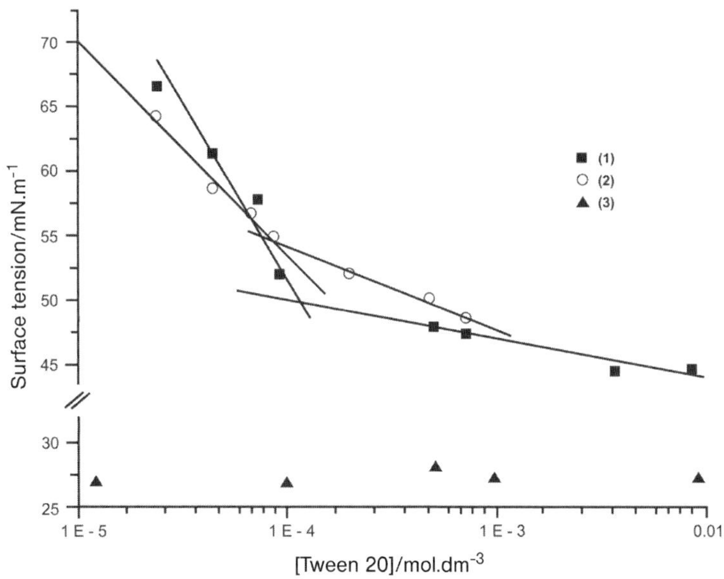

Figure 10.13 Variation of surface tension of aqueous and nonaqueous emulsifier Tween 20 solution with the emulsifier concentration and additive (starch) at 25 °C [126]. (1) 0.5 wt.% starch, in water, CMC $= 1.34 \times 10^{-4}$ mol dm^3; (2) without starch, in water, CMC $= 1.0 \times 10^{-4}$ mol dm^3; (3) cyclohexane (starch insoluble).

plateau but increase with increasing emulsifier concentration. This can result from the interaction between emulsifier and polymer or from the premicelle formation. In the nonaqueous solution (cyclohexane) there is no variation in the surface tension with the CMS concentration. This can result from the low oil solubility of CMS.

Different behavior was observed in the aqueous solution of anionic emulsifier Slovafos 1M. The addition of CMS decreases the critical agglomeration concentration of emulsifier. The slight decrease in CMC (or CAC) can be attributed to the interaction between polymer and emulsifier and the formation of mixed micelles. Most polysaccharides such as hydroxyethylcellulose derivatives usually decrease both the CMC and the surface tension of an aqueous solution on addition of the polymer [142]. When the addition of hydrophobized polysaccharide does not decrease the surface tension and the surface tension was kept unchanged even at the higher concentration means that the hydrophobic core of the polysaccharide aggregates is completely stable and covered by the hydrophilic shell of the polysaccharide skeleton. This is not the case when the surface tension is high and decreases beyond the CMC as in some of the present systems. The very low CAC might indicate the formation of colloidal particles, but the decrease in the surface tension above the CMC can result from the some disorganization.

The value of $(C_2 - \text{CMC})/[\text{CMS}]$ can be used to estimate the amount of emulsifier bound to the polymer chains. For both Slovasol 2430 (Figure 10.12) and Tw 20 (Figure 10.13), the $(C_2 - \text{CMC})/[\text{CMS}]$ molar ratios are about 26.2 $[2.1 \times 10^{-4}\ \text{mol dm}^{-3}/(8 \times 10^{-4}\ \text{mol dm}^{-3})]$ and 4.1 $[0.33 \times 10^{-4}\ \text{mol dm}^{-3}/(8 \times 10^{-4}\ \text{mol dm}^{-3})]$, respectively. These numbers indicate that CMS can bind more hydrophilic Slovasol 2430 than the less hydrophilic Tw 20 at saturation concentration C_2. The aggregation number of nonionic emulsifiers (Tw 20 and Slovasol 2430) alone is ca. 100–150 [143, 144]. In the mixed emulsifier CMS aggregate there are ca. 4–6 CMS molecules for Slovasol 2430 and ca. 30 for Tw 20. These data do not seem real. The shift in the CMC can therefore be attributed to the binding of emulsifier molecules to the CMS skeleton. Furthermore, the cooperative binding of ionic emulsifier (SDS) monomers to PEO was reported to occur when $M_{\text{r,m,PEO}}$ exceeds ca. 3000 [141]. For the hydrophilic polymer with EO segments, the exothermic peak was ascribed to the rehydration of the EO segments into the aqueous phase and these rehydrated segments could form an ion–dipole association with the hydrophilic headgroups of SDS micelles. In the present case, the hydrophilic headgroups of CMS could form an ion–dipole association with the rehydrated EO segments of nonionic emulsifiers. The relative molecular mass of PEO chains is ca. 1000 for Slovasol 2430 and 680 for Tw 20. This might be one of reasons why the loosely associated aggregates are formed and the differences in C_2 values for both emulsifiers appear.

Ultrasonification of CHP (pullulan containing 1,6-cholesterol groups per 100 glucose units) (0.1 wt.%) initiate intermolecular aggregation and provides relatively monodisperse particles [145]. The CHP self-aggregates eluted earlier than the parent pullulan. The apparent molecular weight of the polymer aggregates estimated by size-exclusion (column) chromatography (SEC) is higher than that of the parent pullulan. The M_w/M_n of the CHP self-aggregate in water was 1.17, whereas that of CHP in dimethylformamide (DMF) was 1.65. This was attributed to the formation of relatively monodisperse aggregates in water. The root mean square radius of gyration (R_G) of the CHP self-aggregates was 16.5 nm on average. That the aggregate was colloidally very stable was revealed from evidence that the size of the CHP self-aggregates did not change at all, even after keeping for a long time at room temperature. When aggregate of the palmitoyl group-bearing pullulan (OPP) was kept at room temperature, an increase in the turbidity of the sample suspension was observed and precipitation took place after 24 h. The OPP self-aggregate was colloidally less stable than the CHP self-aggregate [146]. Hence the structure of hydrophobic moiety of the polysaccharides influences the colloidal stability of the aggregates. The average hydrodynamic radius (R_H) of the CHP self-aggregates measured by dynamic light scattering (DLS) was 13.3 nm ($D_{dif} = 1.85 \times 10$ cm s^{-1}) with the formation of relatively monodisperse particles. Spherical particles with relatively uniform size (diameter = 25 nm) were observed in negatively stained electron microscopy of the aqueous CHP solution. The aggregation number determined SLS was approximately 13; the weight-average molecular weight of the self-aggregate was 7.6×10^5, R_G was 16.8 nm and the second virial coefficient (A_2) was 2.6×10^{-4} mol mL g^{-2}. The value of R_G estimated by SEC was almost identical with that obtained by SLS. The results show that one CHP self-aggregate ($M_w = 7.6 \times 10^5$) consists of approximately 13 CHP molecules ($M_w = 5.8 \times 10^4$).

Kato et al. [147] investigated pullulan having different molecular weights by SLS and proposed empirical equations for R_G and A_2 as a function of the weight-average molecular weight (M_w):

$$R_G = 1.47 \times 10^{-2} {M_w}^{0.58} \tag{23}$$

$$A_2 = 5.42 \times 10^{-3} {M_w}^{-0.26} \tag{24}$$

If R_G and A_2 were calculated by using Eqs. (21) and (22) on the assumption that the M_w of pullulan is now 7.6×10^5, the following values are obtained: $R_G =$ 37.9 nm and $A_2 = 1.6 \times 10^{-4}$ mol mL g^{-2}. R_G is almost double the values observed for the CHP self-aggregates. This means that the polysaccharide chain of the CHP self-aggregates must be more compact in water compared with that of the parent pullulan. This can be attributed to the partial dehydration of chain moieties of polysaccharide close to the hydrophobic core of micellar aggregate.

The structural change on dilution of the CHP self-aggregates in water was investigated by fluorimetry in the presence of protein nucleic acid (PNA) as the fluorescent probe [145]. PNA strongly emits in a polar solvent or within a hydrophobic environment, whereas it is fairly quenched in polar media [148]. When PNA was with the CHP self-aggregates, however, the emission maximum of PNA shifted to lower wavelength and the intensity increased drastically as a function of the concentration of CHP. A clear break point was observed for changes in both the emission maximum and the intensity at a concentration around

Table 10.6 Critical micelle concentration (H_0) of various nonionic amphiphiles [145][a].

Nonionic amphiphile	$H_0 \times 10^3$ (g hydrocarbon dm^{-3})	Ref.
1. $CH_3(CH_2)_{11}O(CH_2CH_2O)_8H$	18	[142]
2. $CH_3(CH_2)_{11}O(CH_2CH_2O)_{18}H$	14	[142]
3. $CH_3(CH_2)_{11}O(CH_2)_2CHO(CH_2CH_2O)_{15}H$	0.04	[150]
4. Chol–O–$(CH_2CH_2O)_8H$	0.47	[149]
5. Chol–O–$(CH_2CH_2O)_{25}H$	1.4	[149]
6. Hydroxyethylcellulose ethers with C_{12}–C_{24}-alkyl chains	20–40	[142]
7. OPP	0.15	[146]
8. CHP	0.05	[146]

a) The H_0 values for Nos. 1–6 were calculated by using Landoll's equation from data given in the literature. For 7 and 8, the H_0 values were calculated from data obtained by using PNA [145, 150].

0.01 mg mL^{-1} (0.001 wt.%). The break point was assumed to correspond to the critical concentration where the intermolecular aggregation of CHP occurs.

In order to compare the critical concentration of the different amphiphiles, the parameter H_0 (g of hydrocarbon dm^{-3}) was used. It indicates the concentration of the hydrocarbon moiety of the polymer in solution (Table 10.6). A smaller H_0 value means the formation of an aggregate at a lower concentration. For example, the H_0 value of hydroxyethylcellulose ethers bearing a long alkyl chain (C_{12}–C_{24}) is close to the those of nonionic emulsifier based on poly(ethylene glycol) derivatives (Table 10.6; see samples 1 and 2) [142]. On the other hand, the H_0 value of CHP is 10 or 100 times smaller than those of cholesterol derivatives of poly(ethylene glycol) (samples 4 and 56) [149]. Its value was closer to that of poly(ethylene glycol) bearing two alkyl chains (3), which forms bilayer structures rather than micelles [150]. These data suggest that the cholesterol moiety is a more powerful hydrophobic pendant than the palmitoyl moiety for forming micellar aggregates. The H_0 values of 7 and 8 were smaller than those of hydroxyethylcellulose bearing a long alkyl chain (sample 6). Therefore, the rather flexible skeleton of pullulan may associate and form a compact aggregate more easily compared with the relatively rigid skeleton of cellulose [147].

Most amphiphilic polymers such as hydroxyethylcellulose derivatives usually decrease considerably the surface tension of an aqueous solution with an increase in the polymer concentration [142]. When the OPP concentration was increased to 0.5 mg mL^{-1}, in fact, the surface tension of the solution significantly decreased to 57.0 dyn cm^{-2} at 23 °C. However, the addition of CHP in water did not decrease the surface tension of water at all. The surface tension remained unchanged even at the higher concentration: 74 dyn cm^{-2} at 0.145 mg mL^{-1} CHP.

This means that the hydrophobic core of the CHP aggregates is completely and stably covered by the hydrophilic shell of the polysaccharide skeleton. The very low critical aggregation concentration and the surface inactivity indicate that colloidal stable nanoparticles are certainly formed above the critical concentration. The existence of microdomains which consist of both the rigid core of hydrophobic cholesterol and the relatively hydrophilic polysaccharide shell was suggested on the basis of both the line broadening of the proton signal of the cholesterol moiety of CHP ($\delta = 0.6$–2.4 ppm) in the ^{1}H NMR spectrum and incorporation of several hydrophobic fluorescent probes in the CHP self-aggregates. The CHP self-aggregates strongly complexed with hydrophobic and less hydrophilic fluorescent probes similarly to the case of cyclodextrin.

Optical microscopy showed that the diameter of the corn starch granules in the water dispersion (at concentration 5 g L^{-1} and temperature 40–60 °C) increased due to swelling [151]. At 70 °C, however, most granules are disrupted. This temperature may be taken as the gelatinization temperature [T(G)] for this sample. At 55 °C, no birefringence is observed for the same starch–water dispersion. Some authors showed that crystallinity still decreased after all the granules had lost birefringence, which means that birefringence measurements provide only an approximation of the final melting point of the ordered regions [152].

The melting temperature (T_m) of corn starch expressed as the peak temperature of the DSC melting endotherm varied with the water content as follows [151]:

$$T_m(^\circ\text{C})/\text{water content }(\%)\text{: } 168/0,\ 126/10,\ 123/20,\ 120/30,\ 109/40,\ 108/50,\ 72/60 \qquad (25)$$

For dispersions with water contents in the range 0–50%, melting occurred at much higher temperatures than for those at 60% water content. These results are useful for an estimation of the thermal conditions necessary for processing starch, without degradation. Since the loss of crystallinity during extrusion is caused by the action of heat and intense shear forces in the presence of plasticizers, milder thermal conditions could be used.

Starch samples with higher degrees of crosslinking, cld 0.5 and 0.75, displayed individual particles [92]. They can be described as suspensions of deformable particles suspended in a continuous suspending medium constituted of polymer chains plus water. Depending on the amount of solvent available, the particles swell in water at room temperature and their size varies with the degree of crosslinking. It is usually possible to define a critical concentration C_{crit} for suspensions which corresponds to the space-filling concentration, that is, the concentration at which the system appears homogeneous. It has been demonstrated that this critical concentration is close to the inverse of the swelling power, in the case of starch suspension [153]. Indeed, the higher the extent of crosslinking, the lower is the swelling power and, hence, the higher is the space-filling concentration.

Starch in a mixed solvent of water–DMSO produced stable dispersions [154, 155]. However, increasing shear time during sample preparation led to a decrease in shear moduli. Increasing concentration led to a change in rheological behavior from Newtonian liquid to semidilute solution for high amylose starch and from semidilute solution to viscoelastic solid for common corn and waxy starch. The amylopectin component of starch has been reported to cause shear thickening and flow-induced incipient phase separation at shear rates greater than about 20 s^{-1} [63]. The effects of higher amylopectin content for the same starch concentration were higher shear viscosity and even gel formation. This pointed to the formation of networks with increasing amylopectin content.

Table 10.7 summarizes the kinetic and colloidal data for the miniemulsion polymerization of butyl acrylate in the presence of carboxymethylated starch (CMS). It shows that the polymer particles are formed during the polymerization. This indicates that (grafted) radicals generated in the aqueous phase enter the hydrophobic polymer [poly(butyl acrylate) (PBA)] particles. Furthermore, the fast polymerization favors the polymerization in the polymer particles.

The rate of polymerization is observed to decrease on addition of starch and the decrease is much more pronounced at a low level of Tw 20. The chain transfer to starch increases the fraction of less reactive hydrophilic radicals which partly fail to enter the hydrophobic polymer particles. Furthermore, the polysaccharides are known to act as a reducing agent and so they can deactivate the initiating

Table 10.7 Variation of kinetic and colloidal parameters in the sterically stabilized miniemulsion polymerization of butyl acrylate (BA) with carboxymethylated starch (CMS) and Tween 20 (Tw 20) concentration [127][a)].

[Tw 20] × 10^2 (mol dm^{-3})	WR	$R_{p,max}$ × 10^4 (mol dm^{-3} s^{-1})		Conversion (%)		$D_{p,f}$ (nm)		$N_{p,f}$ × 10^{-16} (dm^3)		D_w/D_n
		(1)	(2)	Max.	F	(1)	(2)	(1)	(2)	
0.41	0.25	4.8	10	21	44	785	335	0.063	1.6	4.7
0.81	5	7.6	12	3	62	797	310	0.085	2.0	3.5
1.22	7.5	10.9	14.5	5	88	388	280	1.05	2.8	1.14
1.63	10	14.5	18	9	90	252	250	3.9	4.0	1.2
2.44	15	18.1	30	11	76	173	200	10.1	7.6	1.07

a) Recipe: 100 g water, 40 g BA, 0.2 g CMS, [APS] = 1×10^{-3} mol dm^{-3}, temperature = 60 °C. (1) with CMS; (2) without CMS. WR, weight ratio Tw 20 and CMS; $R_{p,max}$, the maximal rate of polymerization; $D_{p,f}$, the diameter of final polymer particle, $N_{p,f}$, the number of final polymer particles; PSD_{rel}, the relative particle size distribution; F denotes the final value and Max. the maximal value of R_p; D_w/D_n, particle size distribution.

(primary and oligomeric) radicals. The starch molecules and micelles compete for the initiating radicals. The larger the number of micelles, the larger is the number of radicals entering the micelles or polymer particles. The stronger depression of the polymerization rate with starch at low emulsifier concentration increases the reaction exponent x ($R_{p,max} \propto [\text{Tw } 20]^x$, $x = 0.77$ with starch, $x = 0.57$ without starch).

The stabilization of polymer particles by of emulsifier is related to the size and number of polymer particles. At low emulsifier concentration the generation of large polymer particles and their interaction with starch induce particle agglomeration. This is also accompanied by a broad particle size distribution. At high emulsifier concentration, in contrast, an increase in the number of polymer particles appears (starch somehow favors the formation of larger number of particles) and the polymer dispersion formed is nearly monodisperse (Table 10.7). The synergistic effect is documented by a large value of the exponent y, that is, a strong dependence of the particle number on emulsifier concentration:

$$N_{p,f} \propto [\text{Tw } 20]^{y=3.1} \text{ (with starch) and}$$
$$N_{p,f} \propto [\text{Tw } 20]^{y=0.64} \text{ (without starch)} \qquad (26)$$

The variation of the colloidal and rheological properties of polymer dispersions in the emulsion polymerization of vinyl acetate (VAc) with the concentration and type of starch are summarized in Tables 10.8 and 10.9 [126]. The addition of a small amount of CMS causes a decrease in the viscosity of the polymer dispersion. Further addition of CMS strongly increases the viscosity and at a certain concentration of CMS the polymer dispersion becomes solid. Distinct polymer particles were formed in the emulsion system without CMS. The presence of CMS depresses the formation of distinct polymer particles, that is, the light scattering measurements (hydrodynamic size) did not confirm the formation of polymer particles.

Table 10.8 Preparation of poly(vinyl acetate) dispersions [126][a)].

Sample	Sloviol (g)	Water (g)	CMS (g)	Viscosity (mPa)	D (nm)
1	44.1	30	0	340	430
4	36.75	33.97	3.38	117	–
6	29.4	37.95	6.75	140	–
3	22.05	41.9	10.13	958	–
7	14.7	45.9	13.51	Solid	–
5	7.35	49.9	16.88	Solid	–

a) Recipe: 69.6 g VAc, 1.5 g $NaHCO_3$, 14.3 g water, 0.225 g APS.

Table 10.9 Preparation of poly(vinyl acetate) dispersions [126][a].

Sample	Sloviol (g)	NWS (g)	H_2O (g)	Viscosity (mPa)	D (nm)
1	44.1	0	30	340	430
2	36.75	5	30	4936	412[b]
3	40.43	2.5	30	1814	470[b]
4	36.75	0.75	37.35	913	472
5	29.4	1.5	44.7	–	461
6	22.05	2.25	52.05	4790	508[b]

a) See footnote to Table 10.8.
b) Agglomerates.

The emulsion polymerization of VAc in the presence of native wheat starch (NWS) leads to the formation of a polymer dispersion with distinct polymer particles (Table 10.9). The addition of a larger amount of NWS induces particle agglomeration and the formation of coagulum. The viscosity of the polymer dispersion increases with increasing amount of NWS.

The microemulsion polymerization of butyl acrylate initiated by APS was followed in the presence of unsaturated galacturonides (in H form, UGH; in K form, UGK) [125]. These data show that the addition of UGK leads to a decrease in polymerization rate and the appearance of limiting conversion (Table 10.10). A final conversion close to 90% conversion was reached with UGK. The presence of UGH led to the appearance of limiting conversion in the range 75–90%. The dependence of the maximum rate of polymerization on conversion is described by a curve with a maximum at ca. 30–40% conversion. In the absence of UGH/K a maximum rate was observed at ca. 20–30% conversion [156, 157]. Thus, the addition of UGH/K shifts the maximum rate (or the monomer–saturation interval) to higher conversions. This is discussed in terms of depressed radical entry rate and chain growth due to which the gel effect is shifted to higher conversions.

In the presence of UGH/K, a decrease in the rate of polymerization is observed and the decrease is more pronounced in runs with UGH (Table 10.10). In the runs with UGK a decrease in the rate was observed on addition of a very small amount of UGK. In the range of UGK used the rate of polymerization is nearly constant. The variation of the maximum rate of polymerization with the [UGH] is as follows (with APS):

$$R_{p,\max} \propto [\mathrm{UGH}]^{-0.28} \tag{27}$$

The retardation effect of UGH was attributed to the degradative chain transfer events, the exit of transferred radicals from particles and the decrease in the monomer concentration in particles (the higher the number of particles, the lower is the monomer concentration in the polymer particles; see later). This be-

Table 10.10 Variations of kinetic and colloidal parameters of the microemulsion polymerization of butyl acrylate initiated by APS or AIBN with the additive concentration [125][a].

Additive type	wt.% (per water)	$R_{p,max} \times 10^3$ (mol dm^{-3} s^{-1})	D (nm)	$N \times 10^{18}$ (dm^3)	Conv.$_{final}$ (%)	$M_w \times 10^{-5}$
CRK (APS)	0	6.3	41	2.45	98	9.56
	0.4	5.6	33	4.85	98	9.32
	0.8	5.6	35	3.9	96	9
	1.2	5.7	43	2.19	97	8.13
	2.0	5.6	46	1.79	97	6.65
[b]	1.2	2.8	56	0.75	76	18.9
	1.2	2.7	55	0.74	74	18.5
CRH (APS)	0.4	5.5	33	4.7	90	5.93
	0.8	5.0	28	6.7	85	4.04
	1.2	4.5	24	9.42	75	1.6
	2	3.4	23	13.9	81	0.43
	2	3.3	22	12.5	77	0.16
	2					0.4
CRK (AIBN)	0	2.1	41	2.3	87	9.32
	1.2	1.9	43	2.1	76	8.57
CRH (AIBN)	1.2	0.92	33	1.74	36	0.79

a) Recipe: 100 g water, 20 g SDS, 10 g BA, [APS] = [AIBN] = 1.4×10^{-4} mol dm^{-3} (related to the whole system).
b) 10 g SDS.

havior was attributed to the close packing of emulsifier and coemulsifier (UGH), which generates a barrier to radical entry into emulsified monomer droplets. It was reported that in the microemulsion polymerization of BA the rate of polymerization and the particle size decreased with increasing [SDS] in which BA is supposed to act as a coemulsifier [157]. In the runs with UGK the particle size decreased but the rate of polymerization, in contrast, increased with increasing [SDS]:

$$R_{p,\max} \propto [\mathrm{SDS}]^{0.5-1.0} \tag{28}$$

The aqueous phase (co)polymerization of UGK generates aqueous phase oligomeric radicals with a low entry efficiency. Increasing [SDS] or micelle concentration increases the radical entry efficiency and also the rate of polymerization.

The average particle size was found to decrease with increasing [UGH] and to increase with increasing [UGK]. The particle concentration varied with increasing [galacturonide] as follows:

$$N \propto [\mathrm{UGH}]^{0.66} \quad \text{and} \quad N \propto [\mathrm{UGK}]^{-0.64} \tag{29}$$

The main location of a partly water-soluble UGH (coemulsifier) is the micellar region. This favors the formation of a larger number of micelles and polymer particles. UGH as a radical chain transfer is supposed to act also as a short-stopper. The depletion of monomer-swollen micelles therefore falls, which means that a larger number of monomer-swollen micelles can exist longer and so a larger number of particles can be formed [158]. In contrast, the aqueous phase polymerization of UGK generates a water-soluble polymer initiating particle agglomeration. The presence of polymer derived from UGK is supposed to initiate the particle agglomeration and the formation of larger particles. Hence the difference between these galacturonides [dissociated (UG^-) and undissociated (UGH)] results from the difference in their surface activity, the attraction and repulsion interactions and the location in the reaction system.

Variations of the molecular weight parameters with the initiator, galacturonide and emulsifier concentration are summarized in Table 10.10. The molecular weight varied with [UGK] or [UGH] in the presence APS as follows:

$$M_\mathrm{w} \propto [\mathrm{UGK}]^{-0.16} \quad \text{and} \quad M_\mathrm{w} \propto [\mathrm{UGH}]^{-1.3} \tag{30}$$

These data show that both galacturonides decrease the molecular weight and the decrease is much more pronounced in runs with UGH. It is assumed that UGK is mostly located in water whereas UGH is located in the interface zone (as a coemulsifier). The location of UGK in water may influence the oligomer formation and the radical entry events. However, UGH is supposed to act as a coemulsifier and chain transfer agent within the polymer particles. UGH therefore favors a rise in the particle number and a fall in the molecular weight, as was observed. The same trend can be found in the runs with AIBN, i.e. UGH strongly decreases the molecular weight of the polymer whereas UGK is inactive. The following data also support a slight influence of UGK on the M_w:

$$M_\mathrm{w} \propto [\mathrm{SDS}]^{-0.95} \text{ (without UGK)} \quad \text{and} \quad M_\mathrm{w} \propto [\mathrm{SDS}]^{-0.91} \text{ (with UGK)} \tag{31}$$

The uronic acid content (4.4 and 4.2%) in the copolymers PBA–carbohydrate residue H/K (PBA–UGH and PBA–UGK) and IR spectra (at 3400 cm^{-1}, –OH groups) confirmed the presence of carbohydrates in the polymer chains.

10.4 Nanocomposites, Blends and Their Properties

Processing of starch–water–glycerol mixtures in a single-screw extruder produced films that, although clear, flexible and apparently homogeneous, showed heterogeneities, probably resulting from incomplete melting of starch granules.

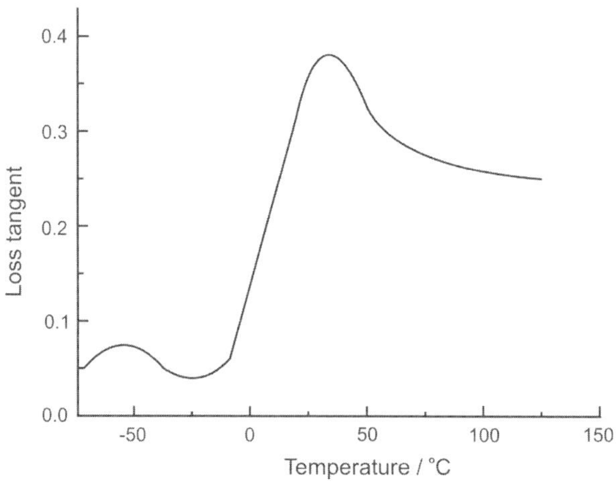

Figure 10.14 Temperature dependence of the loss tangent (tan δ) measured at 10 Hz for a thermoplastic starch sample obtained with 33% (w/w) total plasticizers [151].

Transparent, flexible and homogeneous films were obtained by extrusion in a twin-screw extruder. The mechanical properties of these films varied with the water content. At 8% water content, the material is brittle and at 18% water content, although not elastic, the material reaches a maximum value of strain [151].

Dynamic mechanical properties are reported in Figure 10.14 for a thermoplastic starch sample with 15% glycerol and 18% water, in terms of the temperature dependence of the loss tangent (tan δ) [151]. Two distinct transitions are observed in the tan δ vs. T curve. The first transition has a maximum at around −55 °C and the second transition at around 40 °C. At intermediate plasticizer levels (between 20 and 35% total contents of water and glycerol), two glass transitions were observed and phase separation was suggested to occur. Dynamic mechanical analysis (DMA) carried out for such mixtures showed two loss peaks, attributed to the transitions of a starch-poor phase (lower temperature) and of a starch-rich phase (higher temperature) [159].

The rheological properties of water–DMSO dispersions varied with starch composition and concentration [154, 155]. The limit of strain in the linear viscoelastic region was determined to be between 0.15 and 0.2 for high-amylose corn starch, 0.15 for common corn starch and 0.1 for waxy corn starch. The loss modulus was greater than the storage modulus in the frequency range 0.1–100 rad s^{-1} for all samples except 8% concentration waxy corn starch, where a crossover was observed at a frequency of 1.58 rad s^{-1}. High-amylose corn starch at a concentration of 2% (w/v) was essentially a Newtonian liquid with a viscosity of 18 mPa s. The applicability of Rouse theory to high-amylose starch data was limited by the presence of branching and polydispersity, which resulted in the underprediction of elastic moduli.

Table 10.11 Slope of log G' and log G'' versus log ω curves [155][a].

Starch concentration (% w/v)	High-amylose corn starch		Common corn starch		Waxy corn starch	
	G'	G''	G'	G''	G'	G''
2	1.29	0.89	0.76	0.69	0.63	0.57
4	1.08	0.9	0.65	0.67	0.59	0.56
6	0.99	0.87	0.65	0.63	0.57	0.56
8	0.99	0.87	0.61	0.63	0.49	0.54

a) Each starch sample dissolved in aqueous DMSO (90% DMSO, 10% water). In dynamic measurement the storage (G') and loss (G'') moduli were measured as a function of oscillation frequency.

The rheological properties of common corn starch changed from a viscoelastic liquid at 2% concentration to a near-critical gel at 8% concentration. The proportion of amylopectin in starch dictated whether "gel-like" viscoelastic behaviors were observed. A power-law relaxation was observed for all concentrations of common corn starch, with the slope of the storage (G') and loss (G'') moduli curves decreasing with an increase in concentration. Values for the slope of log G' and log G'' versus log ω curves are summarized in Table 10.11. Power-law behavior was also observed for waxy corn starch, with the slope of the G' and G'' curves decreasing from 0.63 to 0.54, respectively, at 2% concentration, to 0.49 and 0.54, respectively, at 8% concentration. Waxy corn starch was a viscoelastic liquid at 2 and 4% concentration, a near-critical gel at 6% concentration and a viscoelastic solid at 8% concentration. Rheological properties of 8% concentration common corn starch and 6% and 8% concentration waxy corn were sensitive to variations in the preparation method and network rupture effects were observed at a strain magnitude >2. Extensional thinning was observed for 2% concentration waxy corn starch. Extensional viscosity data for 2% concentration high-amylose corn starch were dominated by fluid mechanical and instrument effects and limited conclusions could be drawn from the extensional data.

The power-law relaxation was discussed in terms of the formation of self-similar clusters near the gel point. The slopes of the G' and G'' curves are lowest for waxy corn starch and highest for high-amylose corn starch. This trend was explained as due to the high molecular weight and high branching of the amylopectin compared with amylose. Waxy corn starch consists of approximately 99% amylopectin and thus has a greater proportion of high molecular weight molecules than either common corn starch (~75% amylopectin) or high-amylose corn starch (~30% amylopectin). Thus, the amylopectin content in the starch increased in the following order:

$$\text{high-amylose starch} < \text{common corn starch} < \text{waxy starch} \quad (32)$$

The formation of self-similar clusters gives rise to power-law relaxation in common corn and waxy corn starch, in contrast to the semi-dilute solution behavior of high-amylose corn starch.

The relaxation exponent m in the equation $G(t) = c_p t^m$, where G is the modulus and c_p the prefactor for common and waxy corn starch samples, had values summarized in Table 10.12 [160].

The exponent m has a value in the range $0 \leq m \leq 1$ depending on the molecular composition and crosslinking conditions and the prefactor c_p is a measure of the strength of the critical gel [161]. As expected, c_p increases with an increase in the concentration of both common and waxy corn starch. For a chemical crosslinking system, the exponent m decreases and the gel strength c_p increases as the stoichiometric ratio of crosslinker to starch increases [162]. Entanglements among the precursor polymer molecules results in a decrease in the value of m. Thus, the power-law behavior observed for starch in water–DMSO was attributed to the formation of physical crosslinks. Local helical structures, microcrystallites and nodular domains have been proposed as the interactions that give rise to physical gelation. Starch has an abundance of hydroxyl groups and the phenomena of retrogradation and syneresis in starch–water systems have been attributed to hydrogen bonding. The power-law behavior of starch–water–DMSO systems could be attributed to the formation of helical structures leading to physical crosslinks.

When powders are compressed, it is hypothesized that the aggregates of the polysaccharide chains come closer together and their ability to resist an external force could be related to this new structure. This is supposed to lead to molecular rearrangements and possibly to more extended hydrogen association [104]. The crushing strength of crosslinked high-amylose starch (CLHAS) materials has been shown to depend on the cld [106]. The presence of water and its influence on solid carbohydrate structure are often reflected in their spectra [94, 163, 164]; a variation of the 1646 cm^{-1} band as a function of cld was found and correlated with the hydration state of CLHAS powders. FTIR studies of conformational

Table 10.12 Model parameters for common corn and waxy starch [160].

Concentration (%)	Common corn starch		Waxy corn starch	
	c_p	m	c_p	m
2	0.05	0.78	0.21	0.63
4	0.23	0.68	0.7	0.59
6	0.43	0.66	1.41	0.57
8	0.99	0.61	0.61	0.48

changes due to the retrogradation of starch–water systems during storage [165] indicated that the 1300–800 cm^{-1} region is sensitive to the conformation of the polysaccharides. Since CLHAS films represent, in fact, another type of dry structure, it was interesting to observe the evolution of the 1300–800 cm^{-1} region in powder and film forms versus the cld and to correlate this with the morphological transitions from B- to V-type helix observed by X-ray diffraction. The X-ray and FTIR analysis correlated with dissolution kinetics and mechanical hardness of the dry tablets can generate interesting information on the structure–properties relationship in CLHAS matrices.

Walenta et al. used wide-angle X-ray scattering (WAXS) to investigate the relationship between the supermolecular structure (crystalline polymorphs, degree of crystallinity, crystallite dimensions) of separated amyloses and materials extruded from them and measured the mechanical properties of the extruded materials [166]. Pure amyloses of crystal type V_A with various molecular weights and molecular weight distributions have been extruded with a suitable plasticizer system (water–urea–glycerol) to give clear and homogeneous films with B-type crystals. This means that in the extrusion process the original crystalline structure has been destroyed completely and a new crystalline order was established by recrystallization. V_H crystals were observed in one case where stearic acid had been added to the premix. Compared with the starting materials, the crystallinities of the extruded films are enhanced, whereas the crystallite dimensions are drastically reduced.

The group of amyloses with broad molecular weight distributions (samples A1–A5) have crystallinity values ranging from 33 to 40%, whereas in the case of the narrower chain length distributions (samples A6 and A7) there is a tendency for lower crystallinity (Table 10.13) [166]. The lattice distortions of the amylose crystallites are generally reduced compared with wrinkled pea starch (WPS)

Table 10.13 Weight-average molecular weight (M_w), molecular weight distribution (M_w/M_n), crystalline polymorphs, degree of crystallinity (x_c) and crystallite dimensions (D_{hkl}) of wrinkled pea starch (WPS) and amyloses (A1–A7) separated from it [166].

Sample	$M_w \times 10^5$	M_w/M_n	Crystalline polymorph	x_c (%)	D_{hkl} (nm)
WPS	122.2	–	B(V_H)	31	8.1
A1	3.63	121	V_A	33–40	17.5–24
A2	1.22	81			
A3	1.25	102			
A4	1.84	115			
A5	1.86	124			
A6	1.97	10	V_A	33	22.9
A7	0.58	6	V_A	29	25.3

sample and vary only slightly. The lowest disorder parameter was found for amylose sample A6 with a narrow molecular weight distribution and a high M_w value. The crystallite dimensions of the amyloses generally far exceed the values obtained for WPS.

The ordered regions E1–E7 films of A1–A7 samples after extrusion are crystallized in the B polymorph, despite the fact the starting amyloses were crystallized in the V_A modification (Table 10.14). The occurrence of B-type crystallinity is not uncommon for thermoplastic starch materials and has been reported, e.g., for glycerol-plasticized, compression-molded potato starches [167]. Compared with the staring amyloses, crystallinities are enhanced and the lateral crystallite dimensions are distinctly smaller after the extrusion process. Although the crystallites are smaller (the peaks broader), there is a greater weight fraction of crystallites present in the extruded materials. Comparing extruded films E1 and E2, the higher molecular mass of amylose A1 results in a somewhat lower degree of crystallinity. Furthermore, the sample produced from the starting amylose with lower average molecular weight (sample E7) exhibits a higher degree of crystallinity and enlarged crystallite dimensions.

As found from samples E3, E3a and E5, the addition of up to 15 wt.% amylopectin does not change the supermolecular structure significantly. However, as shown by the results for sample E4, the composition of the plasticizer system in conjunction with the extrusion conditions clearly affects the structure of the extruded products. In this case, in addition to the crystallinity of the B type with slightly enhanced crystallite dimensions, very much larger crystallites of the V_H modification are also formed. This could be due to the addition of stearic acid, which is known to act as a complexing agent, especially for amylose [168].

Table 10.14 Parameters for the supermolecular structure (from WAXS) and averages for strength σ, elongation ε and Young's modulus E of extruded amylose films [166][a].

Sample	x_c (%)	D_{hkl} (nm)	σ (MPa)	ε (%)	E (MPa)
E1	47	4.9	36.5	3.6	1800
E2	53	4.9	30.1	7.3	1180
E3	53	5.3	13.3	2.1	860
E3a	53	5.0	13.3	2.2	800
E4	40	5.5	11.2	1.8	1120
E5	51	4.8	16.4	2.6	1010
E6	43	4.6	18.2	4.3	860
E7	50	5.3	22.2	2.9	1440

a) Crystalline polymorph B – in all samples, film E4 B and V_H crystalline polymorph.

Films E1 and E2 exhibit the highest mechanical property values of all the samples tested (Table 10.14). Compared with samples E1 and E2, the mechanical properties of samples E6 and E7, produced from amyloses with a narrow molecular weight distribution, showed significantly reduced strength values. Compared with pure amylose sample E3, the addition of 7.4 wt.% (sample E3a) does not change the mechanical properties significantly. The addition of 14.7 wt.% amylopectin (sample E5) results in improved mechanical properties, which are, however, still clearly lower than the values for the pure amylose films E1 and E2, even though sample E2 originates from a starting amylose with a lower molecular weight. The changed composition of the plasticizer–melt flow accelerator system (sample E4) leads to improved processability of the material but not to better mechanical properties. In view of the otherwise rather similar sample E2, loss in strength and elongation are attributed to the formation of the V_H crystal polymorph.

Thus, the molecular weight distribution of the pure amylose samples and the addition of amylopectin (synthetic mixtures of amylose and amylopectin) do not substantially influence the mechanical properties. However, the results clearly demonstrate that increased average molar mass enhances the mechanical properties of the extruded films. The measured properties of the extruded starch films do not differ greatly but, surprisingly, films with amylopectin as the starting material exhibit the highest modulus and strength (Table 10.15).

The samples with 24% plasticizer (6.2% urea + 9.4% glycerol + 10.4% water) were observed to be very brittle. As expected, with increasing plasticizer content the strength and modulus increase and the elongation at break decreases (Table 10.16). A significant change in mechanical properties is observed between 26 and

Table 10.15 Mechanical properties strength σ, elongation ε and Young's modulus E and structural parameters of starch films extruded from different starting starch polymers [166].

Starch type[a)]	σ (MPa)	ε (%)	E (MPa)	Crystalline polymorph	x_c (%)
Potato starch	22	11.7	880	B	29
Maize starch	20.3	9.1	800	A	33
Wheat starch	22.1	7.6	1140	A	32
AMS	21.5	10.4	1015	B	27
WPS	17.7	4.8	740		
AmPn	23.4	5.4	1200	A	35
WPS–AmPn, 9:1	20	4.6	940		
WPS–AmPn, 4:1	17.6	5.7	730		

a) WPS, wrinkled pea starch; AMS, amylomaize starch (Hylon VII); AmPn, amylopectin.

Table 10.16 Mechanical properties strength σ, elongation ε and Young's modulus E and structural parameters (determined by WAXS) of starch films extruded from wheat starch with different plasticizer content [166][a].

Plasticizer content (%)	σ (MPa)	ε (%)	E (MPa)
26	18.3	3	1890
28	11.4	8	895
30	7.5	60	440
32	6.9	64	325
34	6.2	74	260

a) Crystalline polymorphs – V_H.

30% plasticizer content. It should be mentioned that above 28% the strain at break not longer coincides with the strain at maximum force and the strain at maximum force values are given.

Starches display their expected crystal type (Table 10.15). The crystallinities vary in a small range between 29 and 35%. The lateral crystallite sizes have values between 5 and 9 nm as a function of starch origin and crystallographic direction. The crystalline V_H polymorph with slightly enhanced crystallinity was found as the plasticizer content increased (Table 10.16). Remarkably, the crystallite dimensions increased substantially with increasing plasticizer content in the range 26–34%.

In the native starting starch polymers, the crystalline A and B polymorphs with the molecules arranged as double helices were found. Amylose produced from WPS exhibited the single-helix V_A polymorph whereas in amylopectin derived from maize starch the crystalline A type was identified. The crystallinities of the native starches were found to be roughly in the same range as those separated from amylose and amylopectin. However, the crystallite dimensions of native starches of different type and of amylopectin are distinctly smaller than those of amylose. After extrusion of native A- and B-type starches, the V_H crystal structure was generally detected. The extrusion products from pure amylose are generally crystallized in the B polymorph. In the case of amylopectin, the starting A polymorph was preserved to a large extent after extrusion but minor transitions to the B polymorph are also possible. The formation of a crystal structure in the extruded product which is different from the crystal structure of the starting starch polymer can be explained by initial melting and a subsequent recrystallization process. Under certain extrusion conditions, residual structures of the starting material can remain unchanged, owing to an insufficient destructuring process. In this respect, amylopectin is a special case with the A polymorph preserved during extrusion or with a rearrangement of the double helices to the B type. It is probable that the original double-helix structure of pure amylopectin cannot be destroyed like that of A- and B-type starches under the same extrusion conditions. Obviously, under the given extrusion conditions, the crystal structure type of the starting material governs the crystal type of the extruded products.

With regard to mechanical properties, the results clearly indicate that a high molecular weight of amylose leads to improved strength and modulus of the amylose films. Amylopectin, with its high molecular weight, also produces extruded films with comparatively good mechanical properties. The influence of molecular weight on mechanical properties is well known from synthetic polymers such as polyethylene [169] and from cellulose materials [170]. For extruded potato starch, an increase in the elongation and tearing energy with increasing molar mass was observed by van Soest et al. [50].

The highest crystallinity was found for the extruded amylose films and the lowest for the potato starch films (Tables 10.14–10.16). The crystallinities of the extruded products did not differ greatly and it was not possible to establish any dependence of mechanical properties on the degree of crystallinity as a single parameter. On the other hand, the crystallite dimensions of the extruded films varied over a wide range. The extruded amylose films of B-type exhibited sizes from 3 to 7 nm, whereas the V_H-type crystallites of the native starch films reached dimensions up to 35 nm. Obviously, the formation of larger ordered regions is favored in the case of single-helix V_H–type crystallization.

With constant plasticizer composition and content, the crystallite dimensions of films made from the different native starches increased substantially with increase in temperature. The increase in crystallite size occurs with virtually unchanged crystallinity, indicating a reduced number of nuclei and higher crystallization rate with increasing extruder temperature. In the experimental series with varied extruder temperatures, a correlation was found between increasing crystallite dimensions and improved strength and modulus for potato and maize starch (Table 10.16). This is in contrast to results by van Soest and Kortleve [171], who reported a sharp increase in elongation and a gradual decrease in modulus for compression-molded potato starch above 160 °C. Above the same temperature the authors found significant starch molecular breakdown to which the drop in properties is likely to be related. In the case of extrusion, the time the material experiences the high temperatures is much shorter than with compression molding and thus chain degradation should be less severe. At the same time, the positive effects of high temperature on chain mobility are likely to allow for the formation of large crystallites.

Generally, starch in its granule form is unsuitable for most uses in the plastics industry, mainly due to processing difficulties during extrusion or injection molding. For this reason, a technology has been developed in which a mixture of native starch, plant fibers, food additives and water is co-extruded and injected into molds. After demolding and humidity equilibration, a stable and flexible material (dispersion) is obtained [172]. Injection molding of starch can only take place in the presence of large amounts of water [173], which acts as a plasticizer, allowing starch to melt under milder temperatures and shear stress conditions.

After removing the excess amount of water, however, the material becomes brittle, having high tensile strength (about 30 MPa) but very low elongation at break (4%). Recent investigations have shown that it is possible to produce drug delivery containers from starch and gelatin in the presence of water by injection machines [174, 175]. If, instead of water, glycols are used as plasticizers, a thermoplastic material can also be produced. However, even in this case, the mechanical properties of the materials produced are very poor (especially in tensile strength), depending on the kind of plasticizer used [176]. Glycerol is the most effective plasticizer but still cannot prevent the degradation of starch macromolecules during plasticization. It was found that the decomposition depends on the amount of glycerol and on the temperature used [177]. At high glycerol amounts (about 43 wt.%), the depolymerization diminishes and it is very small at temperatures between 130 and 150 °C.

Another approach that has been considered to improve the mechanical properties is the use of different additives such as fibers as reinforcement for thermoplastic starch. The use of natural fibers to reinforce thermoplastic starch and other biodegradable materials is a new approach. Unlike biodegradable polyesters, when natural fibers are mixed with polysaccharides (thermoplastic starch and its blends or cellulose derivatives) their mechanical properties become notably improved. This has been attributed to the chemical similarity of polysaccharides and plant fibers, providing good compatibility between them [178]. An initial insight into the use and characteristics of pulp fibers in starch-based composites was performed with cellulosic fibers from *Eucalyptus urograndis* pulp and thermoplastic starch [179]. Pulp readily works as a reinforcement even in relatively low quantities, since 16% produced a significant increase in the modulus and tensile strength (Table 10.17). The modulus and tensile strength show 156 and 120% increases, respectively, while elongation was reduced from 31 to 11%. The moisture content of samples was 22% for the thermoplastic starch and 15% for the composite. Furthermore, the moisture sorption was dramatically reduced

Table 10.17 Tensile test results of thermoplastic starch and its composite with 16% of fiber and sorption of moisture from samples conditioned in 43 and 100% relative humidity (RH) at 25 °C [179][a].

Sample[b]	E (MPa)	ε (%)	UTS (MPa)	43% RH	100% RH
(1) Thermoplastic starch	125	31	5	9	65
(2) Composite	320	11	11	5	34

a) E, Young's modulus; ε, elongation; UTS, ultimate tensile strength.
b) (1) Corn starch (28% amylose) premixed with 30% w/w of glycerin;
(2) mixture of corn starch, glycerin and fibers (16%).

with the incorporation of fiber. These results are ascribed to the fact that starch is more hydrophilic than cellulose and the fibers adsorb part of the glycerin. This results in a less hydrophilic matrix, since plasticized starch is increasingly sensitive to water uptake the higher the glycerin content is [180].

The results of DSC experiments indicate an interaction between the fiber and the glycerin, causing a reduction in the glass transition temperatures (T_g) of the matrix in the composite and a reduction in the water sorption of the composites in comparison with the pure matrix. Two transitions were detected in both materials. One of these transitions, at higher temperatures, occurs at 2 °C for thermoplastic starch and at 17 °C for its composite. The high-temperature transitions are the vitreous transition, determined as ${T_g}^{1/2}$. The difference in ${T_g}^{1/2}$ for the thermoplastic starch and the composite was attributed to fiber interaction with the plasticizer, since the matrix in the composite is less plasticized than the pure matrix. The transitions at lower temperatures, i.e. −55 and −45 °C, were attributed to the main plasticizer itself [181]. The behavior of the mass loss curve is similar in the thermoplastic starch and the composite, while the onset of decomposition occurs at 320 °C in both. Mass loss, at the onset temperature, is 30 and 23% for thermoplastic starch and its composite, respectively. This difference is due to the differences in the equilibrium moisture content of each sample, i.e. 22% for thermoplastic starch and 15% for its composite.

The thermomechanical behavior of hydrophobized starch products was studied by Aburto et al. [88]. The starch and amylose esters produced are mostly amorphous thermoplastic materials with a measurable glass transition temperature (T_g), especially those with a high degree of substitution [88]. This is due to the loss of crystallinity of the starch and amylose after esterification. The T_g of granular starch, estimated from extrapolation data, is about 210 °C [182]. The DSC data of studied starch esters are presented in Table 10.18. In general, an increase in the side-chain length causes a small depression of T_g (for the same degree of

Table 10.18 Properties of starch and amylose esters [88].

Starch ester	T_g (°C)	Tensile strength (MPa)	Elongation at break (%)
OCST 1.8	68	3.1	9
OCST 2.7	−50/40	0.7	380
DODST 2.7	−56/25	0.7	1500
OCDST 1.8	–	3.7	9
OCDST 2.7	–	1.9	10
OCAM 0.54	–	1.8	9
OCAM 2.7	−52	1.2	600
DODAM 2.7	−47	1.1	1550
OCDAM 2.7	–	3.3	19

substitution). This trend is in accordance with the other data where the T_gs of lower starch esters are reported to drop from 65 °C for starch butyrate to 50 °C for starch hexanoate [91]. This behavior could be explained by the increase in the free volume of the polymer, which is caused by the introduction of bulky flexible side-chain groups. The loss of hydrogen bonding interactions, which often stiffen the macromolecular chains, also contributes to a reduction in T_g. The internal plasticization due to the esterification is so effective that the T_g drops to very low temperatures (below −50 °C). This is very close to the T_g of thermoplastic starch (−38 °C) containing 25 wt.% glycerin as external plasticizer. It must be noted that some native starch esters seem to have two glass transition temperatures. These probably correspond to amylose and amylopectin esters, respectively, which, as their unmodified raw materials, remain incompatible after esterification. Since amylopectin is a branched molecule, it has a higher T_g than the linear amylose molecule [183] and this must also apply to their esters. The loss of starch crystallinity after its esterification can be seen more clearly in the octanoated starch with degree of substitution 1.8 (OCST1.8). It shows a weak and broad melting peak at 174 °C, a sign of imperfect crystallization. Unmodified potato starch is estimated to have a degree of crystallinity between 20 and 28% and a melting peak between 220 and 230 °C. The same behavior also appears in the octadecanoated (C_{18}) esters. In none of these esters was a glass transition recorded, but all show a large melting peak around 32 °C. Since this peak does not depend on the degree of substitution and the polymer type, it must be attributed to crystallization of the long C_{18} side-chains. In fact, this melting peak is very close to the melting points of octadecane (28–30 °C) and methyl stearate (40–42 °C).

In octanoated (OCSt) and dodecanoated (DODSt) starch esters, thermogravimetric analysis (TGA) studies indicated that they have higher thermal stability than unmodified starch [184]. Amylose esters behave similarly. Pure amylose is stable up to 290 °C. The maximum decomposition rate appears at 330 °C. The amylose esters appear to be more stable since their decomposition starts at higher temperatures. Comparing the thermal stabilities of starch and amylose esters, no significant differences appear between them. This greater thermal stability of the esters is probably due to the lower amount of remaining hydroxyl groups after esterification. It has been reported that the main decomposition mechanism of starch is the dehydration reaction between starch molecules [185]. Thus, thermal stability increases with the degree of substitution since lower amounts of hydroxyl groups remain. Decomposition was observed to take place in two stages. In the first stage, the weight loss is about 55–63% of the initial weight. The second stage starts above 400 °C and ends at about 525 °C, giving an ash residue. This second decomposition stage may be attributed to the methylene groups of the side esters since it does not appear in pure amylose.

The internal plasticization provided by the bulky side ester groups also has a profound effect on their ability to form films, compared with pure starch or amylose, for which films are almost impossible to prepare. The esters with a high degree of substitution can more easily form flexible film. In contrast, film of esters with a low degree of substitution can be prepared only with great difficulty and are very brittle. Starch and amylose esters with a high degree of substitution behave like typical thermoplastic materials, showing poor tensile strength and high elongation at break. The octanoated and dodecanoated starch or amylose esters with ds 2.7 have about the same tensile strength but the elongation at break increases with increasing side-chain length. Obviously, the bulkier groups are more effective internal plasticizers. The degree of substitution in the above esters also plays an important role in the final properties. The tensile strength decreases as the ds becomes higher, whereas the opposite trend is observed for the elongation at break. It seems that the replacement of only a small fraction of the hydroxyl groups cannot provide a drastic plasticization and the final product retains some of the mechanical properties of the unmodified starch. The picture is different in octadecanoated esters, which behave more like brittle materials. Octadecanoated starch with a lower degree of substitution (ds = 1.8) has higher tensile strength and lower elongation at break than the above-mentioned esters. The elongation at break remains relatively low, even for high degrees of substitution. Such a trend was not observed in octanoated and dodecanoated esters. This sudden reversal in behavior must be attributed to the partial crystallization of C_{18} side-chains, as demonstrated by DSC measurements. Crystalline materials usually show high tensile strength and low elongation at break. Hence it can be said that there is an optimum side-chain length that provides an effective plasticization.

Comparing the corresponding starch and amylose esters, amylose esters have slightly higher tensile strength and also elongation at break. The native starch used contains mainly amylopectin, which is a branched macromolecule and, as such, it has a lower ability to form chain entanglements, compared with the linear amylose macromolecule. It has been found that cast amylose films are more flexible than films prepared by amylopectin [186]. This trend must also apply to their esters. Indeed, it was found that it is easy to prepare amylose triacetate films but preparation of films from acetylated amylose–amylopectin mixtures containing more than 60% amylopectin is not possible, as the resulting films are too brittle [187]. Such a phenomenon was not observed in starch esters, even though they contain a higher proportion of amylopectin (81%). Obviously, the long fatty side groups are more effective plasticizers than acetates. The tensile strength of starch and amylose esters is lower than those mentioned for destructurized starch, which lies between 20 and 30 MPa [182, 188]. The lower values are due to the internal plasticization effect of the bulky fatty ester groups. Because of this plasticization, they have a higher elongation at break compared with that of extruded starch (about 4%). The mechanical properties of the above esters appeared to have similar behavior with starch plasticized with glycols and especially glycerin, which acts as an external plasticizer. The properties of the plasticized starch depend on the glycol type and also its concentration. In the case of starch and amylose esters, properties can be easily adjusted by changing the chain size and/or the degree of substitution. One of the advantages of polysaccharide esters is that the plasticizing groups are covalently bonded and cannot migrate. This leads to stable mechanical properties throughout the service life of the material. In contrast, in plasticized starch it is possible to have some loss of the plasticizer molecules through migration or evaporation, which leads to an alteration of its mechanical properties.

10.5 Biodegradability

A critical issue concerning the usability of the newly synthesized amylose esters is their biodegradability [88]. The biodegradation rate of the starch and its esters can be followed by determining their weight loss when they were exposed to activated sludge. The biodegradability of the several starch esters appears to be rather low since the weight loss (Wl) does not exceed 6% within the time period studied:

$$\text{Wl(\%)/starch: } 5.8/\text{OCST1.8}, 5.3/\text{OCDST1.8}, 5.2/\text{OCST2.7}, 4.9/\text{DODST2.7}, 3.5/\text{OCDST2.7} \quad (33)$$

$$\text{Wl(\%)/amylose: } 5.5/\text{OCAM2.7}, 5/\text{DODAM2.7}, 4/\text{OCDAM2.7} \quad (34)$$

The amylose esters show slightly higher biodegradability than their native starch counterparts, probably due to the linear character of the amylose molecule. It was observed by Bhattacharya et al. [189] that starch (with 70% amylose)–styrene maleic anhydride blends were slightly more biodegradable than those containing only amylopectin. Comparing the different esters, it can be said that those with lower degrees of substitution show higher biodegradability. Octanoated amylose with low degree of substitution has the highest biodegradation rate. Similar findings have been reported for starch acetate [89]. It seems also that the biodegradation rate increases with shorter side-chain length. Obviously, the bulky groups introduced by esterification interfere with the biodegradation process, possibly by inhibiting the catalytic action of amylases, which are responsible for the biodegradation of the starch.

Scanning electron microscopy (SEM) photographs of the exposed samples (octanoated amylose sample with degree of substitution 2.7, before and after 3 weeks of exposure) seem to corroborate the findings from the weight loss measurements. The film surface becomes progressively rough as time passes due to starch removal. It is evident that there is some material consumption, but this happens only in small areas of the film surface and, thus, only a small weight loss was detected within this exposure time. More interestingly, microbial colonies were accumulated on some films. Starch removal seems more intense around these areas of increased microbial population. This is an indication that starch removal is mainly attributable to microbial activity, although there may be other secondary factors, such as mechanical abstraction and starch dissolution. The same behavior appears for loss measurements and SEM photographs. The tensile strength decreases only slightly with exposure time. It must be noted, however, that the tensile strength of the unexposed samples is already rather low, so the differences might not be so pronounced as they lie within experimental error. Only amylose C_{18}-ester shows a significant decrease (25%) in tensile strength. This is probably due to the fact that this ester has a significant tensile strength (4 MPa) and the differences are more pronounced.

The modified starch with the lower degree of substitution is affected to a greater extent (two samples of octadecanoated starches with degrees of substitution 1.8 and 2.7, respectively, after 3 weeks of exposure to activated sludge). This finding is in good agreement with the biodegradation results for octanoated starch esters during soil burial [189]. The mechanical properties of biodegraded esters seem to corroborate the findings of weight at break following a similar trend. Again, there is a small reduction, especially in samples with high elongation at break. Most notably, the DODSt2.7 ester shows a large decrease in elongation at break.

Bacteria consuming starch use enzymes, such as α,β-amylases, which act through complex formation in an active site close to the ether bond formed between two α-D-glucopyranose groups and finally lead to its breaking [190, 191]. Since most of these bonds are shielded in the starch esters by the bulky ester groups, the above complexes are more difficult to form. It must be noted, however, that many microorganisms also produce enzymes called esterases, which are able to break ester linkages. Hence the actual biodegradation process may involve all the above-mentioned enzymes. To verify the above assumption, octanoated starches with different degrees of substitution (0.54, 1.8 and 2.7) were exposed to enzymatic hydrolysis using α-amylase–lipase mixtures (4 days, $C_{\text{amylase}} = 40$ mg mL^{-1} and $C_{\text{lipase}} = 4$ mg mL^{-1}):

$$\text{\%glucose elimination/starch: 60/raw starch, 5/OCST0.54,} \quad \text{OCST1.7} \approx \text{OCST2.7/4} \tag{35}$$

Pure starch seems to be easily hydrolyzed irrespective of the environment. Esterified starches, however, show only limited hydrolysis, in agreement with weight loss measurements observed. Samples exposed to a higher concentration of lipase show a greater extent of hydrolysis. Also, the differences between the esters with different degrees of substitution are more visible than in the case where only 4 mg mL^{-1} lipase were used. This proves that the presence of lipase has a beneficial effect on hydrolysis because it leads to cleavage of the ester groups and permits easier attack by α-amylase. The enzymatic hydrolysis experiments further confirm that the degradability of starch decreases with increasing degree of substitution, as was found from exposure to activated sludge.

The disintegration state of starch granules determines the accessibility of amylose and amylopectin to starch-degrading enzymes and the kinetics of the degradation process [192]. This is used as an analytical tool for quantifying the extent of starch gelatinization [193].

10.6
Starch–Additive Complexes

It is well known [194, 195] that heavy metal ions such as Pb(II) and Cu(II) ions released into the environment affect ecological life owing to their tendency to accumulate in living organisms and are highly toxic when absorbed into the body. Various methods such as ion exchange, reverse osmosis and electrodialysis techniques have been developed for the removal and recovery of heavy metal ions from sewage and industrial waste water [196]. In spite of their removal effectiveness, they are often expensive. Traditional chemical precipitation can be envisaged, but the generation of precipitated bulky hydroxides and colloidal particles is a major disadvantage. The search for an effective and economic method of removing toxic heavy metal ions requires the consideration of unconventional materials and processes. In this respect, many natural polysaccharides and their derivatives containing various functional groups may have some potential. In the last decade, chitin and its derivatives [197, 198], modified cellulose [199, 200] and modified starch ethers [201, 202] have been studied with respect to their ability to remove heavy metal ions from aqueous solutions.

Several complexes formed by starch and salts of metals from transition groups have been described [203, 204]. Potato starch with random phosphoric acid moieties bound to amylopectin formed metal complexes with involvement of these moieties [205]. A group of metal starch derivatives have been reported, including iron(III) [206], titanium(IV) [207] and lanthanum [208] atoms bound covalently to the hydroxyl oxygen atoms. They were prepared by "rusting" reduced iron powder in starch gel and by reacting starch with corresponding metal alkoxides [Ti(IV) and La]. It was proven [209, 210] that contrary to many opinions [211–213], alkali metal (group IA) salts interact with starch with involvement of their anions rather than cations. On the other hand, in solutions of the salts of metals from higher non-transition groups also cations interact with starch [214].

Interactions between salts of metals from transition groups and starch have been investigated on $CuSO_4$, silver(I) halides [215], $ZnCl_2$ [216], HgI [217] and titanium carboxylates [218]. Several Fe(III) complexes have also been reported [219–221]. In several cases, the complexes described were sorption complexes in which amylopectin rather than amylose was indicated as the place of sorption.

The problem of penetration of the interior of starch granules by either salts or ions was discussed in more detail by Ciesielskia et al. [222]. In their paper, evidence is given for the ligation of selected cations of metals from transition groups with starch. The metal cations used were paramagnetic. In every case the EPR spectra were run for either granular or gelatinized corn starch, waxy corn starch and potato starch, in addition to plain and gelatinized amylopectin. In the first series of experiments, solid polysaccharides were thoroughly blended with solid hydrated salts. In this series of experiments no evidence could be found for interaction of polysaccharide with metal salts when the components were blended in the solid state. They could be considered as secondary control samples. The pattern of the EPR spectra showed that there were clear interactions between polysaccharides and metal ions under investigation. Sharp signals in the spectra of pure salts, particularly of $CoSO_4$, should be ascribed to defects in the solid salt lattice rather than to unpaired spins in the corresponding metal ions. Corresponding signals in the spectra of samples of polysaccharides blended with aqueous solutions usually turned into typical spectra of ligated metal ions. This effect was particularly pronounced when metal ions were ligated by pasted polysaccharides.

Granular, i.e. non-pasted, polysaccharides could form complexes either by sorption of metal ions on the granule surface or by penetration of ions into the interior of starch granules. Since anions are capable of such penetration rather than cations, such complexes could have the character of ion pairs with negatively charged salt anion containing granules and assisting cation. Sharp signals in the spectra of Cr(III) and Fe(III) sulfates suggested that these salts crystallized in starch, forming defected lattices responsible for such a pattern of the spectra. Inspection of g-factors indicated that this might happen also in the case of cobalt salt with potato starch, whereas corn starch and waxy corn starch ligated the cobalt cation. In other cases, the g-value exceeded 2.2. It should be noted that contrary to potato starch [223], corn starch and waxy corn starch had their outer shell perforated by micropores [224].

Gelatinized polysaccharides formed high-spin complexes with all metal cations under study. Values of the g-factor in the relevant spectra of the samples prepared from solid polysaccharides and salt solutions suggested changes in the coordination sphere of the central metal atoms. Whenever splitting of the signal was observed [some Co(II), Cr(III), Cu(II) and Fe(III) complexes] higher g-parallel and lower g-perpendicular indicated unpaired electrons localized in the d_{z^2} orbital and tetrahedral symmetry around this central atom. Ligation of Mn(II) atom was manifested by a vanishing of the original splitting of the signal of Mn(II) in the spectrum of pure salt. A strong decrease in the signal of the Mn(II) ion in the sample prepared from granular potato starch indicated that that ion weakly sorbed on the surface of that kind of granules. Changes in the intensity of the signal of Fe(III) in the spectrum of granular corn starch could be interpreted in a similar manner.

Documented by varying EPR spectral patterns, changes in the coordination sphere of the central atoms upon their contact with pasted polysaccharides could originate from changes in the hydration of the central atom. Thermogravimetric and differential scanning calorimetric analyses (TGA and DSC) of these complexes showed that except for the Cr(III) complex there was only a residual amount of water either in the coordination sphere of the central atoms or/and in the starch matrix (Table 10.19). Chromium sulfate octadecahydrate lost all water molecules up to 110 °C [225], whereas the weight loss from the starch complex of Cr(III) began at 129 °C. Hence one might accept that starch was able to repulse all water molecules from the coordination sphere of that central atom. Copper sulfate pentahydrate dehydrated into monohydrate up to 110 °C [226] in order to turn into the anhydrous salt at 250 °C. Since the weight loss from the complex began at 188 °C, only one water molecule could remain in the coordination sphere of Cu(II) in the starch complex. Hydrated Co(II), Fe(III) and Mn(II)

Table 10.19 Results of thermogravimetry (TG), differential thermogravimetry (DTG) and differential scanning calorimetry (DSC) analyses of pasted potato starch and its metal complexes [225].

Sample	TG		DTG	DSC
	Temperature (°C)	Total weight loss (%)	Temperature (°C)	Temperature (°C)
Potato starch	34	0.0	71	70 (*endo*)
	243	18.5	285	260 (*endo*)
	307.5	61.6		282 (*endo*)
				340 (*exo*)
Co(II) complex	178	2.8	231	181 (*exo*)
	262	40.5	399	297 (*exo*)
	356	62.1		402 (*exo*)
Cr(III) complex	129	0.5	155	142 (*endo*)
	244	32.6	384	213 (*exo*)
	358	79.7	396	359 (*exo*)
Cu(II) complex	155.5	2.6	207	188 (*endo*)
	256	41.0	329	248 (*exo*)
	328	97.0		280 (*endo*)
Fe(II) complex	132	0.2	180.5	161 (*exo*)
	240	32.5	373	380 (*endo*)
Mn(II) complex	176	1.2	243	239 (*exo*)
	371	53.4	434	396 (*exo*)

sulfates lost all but one of their water of crystallization up to 71, 140 and 100 °C, respectively. Full dehydration took place above the temperatures noted for the beginning of the weight loss of the starch complexes with these metals [226] (Table 10.19). Therefore, also in these complexes retention of a single water molecule in the coordination sphere of these metals could be assumed. It should be noted that in the thermogram (differential thermogravimetry, DTG) of pasted native potato starch after removal of water reflected by the endothermic peak centered at 71 °C, the subsequent effect at 260 °C associated with glassy transition followed by decomposition [227] was also endothermic. In the corresponding thermogram of the complex with Cr(III) the first effect which began at 129 °C (TG) and centered at 149 °C (DTG) could correspond to the elimination of water from the coordination sphere of the central atom. In the spectrum of the copper complex the first thermal effect was endothermic and could be related to decomposition of starch without participation of the central metal atom. The first exothermic effects in the thermograms of Co(II), Fe(III) and Mn(II) complexes suggested the beginning of either redox reactions or metal ion-catalyzed air oxidation in these steps of starch decomposition.

These data should be related to the presence of metal atoms bound to given polysaccharides by coordination, sorption and, in the case of potato starch, covalently to phosphoric acid moieties where they exchanged atoms residing therein in original starch. Each polysaccharide possessed its own specific affinity to the same metal ion and there was different affinity among metal ions to a given ligand. The total amount of a given metal atom trapped was higher for solid polysaccharides, showing the importance of surface sorption and, in the case of granular starch, penetration of ions into starch granules. Formation of complexes by granular starch was additionally confirmed by means of SEM. Starch granules soaked in water for several hours swelled and cracked and/or leached the content of their interior into water.

Leaching from starch granules was demonstrated by Gallant et al. [55]. Such exudations could be removed from the surface of granules by vigorous agitation of suspensions. After such treatment, the surface of granules remained smooth. In a further paper the authors reported [210] that NaCl present in such suspensions retarded cracking of granules. Ciesielskia et al. did not observe under SEM any cracked starch granules after soaking them in solutions of transition metal sulfates [222]. Prior to separation of granules from suspensions they were vigorously agitated and washed several times with water, then dried. There could be visibly abundant, not removed exudations on the granule surface. Starch granules were treated by soaking in water and in $CuSO_4$ solution and then agitated prior to isolation. After soaking and washing, the starch granules retained a pale color of salts. Thus, after such treatment salts still resided on the granule surface. Therefore, the assumption that observed, poorly removable exudations on the surface of granules might be starch–metal complexes was sound.

Zhang and Chen [228] explored the adsorption behavior of new starch graft copolymers containing tertiary amine groups (Table 10.20), which were prepared by grafting dimethylaminoethyl methacrylate (DMAEMA) on to commercial crosslinked starch using potassium permanganate–sulfuric acid initiating, from

Table 10.20 Preparation of water-insoluble starch graft copolymers [228][a)].

Sample	DMAEMA (mol dm^{-3})	Grafting (%)
A	0.13	24
B	0.25	60
C	0.76	40

a) Other conditions: [starch] = 2.0 g per 100 cm^3; [$KMnO_4$] = 2.0×10^{-3} mol dm^{-3}; [H_2SO_4] = 5.0×10^{-2} mol dm^{-3}; grafting temperature, 45 °C; grafting time, 5 h.

aqueous solutions towards lead(II) and copper(II) ions. For this purpose, various factors affecting the adsorption, such as treatment time, initial pH of the solution, metal ion concentration and grafting percentage of the starch graft copolymers, were investigated by a batch technique.

The adsorption capacity of both Pb(II) and Cu(II) ions on water-insoluble starch graft copolymer increases with the treatment time during the first 2 h and then levels off towards the equilibrium adsorption capacity. In contrast, the adsorption rate of Cu(II) ions is higher than that of Pb(II) ions ($Q_{abs} = 0.5$ mmol g^{-1} in 2 h) (sample A). This may be due to the higher complex formation rate between Cu(II) ions and the tertiary amine groups on the surface of the starch graft copolymer. Chan and Wu [202] studied the adsorption kinetics of Cu(II) ions by water-insoluble starch ethers containing tertiary amine groups and also found that the adsorption reached equilibrium in about 2 h.

The molar adsorption capacity was calculated from the following expression:

$$Q_{abs} = (C_i - C_f)V/m_{rm} \quad (36)$$

where C_i and C_f (mmol mL^{-1}) are the initial and final concentrations of the metal ions in the adsorption medium, respectively, and V (mL) and m_{rm} (g) are the volume of the reaction medium and the amount of the starch graft copolymer, respectively.

Due to the protonation and deprotonation of the tertiary amine groups on the surface of the starch graft copolymer, its adsorption behavior for metal ions can be influenced by the pH. It was found that at strongly acidic pH the starch graft copolymer has lower adsorption capacities. This can be explained by the fact that in this case most of the tertiary amine groups are protonated. Then cationic repulsion can occur between metal ion species and protonated graft chains. For Pb(II) ions, the adsorption capacity has a maximum value at pH 6.0. For Cu(II) ions, the adsorption capacity increased with increase in pH, reaching a plateau value at around pH 7.0. A contact time of 2 h and pH values of 6.0 and 7.0 were chosen as the experimental conditions for the determination of adsorption isotherms for Pb(II) and Cu(II) ions, respectively. The adsorbent used is sample B with a grafting percentage of 60%.

To evaluate and compare the saturation capacities of the starch graft copolymer towards two heavy metal ions, the adsorption isotherms were analyzed and fitted using the Langmuir model, written as

$$C_e/Q_{abs} = C_e/Q_{abss} + 1/(KQ_{abss}) \quad (37)$$

where Q_{abs} is the amount of metal ions adsorbed (mmol g^{-1}), C_e the concentration of metal ions at the equilibrium (mmol dm^{-3}), Q_{abss} the capacity at the saturation (mmol g^{-1}) and K the adsorption coefficient (dm^3 g^{-1}). The adsorption behavior of Pb(II) and Cu(II) ions on the starch graft copolymer can be well described by the Langmuir isotherm and the saturation adsorption towards lead(II) and copper(II) ions was found to be 2.09 and 2.12 mmol g^{-1} (dry weight),

Table 10.21 Saturation capacities evaluated from Langmuir isotherms[a].

Metal ion	Q_{abs} (mmol g^{-1})	K (dm^3 g^{-1})
Pb(II)	2.09	0.0414
Cu(II)	2.12	0.0668

a) Adsorption conditions: treatment time, 2 h; pH, 6.0 (Pb^{2+}), 7.0 (Cu^{2+}).

respectively (Table 10.21). Kang et al. [198] investigated the adsorption of Pb(II) and Cu(II) ions by water-insoluble graft copolymer of crosslinked chitosan with acrylonitrile, which has a grafting percentage of 170%, and found that the maximum adsorption capacity towards Pb(II) and Cu(II) ions is 0.35 and 0.42 mmol g^{-1}, respectively. The adsorption coefficient K (dm^3 g^{-1}) was found to be 0.0414 and 0.0668 for Pb(II) and Cu(II), respectively.

For the two heavy metal ions, a higher grafting percentage results in a higher adsorption capacity (samples A, B and C):

$$Q_{abs} = 0.4,\ 0.7 \text{ and } 1.0 \text{ mmol g}^{-1} \text{ for Pb(II) and } 0.5,\ 0.9 \text{ and } 1.2 \text{ mmol g}^{-1} \text{ for Cu(II)} \tag{38}$$

This was attributed to the fact that the starch graft copolymer with a higher grafting percentage has more tertiary amine groups, which increases the adsorption ability towards Pb(II) and Cu(II) ions due to stronger complexation.

It was found that an adsorption time of 2 h is sufficient to reach adsorption equilibrium and that the adsorption equilibrium data correlate well with the Langmuir isotherm equation. For the starch graft copolymer with a grafting percentage of 60% (sample B), the saturation adsorption capacity towards lead(II) and copper(II) ions was found to be 2.09 and 2.12 mmol g^{-1}, respectively. Further, the pH value and the grafting percentage have greater influences on the adsorption of lead(II) and copper(II) ions on the starch graft copolymer.

Different neutral starch derivatives prepared by reacting starch with various mono- and dimethylol derivatives of urea, thiourea and melamine (MMU, monomethylolurea; DMU, diethylolurea; MMTU, monomethylolthiourea; DMTU, dimethylolthiourea; MMM, monomethylolmelamine; DMM, dimethylolmelamine) were reacted with heavy metals (Hg^{2+}, Cu^{2+}, Zn^{2+}, Cd^{2+}, Co^{2+} and Pb^{2+}) [75].

The data (heavy metal sorption on various starch derivatives) show the following:

1. The sorption values depend on the metal ion and starch derivatives.
2. The sorption values on starch derivatives follow the order

$$Hg^{2+} > Cu^{2+} > Zn^{2+} > Co^{2+} > Cd^{2+} > Pb^{2+} \tag{39}$$

which is in accord with the William–Irving series [229] and the finding of a report [230]. It was obvious that Hg^{2+} acquires the highest sorption value on these ligands; hence this metal ion will be used in comparing the sorption behavior of different prepared ligands.

3. The sorption values of monomethylol resin–starch derivatives were higher than those of dimethylol resin–starch derivatives. This may be attributed to the following:
   a) the difference in sorption of a substituted and unsubstituted amide group; the sorption value was higher for unsubstituted groups [231–234] due to lack of steric hindrance;
   b) different available surface areas on each type are also higher for unsubstituted amide groups in the case of monomethylol derivatives and lower in the case of dimethylol resins, due to the higher crosslinking ability of the latter.

4. The sorption values of metal ions on starch derivatives (either mono- or dimethylol resin–starch reaction products) follow the order

   $$\text{thiourea resins} > \text{urea resins} > \text{melamine resins} \qquad (40)$$

   This was due to:
   a) the different abilities of these resins to undergo self-polymerization, which is maximum in the case of melamine resins and minimum in thiourea resins;
   b) the number of donating atoms per resin molecule.

5. Generally, sorption values of mono- and dimethylol resin–starch reaction products follow the order

   $$\text{MMTU} > \text{DMTU} > \text{MMU} > \text{MMM} > \text{DMU} > \text{DMM} \qquad (41)$$

6. The sorption efficiency (%) on starch derivatives increased with increasing nitrogen content, reaching a maximum value and then decreasing. It is known that starch is a polyhydroxy compound and hydrogen bonding greatly affects the available surface area. At lower reaction extent these forces are effective and minimize the available surface area, affecting the sorption (%) of metal via chelation. On increasing the reaction extent, these forces were minimized and the available surface area increased, leading to higher chances for chelation to reach maximum sorption. At a higher extent of reaction crosslinking occurred, which minimized the available surface area, leading to a lower chelation ability.

7. The sorption values of Hg^{2+} (mmol: resin molecule) at maximum sorption efficiency were 1135, 2624 and 2538 for starch–monomethylolurea, starch–monomethylolthiourea and starch–monomethylolmelamine, respectively. This indicates that there are two donating atoms in these starch derivatives in the

case of urea derivatives and three in the case of thiourea and melamine. These atoms are:

a) oxygen and one of the nitrogen atoms in the urea derivatives – the ligand is bonded to two metal atoms, one via an oxygen atom and the other (less strongly) via a nitrogen atom;
b) one sulfur atom and two nitrogen atoms in thiourea derivatives;
c) three nitrogen atoms, only, in melamine derivatives – thiourea and melamine act as a tridentate ligand, while urea acts as a bidentate ligand only in the starch–monomethylol resin derivatives.

It has been reported [77–79] that the introduction of reactive functional groups into the backbone of highly crosslinked starches brings about products (anionic and cationic starches) that are capable of removing heavy metal ions from industrial waste water. All starch derivatives have been used in heavy-metal removal. Khalil and Farag [230] reported that polyacrylamide–starch graft copolymer, carbamoylethylated starch and starch carbamate acquire high efficiency in absorbing heavy metal ions from solutions.

The addition of raw starch to UV-curable formulations was reported to increase the water absorption from 0.17 to 0.59% (Table 10.4) [128]. The starch grafted with GMA reduced the water absorption of the UV-cured film from 0.59 to 0.40%, a 32.20% improvement in the water resistance. The reduction of the water absorption is due to the formation of a protective shell of GMA around the starch particles. This modification by grafting is, however, confined only to the surface of the starch granules, with the bulk of the interior remaining unaffected with the formation of a hydrophobic shell on the surface of the sago starch granule. The shell layer further reduces the intermolecular hydrogen bonding in starch and the affinity for water. When both the native starch and the modified starch samples were boiled in excess water, the native starch underwent gelation and the gel occupied the entire volume with no visible phase separation between the dispersed phase and the dispersion medium. On the other hand, the modified starch remained essentially unaffected after the boiling, retained the particulate integrity and remained discrete after the boiling. The mixture could be centrifuged and the dispersion medium (water) remained as a clear liquid over the unagglomerated particles of modified starch. This showed that the hydrophobic shell caused by the grafting of GMA on to starch allowed the starch particles to survive the boiling test. This core–shell structure is expected to impart adequate dimensional and mechanical stability to the cured film while at the same time allowing an eventual possibility of biodegradation since the core constitutes of predominantly ungrafted starch. The core–shell configuration also promotes good adhesion between starch particles and the matrix resin of the UV-curable coating system (Figure 10.15).

Regarding the interactions of starch with complex-forming emulsifiers, different rheological effects may be obtained. These effects range from viscosity decrease to viscosity increase and gelation, depending on the degree of granule swelling and disintegration and amylose leaching [235]. Similarly, the kinetics of

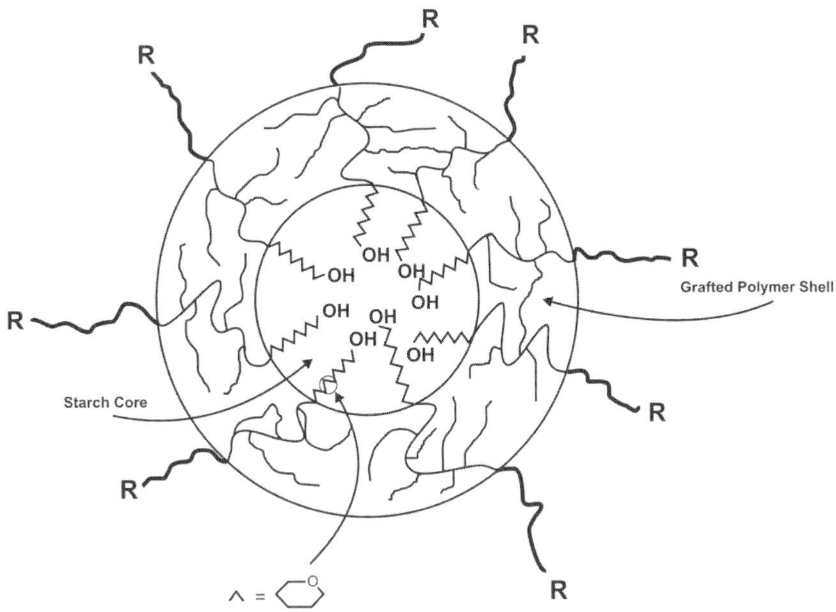

Figure 10.15 Depiction of core–shell structure of starch-g-GMA [128].

starch flavor complexation are determined by the physicochemical properties of the ligand, such as solubility, and also by the supramolecular structure of starch [236]. By combining starch with non-starch hydrocolloids, large differences in the rheological properties of the mixtures result, depending on the microstructure of starch [122].

The formation of a helical complex between amylose and iodine gives rise to the typical deep color of starch dispersions stained with iodine and forms the basis for the quantitative determination of amylose content. In I_2–KI solutions, polyiodide ions such as I_3^- and/or I_5^- interact with amylose, forming single left-handed V-type helices [237]. A helix consists of six anhydrous glucose residues per turn with a pitch of 0.8 nm and a hydrophobic helical cavity of diameter 0.5 nm. Amylose also forms a V-helix complex with the hydrocarbon portion of monoglycerides and fatty acids [238].

10.7 Conclusions

Starch is an inexpensive material in comparison with most synthetic plastics and is readily available. It is produced by nearly all green plants to store energy and consists of amylose and amylopectin. The native starch granule is heterogeneous both chemically (e.g. amylose and amylopectin) and physically (e.g. crystalline and noncrystalline regions). Starch is the most promising raw material for the production of biodegradable plastics. The presence or absence of crystalline order is often a basic factor underlying starch properties. When starch is heated in excess water, the crystalline structure is disrupted (due to breakage of hydrogen bonds) and water molecules become linked by hydrogen bonding to the exposed hydroxyl groups of amylose and amylopectin. The swelling power and solubility provide evidence of the magnitude of the interaction between starch chains within the amorphous and crystalline domains. The extent of this interaction is influenced by the amylose/amylopectin ratio and by the characteristics of amylose and amylopectin in terms of molecular weight/distribution, degree of length of branching and conformation. Starch is not a true thermoplastic but in the presence of plasticizers (water, glycerin, sorbitol, etc.) at high temperatures and under shear it readily melts and flows, allowing for its use as an injection, extrusion or blow molding material, similar to most conventional synthetic thermoplastic polymers.

The possible use of starch as a thermoplastic, biodegradable, non-food material depends on its attainable properties. To fulfill the various demands for the functionality in different starch products, industrially processed starch is modified enzymatically, physically or chemically. This allows other potential uses in many different industries. Chemical substitution and chemical crosslinking are the main types of modifications that are carried out. In most chemical modifications of starch, usually referred to as chemical derivatization, the granule form is maintained and the hydroxy groups are partially substituted, yielding starch ethers and esters, and also anionic and cationic starches. Chemically modified starches with improved properties are gaining increasing importance in industry, not only because they are inexpensive, but also mainly because the polysaccharide portion of the product is biodegradable. Their applications relate to agriculture, industry, medical treatment and sanitation, etc., which make them important polymeric materials in the fields of dehumidification, dehydration, water preservation and water absorption. For the last few decades, chemical modification of starch by graft copolymerization of vinyl monomers on to it has been a subject of both academic and industrial interest. For example, chemical crosslinking is particularly used to inhibit granule disruption. Depending on the extent of crosslinking, granule swelling will be more or less important, leading to a wide range of rheological properties. The properties of crosslinked starch suspensions are very similar to those of closed-packed microgel dispersions and can be described on a similar basis. Starch samples with higher degrees of crosslinking display individual particles.

Grafting has also been used as an important technique for modifying the physical and chemical properties of polymers. Graft polymerization originates from the formation of an active site at a point on a polymer chain other than its end and exposure of this site to a monomer. Most graft copolymers are formed by radical polymerization. In many instances, chain transfer reactions are involved by the abstraction of hydrogen atoms. Starch has also been modified by grafting with vinyl monomers (e.g. methyl acrylate) on to the starch backbone, yielding thermoplastic materials that can be injection molded or extruded into films with properties similar to low-density polyethylene. Several synthetic polymers have been used to improve the mechanical properties of thermoplastic starch, such as ethylene–acrylic acid copolymer and ethylene–vinyl alcohol copolymer. The grafting efficiency of starch by copolymerization varies with the type of monomer (the water-soluble or oil-soluble unsaturated monomers) and starch. The grafting of hydrophilic starch particles by hydrophobic monomers is confined only to the surface of the starch granules, with the bulk of the interior remaining unaffected with the formation of a hydrophobic shell on the surface of the sago starch granule. The shell layer further reduces the intermolecular hydrogen bonding in starch and the affinity for water. The hydrophobic shell caused by the grafting of hydrophobic monomers on to starch strongly decreased the interaction with water and allowed the starch particles to survive the boiling test. This core–shell structure is expected to impart adequate dimensional and mechanical stability to the cured film while at the same time allowing an eventual possibility of biodegradation since the core is constituted of predominantly ungrafted starch. When a water-soluble starch is used (carboxymethyl starch) the grafting is more effective because starch molecules (their radicals) dissolved in the reaction medium interact with monomer. The grafting of a water-soluble starch with a water-soluble monomer leads to the water-soluble graft copolymer or its gel. Much effort has been made in recent years to develop biodegradable materials, particularly compostable plastics, i.e. plastics that degrade easily under well-defined environmental conditions. These materials may be synthetic, natural or a combination of both. Polysaccharide-based polymer is one of the most promising materials to achieve this objective. Amphiphilic polymer-g-starch derivatives usually decrease considerably the surface tension of an aqueous solution with an increase in the polymer concentration. When the concentration of hydrophobic polymer fraction was increased, the surface tension of the solution decreased significantly. However, the less hydrophobic or hydrophilic polymer did not decrease the surface tension of water at all. This means that the hydrophobic core of the aggregates is completely and stably covered by the hydrophilic shell of the polysaccharide skeleton. The very low critical aggregation concentration and the surface inactivity indicate that colloidal stable nanoparticles are certainly formed above the critical concentration.

Generally, starch in its granule form is unsuitable for most uses in the plastics industry, mainly due to processing difficulties during extrusion or injection molding. Injection molding of starch can only take place in the presence of large amounts of water, which acts as a plasticizer, allowing starch to melt under milder temperatures and shear stress conditions. If, instead of water, glycols are used as plasticizers, a thermoplastic material can also be produced. However, even in this case, the mechanical properties of the materials produced are very poor (especially in tensile strength), depending on the kind of plasticizer used. Glycerol is the most effective plasticizer but still cannot prevent the degradation of starch macromolecules during plasticization. At high glycerol amounts, the depolymerization diminishes and it is very small at temperatures between 130–150 °C. The plasticizer (urea, glycerol, water, etc.) favors an increase in the strength, modulus, elongation, etc. of the starch materials. A significant change in mechanical properties is observed above a critical concentration of plasticizer. The change in the composition of the plasticizer/melt flow accelerator system can improve the processability of the starch material. With regard to mechanical properties, a high molecular weight of amylose leads to improved strength and modulus of the amylose films. Amylopectin, with its high molecular weight, also produces extruded films with comparatively good mechanical properties. Another approach that has been considered to improve the mechanical properties is the use of different additives such as fibers as reinforcement for thermoplastic starch. The use of natural fibers to reinforce thermoplastic starch and other biodegradable materials is a new approach. Unlike biodegradable polyesters, when natural fibers are mixed with polysaccharides (thermoplastic starch and its blends or cellulose derivatives), their mechanical properties become notably improved. This has been attributed to the chemical similarity of polysaccharides and plant fibers, providing good compatibility between them.

It is well known that heavy metal ions released into the environment affect ecological life owing to their tendency to accumulate in living organisms and are highly toxic when absorbed into the body. Various methods such as ion exchange, reverse osmosis and electrodialysis techniques have been developed for the removal and recovery of heavy metal ions from sewage and industrial waste water. In spite of their removal effectiveness, they are often expensive. The search for an effective and economic method of removing toxic heavy metal ions requires the consideration of unconventional materials and processes. In this respect, many natural polysaccharides and their derivatives containing various functional groups may have some potential. Starch and modified starch ethers have been studied with respect to their ability to remove heavy metal ions from aqueous solutions.

Acknowledgments

This research was supported by the Science and Technology Assistance Agency through grants APVT-51-021702 and APVT-20-017304 and the Slovak Grant Agency (VEGA) through the grant 2/4008/04. The author is also indebted to the Alexander von Humboldt Stiftung for support.

List of Abbreviations

AAm	acrylamide
AGU	anhydroglucose units of starch
Am	amylose
APS	ammonium peroxodisulfate
BA	butyl acrylate
CAE	cycloaliphatic diepoxide

CAN	ceric ammonium nitrate
CHP	hydrophobized polysaccharides (pullulan containing 1,6-cholesterol groups per 100 glucose units)
cld	crosslinking degree (or degree of crosslinking)
CLHAS	crosslinked high-amylose starch
CMS	carboxymethyl starch
CMS-g-AAm	carboxymethyl starch graft acrylamide
DB	degree of branching
DLS	dynamic light scattering
DMAEMA	dimethylaminoethyl methacrylate
DMM	dimethylolmelamine
DMSO	dimethyl sulfoxide
DMTU	dimethylolthiourea
DMU	diethylolurea
DODAm	dodecanoated amylose
DODSt	dodecanoated starch
DP	degree of polymerization
ds	degree of substitution
DSC	differential scanning calorimetry
DTG	differential thermogravimetry
FTIR	Fourier transform infrared
GLC	gas–liquid chromatography
GMA	glycidyl methacrylate monomer
HASs	high-amylose starches
HLB	hydrophilic–lipophilic balance
IR	infrared
LDPE	low-density polyethylene
MAP	methylated amylopectin potato starches
MMM	monomethylolmelamine
MMTU	monomethylolthiourea
MMU	monomethylolurea
MP	methylated potato starches
MS	molar substitution
NWS	native wheat starch
OCAm	octanoated amylose
OCDAm	octadecanoated amylose
OCDSt	octadecanoated starch
OCSt	octanoated starch
OPP	palmitoyl group-bearing pullulan
PMMA	poly(methyl methacrylate)
PNA	protein nucleic acid
PSD_{rel}	relative particle size distribution
RE	reaction efficiency
SAXS	small-angle X-ray scattering
SEC	size-exclusion chromatography

SEM	scanning electron microscopy
Slovafos 1M	anionic emulsifier
Slovasol 2430	nonionic emulsifier, alkyl polyoxyethylene ether-type emulsifier: $C_{24}H_{49}O(CH_2CH_2O)_{29}CH_2CH_3$
Sloviol	poly(vinyl acetate)
SLS	static light scattering
StOH	starch
TEM	transmission electron microscopy
TG	thermogravimetry
TGA	thermogravimetric analysis
Tw 85	Tween 85 (nonionic emulsifier, polyoxyethylene sorbitan trioleate, HLB = 11.0)
Tw 20	Tween 20 (nonionic emulsifier, polyoxyethylene sorbitan monolaurate, HLB = 16.7
UGH	unsaturated galacturonide in H form
UGK	unsaturated galacturonide in K form
UTS	ultimate tensile strength
VAc	vinyl acetate
WAXS	wide-angle X-ray scattering
WPS	wrinkled pea starch
WR	weight ratio

List of Symbols

C_{crit}	critical concentration of crosslinked starch derivatives
D	particle diameter
D_{hkl}	crystallite dimensions
$D_{p,f}$	diameter of final polymer particle
E	Young's modulus
(G') moduli	storage
(G'') moduli	loss
ΔH	enthalpy
M_w	weight-average molecular weight
$N_{p,f}$	number of final polymer particles
Q	swelling power
R	ratio
R_G	root mean square radius of gyration
$R_{p,max}$	maximal rate of polymerization,
$T(G)$	gelatinization temperature
T_g	glass transition temperature
T_m	melting temperature
x_c	degree of crystallinity
ε	elongation
σ	strength

References

1 R.L. Reis, A.M. Cunha, P.S. Allan, M.J. Bevis, *Adv. Polym. Technol.* **16** (1997) 263.

2 W.M. Doane, C.L. Swanson, G.F. Fanta, *ACS Symp. Ser.* 476. American Chemical Society, 1992, p. 197.

3 W.M. Doane, C.L. Swanson, G.F. Fanta, in *Emerging technologies for materials and chemicals from biomass*, R.M. Rowell, T.P. Schultz, R. Narayan (eds.), American Chemical Society, Washington, DC, 1992, pp. 197–230.

4 Y.E.M. Burgt, J.B.I.P. Bleeker, P.J.H.C. Mijland, J.P. Kamerling, J.F.G. Vliegenthartart, *Starch/Stärke* **52** (2000) 40–43.

5 Y. Takeda, S. Tomooka, S. Hizukuri, *Carbohydr. Res.* **246** (1993) 267–272.

6 A. Imberty, S. Perez, *Biopolymers* **27** (1988) 1205–1221.

7 A. Imberty, H. Chanzy, S. Perez, A. Buleon, V. Tran, *Macromolecules* **20** (1987) 2634–2636.

8 D. Sievert, J. Holm, *Starch/Stärke* **45** (1993) 136–139.

9 O.B. Wurzburg (ed.), *Modified Starches: Properties and Uses*, CRC Press, Boca Raton, FL, 1986.

10 Y.E.M. van der Burgt, J. Bergsma, I.P. Bleeker, P.J.H.C. Mijland, A. van der Kerk-van Hoof, J.P. Kamerling, J.F.G. Vliegenthart, *Carbohydr. Res.* **312** (1998) 201–208.

11 Y.E.M. van der Burgt, J. Bergsma, I.P. Bleeker, P.J.H.C. Mijland, A. van der Kerk-van Hoof, J.P. Kamerling, J.F.G. Vliegenthart, *Carbohydr. Res.* **320** (1999) 100–107.

12 D. Manners, *Carbohydr. Polym.* **11** (1989) 87–112.

13 Y. Takeda, S. Hizukuri, C. Takeda, A. Suzuki, *Carbohydr. Res.* **165** (1987) 139.

14 R. Hoover, *Carbohydr. Polym.* **45** (2001) 253.

15 G.T. Oostergetel, E.F. van Bruggen, *J. Carbohydr.Polym.* **21** (1993) 7.

16 H.F. Zobel, in *Developments in Carbohydrate Chemistry*, R.J. Alexander, H.F. Zobel (eds.), American Association of Cereal Chemists, St. Paul, MN, 1994, pp. 1–20.

17 C.E. Snape, W.R. Morrison, M.M. Marato-Valer, J. Karkalas, R.A. Pethrick, *Carbohydr. Polym.* **36** (1998) 225–237.

18 M.C. Codet, A. Buleon, V. Tran, P. Colonna, *Carbohydr. Polym.* **21** (1993) 91–95.

19 M. Yamamoto, T. Sano, S. Harada, T. Yasunaga, *Bull. Chem. Soc. Jpn.* **56** (1983) 2643–2646.

20 G. Rappennecker, P. Zugenmaier, *Carbohydr. Res.* 89 (1981) 11–19.

21 P. Tomasik, V. Wang, J. Jane, *Starch* **47** (1995) 91–95.

22 A. Imberty, H. Chanzy, S. Perez, A. Buleon, V. Tran, *J. Mol. Biol.* **201** (1988) 365–378.

23 F. Duprat, D.J. Gallant, A. Guilbot, C. Mercier, J.P. Robin, B. Monties (eds.), *Les Polymères Végétaux*, Gauthier-Villars, Paris, 1980, pp. 176–231.

24 A. Buleon, H. Bizot, M.M. Delage, J.L. Multon, *Starch/Stärke* **34** (1982) 361–366.

25 K. Svegmark, A.M. Hermansson, *Carbohydr. Polym.* **15** (1991) 151.

26 D.J. Stevens, G.A. Elton, *Stärke* **23** (1981) 8.

27 P. Jenkins, *PhD Thesis*, University of Cambridge, 1994.

28 J.W. Donovan, *Biopolymers* **18** (1979) 263.

29 J.K. Jang, Y.R. Pyun, *Starch/Stärke* **48** (1996) 48.

30 B. Vazquez, I. Goni, M. Gurruchaga, J. Areizaga, M. Valero, *Makromol. Chem.* **193** (1992) 2189.

31 V. Giri, P.K. Saboo, R.K. Samal, *J. Appl. Polym. Sci.* **47** (1990) 471.

32 V.D. Athawale, M.P. Padwaldesai, *Eur. Polym. J.* **35** (1999) 1237.

33 A. Dave, U. Vaishnav, R. Desai, A. Shah, B. Ankleshwaria, M. Mehta, *J. Appl. Polym. Sci.* **58** (1995) 853.

34 G.F. Fanta, R.C. Burr, W.M. Doane, *J. Appl. Polym. Sci.* **24** (1979) 2015.

35 D. Trimnell, G.F. Fanta, *J. Polym. Mater.* **3** (1992) 203.

36 I.C. Eromosele, *J. Appl. Polym. Sci.* **51** (1994) 1817.

37 K.M. Mostafa, *J. Appl. Polym. Sci.* **56** (1995) 263.

38 V.D. Athawale, L.L. Vidyagauri, *Carbohydr. Polym.* **35** (1998) 21.
39 M.H. El-Rafe, M.K. Zahran, K.F. El-Tahlawy, A. Hebeish, *Polym. Degrad. Stab.* **47** (1995) 73.
40 V.D. Athawale, S.C. Rathi, *Eur. Polym. J.* **33** (1997) 1067.
41 D. Trimnel, G.F. Fanta, J.H. Salch, *J. Appl. Polym. Sci.* **60** (1996) 285.
42 J. Gao, R. Tian, J. Yu, M. Duan, *J. Appl. Polym. Sci.* **53** (1994) 1091.
43 Y. Cao, X. Qing, J. Sun, F. Zhou, S. Lin, *Eur. Polym. J.* **38** (2002) 1921–1924.
44 F.R. Dintzis, R. Tobin, *Biopolymers* **7** (1969) 581.
45 M.J. Gidley, *Macromolecules* **22** (1989) 351.
46 J.J.G. van Soest, S.H.D. Hullemann, D. de Wit, J.F.G. Vliegenthart, *Carbohydr. Polym.* **29** (1996) 225.
47 J.J.G. van Soest, D. de Wit, J.F.G. Vliegenthart, *J. Appl. Polym. Sci.* **61** (1996) 1927.
48 H.G. Fritz, W. Aichholzer, T. Seidenstucker, B. Widmann, *Starch/Stärke* **47** (1995) 475.
49 C. Bastioli, *Macromol. Symp.* **135** (1998) 193.
50 J.J.G. van Soest, K. Benes, D. de Wit, J.F.G. Vliegenthart, *Polymer* **37** (1996) 3542.
51 G.J.L. Griffin, US Patent 4 125 495 (1978).
52 F.H. Otey, R.P. Westhoff, US Patent 4 337 181 (1982).
53 R.L. Shogren, J.W. Lawton, K.F. Tiefenbacher, L. Chen, *J. Appl. Polym. Sci.* **68** (1998) 2129.
54 R.L. Shogren, G.F. Fanta, W.M. Doane, *Starch/Stärke* **45** (1993) 276.
55 D.J. Gallant, B. Bouchet, P.M. Baldwin, *Carbohydr. Polym.* **32** (1997) 177–191.
56 K. Svegmark, A.M. Hermansson, *Food Struct.* **10** (1991) 117–129.
57 C.G. Biliaderis, T.J. Maurice, J.R. Vose, *J. Food Sci.* **45** (1980) 1669–1674, 1680.
58 J. Owusu-Ansah, F.R. Van de Voort, D.W. Stanley, *Cereal Chem.* **59** (1982) 167–171.
59 E. Jaska, *Cereal Chem.* **48** (1971) 437–444.
60 A.T. Czaja, *Die Mikroskopie der Stärkekörner*, Verlag Paul Parey, Berlin, 1969.
61 K. Autio, K. Poutanen, T. Suortti, E. Pessa, *Food Struct.* **11** (1992) 315–322.
62 J.L. Doublier, *J. Cereal Sci.* **5** (1987) 247–262.
63 F.R. Dintzis, E.B. Bagley, *J. Rheol.* **39** (1995) 1399–1495.
64 H.W. Leach, L.D. McCowen, T.J. Schoch, *Cereal Chem.* **36** (1959) 534–544.
65 A.C. Eliasson, *J. Texture Stud.* **17** (1986) 357–375.
66 J.L. Doublier, *Starch/Stärke* **33** (1981) 415–420.
67 J.W. Mullen, E. Pacsu, *Ind. Eng. Chem.* **34** (1942) 1209.
68 L. Whistler, *Adv. Carbohydr. Chem.* **1** (1945) 279.
69 D. Trimnell, C.L. Swanson, R.L. Shogren, G.F. Fanta, *J. Appl. Polym. Sci.* **48** (1993) 1665–1675.
70 D.R. Patil, G.F. Fanta, *Starch/Stärke* **46** (1994) 142–146.
71 D. Quemada, *Prog. Colloid Polym. Sci.* **79** (1989) 112.
72 V.D. Athawale, S.C. Rathi, *J. Appl. Polym. Sci.* **66** (1997) 1399–1403.
73 J.L. Willet, B.K. Jasberg, L.L. Swanson, *ACS Symp. Ser.* **575** (1994) 50.
74 N.St. Pierre, B.D. Favis, B.A. Ramsay, J.A. Ramsay, H. Verhoogt, *Polymer* **38** (1997) 647.
75 M.I. Khalil, M.G. Abdel-Halim, *Carbohydr. Res.* **324** (2000) 189–199.
76 T. Tamada, Y. Morimoto, M. Hisamatsu, T. Komiya, *Stärke* **39** (1987) 208.
77 R.L. Whistler, J.N. BeMiller, E.F. Paschall, *Starch Chemistry and Technology*, 2nd edn., Academic Press, Orlando, FL, 1984.
78 M.I. Khalil, A. Waly, S. Farag, A. Hebeish, *J. Appl. Polym. Sci.* **43** (1991) 2303–2309.
79 R.E. Wing, W.M. Doane, C.R. Russell, *J. Appl. Polym. Sci.* **19** (1975) 847–854.
80 I. Kuniak, R. Marchessault, Can. Pat. CA. 960 652 (1975).
81 M.I. Khalil, S. Farag, A. Hebeish, *Starch*, **43** (1991) 254–261.
82 M.I. Khalil, S. Farag, S. Abdel-Fattah, *Starch*, **48** (1996) 270–275.
83 M.I. Khalil, S. Abdel-Rahman, S. Farag, S. Abdel-Fattah, *Egypt. J. Appl. Sci.* **10** (1990) 900–916.

84 R.E. Wing, W.E. Rayford, W.M. Doane, C.R. Russell, *J. Appl. Polym. Sci.* **22** (1978) 1405–1416.
85 M.I. Khalil, S. Farag, A. Hashem, *Starch* **45** (1993) 226–231.
86 M.I. Khalil, A. Waly, S. Farag, A. Hebeish, *Starch* **43** (1991) 349–355.
87 M.I. Khalil, S. Farag, *Starch* **46** (1994) 17–22.
88 J. Aburto, I. Alric, S. Thiebaud, E. Borredon, D. Bikiaris, J. Prinos, C. Panayiotou, *J. Appl. Polym. Sci.* **74** (1999) 1440–1451.
89 S. Parandoosh, S.M. Hudson, *J. Appl. Polym. Sci.* **48** (1993) 787.
90 C.L. Swanson, R.D. Westhoff, W.P. Doane, in *Proceedings of the Corn Utilization Conference E′E′*, Columbus, OH, National Corn Growers Association, St. Louis, MO, 1988.
91 A.D. Sagar, E.W. Merrill, *J. Appl. Polym. Sci.* **58** (1995) 1647.
92 I. Dubois, L. Picton, G. Muller, A. Audibert-Hayet, J.L. Doublier, *J. Appl. Polym. Sci.* **81** (2001) 2480.
93 P. Ispas-Szabo, F. Ravenelle, I. Hassan, M. Preda, M.A. Mateescu, *Carbohydr. Res.* **323** (2000) 163–175.
94 Y. Dumoulin, S. Alex, P. Szabo, L. Cartilier, M.A. Mateescu, *Carbohydr. Polym.* **37** (1998) 361–370.
95 A. Buleon, P. Colonna, V. Leloup, *Ind. Aliment. Agric.* **6** (1990) 515–532.
96 J.F. Foster, R.L. Whistler, E.F. Paschall (eds.), *Starch: Chemistry and Technology*, Vol. 1, 1st edn., R.L. Whistler, E.F. Paschall (eds.), Academic Press, New York, 1965, pp. 349–384.
97 V.S.R. Rao, J.F. Foster, *J. Phys. Chem.* **69** (1965) 636–645.
98 U. Funke, M.G. Lindhauer, *Starch/Stärke* **46** (1994) 384–388.
99 J.W. Lawton, *Carbohydr. Polym.* **29** (1996) 203–208.
100 R.P. Veregin, C.A. Fyfe, R.H. Marchessault, *Macromolecules* **20** (1987) 3007–3012.
101 M.J. Gidley, S.M. Bociek, *J. Am. Chem. Soc.* **107** (1985) 7040–7044.
102 J.J. Cael, J.L. Koening, J. Blackwell, *Carbohydr. Res.* **29** (1973) 123–134.
103 M.A. Mateescu, Y. Dumoulin, G. Delmas, V. Lenaerts, L. Cartilier, *Proc. Int. Symp. Control Rel. Bioact. Mater.* **20** (1993) 290–291.
104 V. Lenaerts, I. Moussa, Y. Dumoulin, F. Mebsout, F. Chouinard, P. Szabo, M.A. Mateescu, L. Cartilier, R.H. Marchessault, *J. Control. Release* **53** (1998) 225–234.
105 J. Kost, R. Langer, N.A. Peppas, *Hydrogels in Medicine and Pharmacy*, Vol. 3, CRC Press, Boca Raton, FL, 1987, pp. 96–97.
106 Y. Dumoulin, P. Clement, M.A. Mateescu, L. Cartilier, *STP Pharm. Sci.* **5** (1994) 329–335.
107 B. Gander, R. Gurny, E. Doelker, N.A. Peppas, *Pharm. Res.* **6** (1989) 578–584.
108 C.J. Kim, P. Lee, *Pharm. Res.* **9** (1992) 10–16.
109 I. Moussa, L. Cartilier, *J. Control. Release* **42** (1996) 47–55.
110 B. Conde-Petit, J. Nuessli, S. Handschin, F. Escher, *Starch/Stärke* **50** (1998) 184–192.
111 R.F. Tester, W.R. Morrison, *Cereal Chem.* **67** (1990) 551–557.
112 D.R.M. Prentice, J.R. Stark, M. J. Gidley, *Carbohydr. Res.* **227** (1992) 121–130.
113 A.M. Hermansson, K. Svegmark, *Trends Food Sci. Technol.* **7** (1996) 345–353.
114 G.E. Babcock, R. Tobin, R.J. Dimler, F.R. Senti, *Cereal Chem.* **37** (1960) 645–655.
115 W. G. Bechtel, *Cereal Chem.* **28** (1951) 29–39.
116 W. Banks, C.T. Greenwood, *Starch and its Components*, Edinburgh University Press, Edinburgh, 1975.
117 M.T. Kalichevsky, S.T. Ring, *Carbohydr. Res.* **162** (1987) 323–328.
118 T. Virtanen, K. Autio, T. Suortti, K. Poutanen, *J. Cereal Sci.* **17** (1993) 137–145.
119 H.S. Ellis, S.G. Ring, M.A. Whittam, *J. Cereal Sci.* **10** (1989) 33–44.
120 B. Conde-Petit, Interaktionen von Stärke mit Emulgatoren in wasserhaltigen Lebensmittel-Modellen, *PhD Thesis*, Swiss Federal Institute of Technology (ETH), Zurich, 1992.
121 M. Ott, E.E. Hester, *Cereal Chem.* **42** (1965) 476–484.
122 B. Conde-Petit, A. Pfirter, F. Escher, *Food Hydrocolloids* **11** (4) (1997) 393–399.

123 R.J. Ceresa, *Block and Graft Copolymerization*, Vol. 2, Wiley, New York, 1973, pp. 23–25.

124 M.R. Lutfor, S. Sidik, W.M. Zin, M.Z. Ab Rahman, A. Mansor, H. Jelas, *Eur. Polym. J.* **36** (2000) 2105–2113.

125 I. Capek, P. Capek, *J. Macromol. Sci. Pure Appl. Chem.* **1**, (1998) 147.

126 I. Capek, S. Janickova, P. Capek, J. Kovacik, P. Kostial, *Chem. Pap.* **59** (2005) 48.

127 I. Capek, to be published.

128 T.L. Han, R.N. Kumar, H.D. Rozman, M.A.M. Noor, *Carbohydr. Polym.* **54** (2003) 509.

129 R.N. Kumar, W.C. Woo, A. Abusamah, *J. Coat. Technol.* 71, **896** (1999) 79–88.

130 S.R. Shukla, A.R. Athalye, *J. Appl. Polym. Sci.* **54** (1994) 279–288.

131 Y.E.M. van der Burgt, J. Bergsma, I.P. Bleeker, P.J.H.C. Mijland, J.P. Kamerling, J.F.G. Vliegenthart, *Carbohydr. Res.* **327** (2000) 423–429.

132 J.F. Robyt, *Starch*, 2nd edn., Academic Press, New York, 1984, pp. 87–123.

133 E. Bertoft, *Carbohydr. Res.* **189** (1989) 195–207.

134 M.J. Gidley, *Carbohydr. Res.* **139** (1985) 85–93.

135 J.D. Blake, M.L. Clarke, J. Littlemore, *Carbohydr. Res.* **138** (1985) 161–167.

136 I. Capek, *Trends Photochem. Photobiol.* **7** (2001) 147–157.

137 D.J. Liaw, C.C. Huang, E.T. Kang, *Curr. Trends Polym. Sci.* **4** (1999) 117.

138 F.M. Winnik, *Langmuir* **7** (1991) 905.

139 S.M. Ghoreishi, G.A. Fox, D.M. Bloor, J.F.Holzwarth, E. Wyn-Jones, *Langmuir* **15** (1999) 5474.

140 Z. Zhen, C.H. Tung, *Polymer* **33** (1992) 812–816.

141 S. Dai, K.C. Tam, L. Li., *Macromolecules* **34** (2001) 7049.

142 L.M. Londoll, *J. Polym. Sci., Polym. Chem. Ed.* **20** (1982) 443.

143 M.E. Haque, A.R. Das, S.P. Moulik, *J. Colloid Interface Sci.* **217** (1999) 1.

144 K. Streletzky, G.D.J. Phillies, *Langmuir* **11** (1995) 42.

145 K. Akiyoshi, S. Deguchi, N. Noriguchi, S. Amaguchi, J. Sunamoto, *Macromolecules* **26** (1993) 3062.

146 K. Akiyoshi, S. Yamaguchi, S. Sunamoto, *J. Chem. Lett.* (1991) 1263.

147 T. Kato, T. Katsuki, A. Takahashi, *Macromolecules* **17** (1984) 1726.

148 W.O. Mcclure, G.M. Edelman, *Biochemistry* **5** (1966) 1908.

149 K. Miyajima, T. Baba, M. Nakagaki, *Colloid Polym. Sci.* **265** (1987) 943.

150 T. Kuwamura, *Kogyo Kagaku Zasshi* **64** (1961) 1958.

151 R.C.R. Souza, C.T. Andrade, *Adv. Polym. Technol.* **21** (2002) 17.

152 H. Liu, J.L. Lekievre, W. Ayoung-Chee, *Carbohydr. Res.* **210** (1991) 79.

153 P.A.M. Steeneken, *Carbohydr. Polym.* **11** (1989) 23.

154 B. Kapoor, M. Bhattacharya, *Carbohydr. Polym.* **41** (2000) 323.

155 B. Kapoor, M. Bhattacharya, *Carbohydr. Polym.* **44** (2001) 217.

156 L. Rexová-Benková, O. Markovič, *Adv. Carbohydr. Chem. Biochem.* **33** (1976) 323.

157 I. Capek, O. Potisk, *J. Polym. Sci., Part A* **33** (1995) 1675.

158 I. Capek, *Makromol.Chem.* **190** (1989) 789.

159 P.M. Forssell, J.M. Mikkila, G.K. Moates, R. Parker, *Carbohydr. Polym.* **34** (1997) 275.

160 R.G. Larson, *Rheol. Acta* **24** (1985) 327.

161 F. Chambor, H.H. Winter, *J. Rheol.* **31** (1987) 683.

162 A. Izuka, H.H. Winter, T. Hashimoto, *Macromolecules* **25** (1992) 2422.

163 S.F. Tanner, S.G. Ring, M.A. Whittam, P.S. Belton, *Int. J. Biol. Macromol.* **9** (1987) 219–224.

164 B.J. Goodfellow, R.H. Wilson, *Biopolymers* **30** (1990) 1183–1189.

165 J.J.G. van Soest, H. Tournois, D. de Wit, J.F.G. Vliegenthart, *Carbohydr. Res.* **279** (1995) 201–214.

166 E. Walenta, H.P. Fink, P. Weigel, J. Ganster, *Macromol. Mater. Eng.* **286** (2001) 456.

167 S.H.D. Hulleman, M.G. Kalisvaart, F.H.P. Janssen, H. Feil, J.F.G. Vliegenthart, *Carbohydr. Polym.* **39** (1999) 351.

168 H.G. Bader, D. Goritz, *Starch/Stärke* **46** (1994) 229.

169 J.M. Ward, *Mechanical Properties of Solid Polymers*, Wiley, Chichester, 1983.

170 H.A. Krassig, *Cellulose – Structure, Accessibility and Reactivity*, Gordon and

Breach Science Publisher, New York, 1993.

171 J.J.G. van Soest, P.M. Kortleve, *J. Appl. Polym. Sci.* **74** (1999) 2207.

172 F. Tiefenbacher, *J. Macrol. Sci., Pure Appl. Chem. A* **30** (9, 10) (1993) 727.

173 F.T. Stepto, I. Tomka, *Chimia* **41**(3) (1987) 76.

174 R. Eith, F.T. Stepto, I. Tomka, E. Wittwer, in *Proceedings of the 5th Pharmaceutical Technology Conference*, Harrogate, 1986, p. 178.

175 R. Eith, F.T. Stepto, I. Tomka, E. Wittwer, *Drug Dev. Ind. Pharm.* **12** (1986) 2113.

176 L. Shogren, C.L. Swanson, A.R. Thompson, *Starch/Stärke* **44** (9) (1992) 335.

177 P. Forssell, J. Mikkila, T. Suortti, J. Seppala, K.J.M. Poutanen, *Pure Appl Chem. A* **33** (1996) 703.

178 M. Wollendorfer, H. Bader, *Ind. Crops Prod.* **8** (1998) 105.

179 A.S. Curvelo, A.J.F. de Carvalho, J.A.M. Agneli, *Carbohydr. Polym.* **45** (2001) 183.

180 A. Dufresne, M.R. Vignon, *Macromolecules* **31** (1998) 2963.

181 D. Lourdin, H. Bizot, P. Colonna, *J. Appl. Polym. Sci.* **64** (1997) 1047.

182 R.M. Sala, I.A. Tomka, *Angew. Makromol. Chem.* **199** (1992) 45.

183 M. Goheen, R.P. Wool, *J. Appl. Polym. Sci.* **42** (1991) 2691.

184 S. Thiebaud, J. Aburto, I. Alric, E. Borredon, D. Bikiaris, J. Prinos, C. Panayiotou, *J. Appl. Polym. Sci.* **65** (1997) 705.

185 H. Morita, *Anal. Chem.* **28** (1956) 64.

186 I.A. Wolff, H.A. Davis, J.E. Cluskey, L.J. Gundrum, C.E. Rist, *Ind. Eng. Chem.* **43** (1951) 915.

187 C. Fringant, J. Desbrieres, M. Rinaudo, *Polymer* **37** (1996) 2663.

188 R.L. Shogren, *Carbohydr. Polym.* **19** (1992) 83.

189 M. Bhattacharya, U.R. Vaidya, D. Zhang, R. Narayan, *J. Appl. Polym. Sci.* **57** (1995) 539.

190 D. Bikiaris, E. Pavlidou, J. Prinos, J. Aburto, I. Alric, E. Borredon, C. Panayiotou, *Polym. Degrad. Stab.* **60** (1998) 437.

191 J.F. Robyt, R.L. Whistler, J.M. Bemiller, E.F. Paschall, *Starch: Chemistry and Technology*, 2nd ed., Academic Press, New York, 1984, Chapter IV.

192 M. Lauro, T. Suortti, K. Autio, P. Linko, K. Poutanen, *J. Cereal Sci.* **17** (1993) 125–136.

193 T. Shiotsubo, *Agric. Biol. Chem.* **47**(11) (1983) 2421–2425.

194 B. Volesky, *Biosorption of Heavy Metals*, CRC Press, Boca Raton, FL, 1990.

195 Y. Terashima, H. Ozaki, M. Sekiue, *Water Res.* **20** (1986) 537.

196 B.A. Boto, L. Pawlowski, *Waste Water Treatment by Ion Exchange*, Chapman and Hall, New York, 1987.

197 Y. Yang, J. Shao, *J. Appl. Polym. Sci.* **77** (2000) 151.

198 D.W. Kang, H.R. Choi, D.K. Kweon, *J. Appl. Polym. Sci.* **73** (1999) 469.

199 P.M. Padilha, J.C. Pacha, J.C. Moreira, J.T.S. Campos, C.C. Federici, *Talanta* **45** (1997) 317.

200 R. Saliba, H. Cauthier, R. Gauthier, M. Petit-Ramel, *J. Appl. Polym. Sci.* **75** (2000) 1624.

201 W.C. Chan, J.C. Ferng, *J. Appl. Polym. Sci.* **71** (1999) 2409.

202 W.C. Chan, J.Y. Wu, *J. Appl. Polym. Sci.* **81** (2001) 2849.

203 J.A. Rendleman, *Food Chem.* **3** (1978) 47–62.

204 P. Tomasik, C.H. Schilling, *Adv. Carbohydr. Chem. Biochem.* **53** (1998) 263.

205 W. Leszczynski, *Acta Aliment. Pol.* **11** (1985) 21–34.

206 P. Tomasik, J. Jane, K. Spence, J.W. Anderegg, *Starch/Stärke* **47** (1995) 68–72.

207 S.K. Tyrlik, P. Tomasik, J.W. Anderegg, M. Baczkowicz, *Carbohydr. Polym.* **34** (1997) 1–7.

208 P. Tomasik, C.H. Schilling, M. Refvik, J.W. Anderegg, *Carbohydr. Polym.* **41** (2000) 61–68.

209 F.B. Ahmad, P.A. Williams, *J. Agric. Food Chem.* **47** (1999) 3359–3366.

210 C.Y. Lii, P. Tomasik, W.L. Hung, V.M.F. Lai, *Food Hydrocolloids* **16** (2002) 35–45.

211 A.H.A. De Willigen, P.W. de Groot, *Starch/Stärke* **23** (1971) 37–57.

212 J. Jane, *Starch/Stärke* **45** (1993) 172–175.

213 O.B. Josten, *Starch/Stärke* **35** (1983) 166–169.

214 C.Y. Lii, P. Tomasik, M.T. Yen, V.M.F. Lai, *Int. J. Food Sci. Technol.* **36** (2001) 321–330.

215 N. Itoh, *Bull. Soc. Sci. Photogr. Jpn.* **16** (1966) 24–29.

216 R.T. Hart, Insolubilization of starch adhesieves, British Patent 1 024 881 (1966); *Chem. Abstr.*, **65** (1966) 906.

217 W.A. Samuel, *Chemist Analyst* **37** (1948) 33–35.

218 C.M. Langkammerer, Organic compounds of titanium, US Patent 2 489 651 (1949); *Chem. Abstr.* **44** (1950) 1534.

219 A. Krause, M. Rychlewska, *Kolloid-Z.* **185** (1962) 73–74.

220 D. Shi, *Shipin Yu Fajiao Gongye* **2** (1989) 68–70.

221 L. Zhang, L. Chen, Y. Zhang, *Yingyang Xuebao* **10** (1988) 220–223.

222 W. Ciesielskia, C. Lii, M.T. Yen, P. Tomasik, *Carbohydr. Polym.* **51** (2003) 47–56.

223 P.M. Baldwin, J. Adler, M.C. Davies, C.D. Melia, *Starch/Stärke* **46** (1994) 341–346.

224 K.C. Huber, J.N. BeMiller, *Cereal Chem.* **74** (1997) 537–541.

225 C.L. Rollinson, J.C. Bailar, Jr., *Inorg. Synth.* **2** (1946) 197–200.

226 J.E. Macintyre, *Dictionary of Inorganic Compounds*, Chapman and Hall, London, 1992.

227 J. Kapusniak, W. Ciesielski, J.J. Kozio, P. Tomasik, *Eur. Food Res. Technol. A* **209** (1999) 325–329.

228 L.M. Zhang, D.Q. Chen, *Colloids Surf. A* **205** (2002) 231ff.

229 H.I. Irving, R.J.P. Williams, *Nature* **162** (1948) 746–747.

230 M.I. Khalil, S. Farag, *J. Appl. Polym. Sci.* **69** (1998) 45–50.

231 H.I. Irving, R.J.P. Williams, *Chem. Rev.* **56** (1956) 271–297.

232 M. Furue, K. Sumi, S.N. Nozakura, *J. Polym. Sci., Polym. Lett. Ed.* **20** (1982) 291–295.

233 W.D. Johanston, H. Freiser, *Anal. Chim. Acta* **11** (1954) 201–213.

234 S.M. Nelson, D.W.A. Sharp, MTP International Review of Science, Inorganic Chemistry, Series 1, Vol. 5, Transition Metals, Part 1, Butterworth, London, 1972.

235 B. Conde-Petit, F. Escher, *Food Hydrocolloids* **6** (1992) 223–229.

236 J. Nuessli, B. Conde-Petit, U.R. Trommsdorff, F. Escher, *Carbohydr. Polym.* **28** (1996) 167–170.

237 R.C. Teitelbaum, S.L. Ruby, T.J. Marks, *J. Am. Chem. Soc.* **100** (1978) 3215.

238 M.C. Godet, V. Tran, M.M. Delage, A. Buleon, *Int. J. Biol. Macromol.* **15** (1993) 11.

11
In Vivo Skin Performance of a Cationic Emulsion Base in Comparison with an Anionic System

Slobodanka Tamburic

Abstract

Cationic emulsifiers are a relatively recent addition to the vast range of emulsifiers for personal care products. There are very few data regarding their *in vivo* skin performance. This chapter presents a comparative assessment of skin hydration potential of two creams: a *cationic emulsion*, based on distearyldimonium chloride (distearyldimethylammonium chloride), and an *anionic emulsion*, based on hydrophobically modified acrylic acid polymer. Both emulsions contained the same concentration of emollients (19%), humectants (8%) and preservatives (0.1%). In addition to the vehicles, creams containing 5% herbal extracts (chamomile flower and blackthorn fruit, respectively) in each of the emulsion bases were also evaluated.

A balanced randomized study was performed on lower legs of 10 volunteers, using a Corneometer CM 825. The study design included a negative control (untreated site) and a positive control (10% glycerol), in addition to the six test samples. Changes were followed for 3 h and the results analyzed using an ANOVA test. Higher skin hydration was detected with the cationic emulsion, especially in the initial stages. This applied to both the cationic vehicle and the two cationic creams with herbal extracts, showing that the emulsifier type had a prevailing effect. Further studies are needed to assess the relative merit of the conditioning effect on increased skin hydration.

11.1
Introduction

Cationic materials used in cosmetics and toiletries (typically containing N^+) are normally associated with hair conditioning and are known to impart lubricity, antistatic properties and smoothness to hair fibers. Some classes of cationics have excellent emulsifying properties and good skin substantivity and are suitable for skin care products. The use of cationic emulsifiers has been advocated for more

Colloids and Interface Science Series, Vol. 4
Colloids in Cosmetics and Personal Care. Edited by Tharwat F. Tadros

ISBN: 978-3-527-31464-5

than 20 years [1, 2], but their potential in skin care is still largely unexploited. It is argued that this may be due to their undeserved reputation as irritants, coming from the use of short-chain monoalkyl cationic materials [3]. The results of recent studies have shown that long-chain and dialkyl quaternaries are much milder on the skin than their shorter chain analogues [4] and as a result, modern cationic emulsifiers are finding their way into skin formulations [5]. In his review of the formulation strategies for moisturizers, Barton [6] included cationic emulsifiers, stating that some of them exhibit antimicrobial activity, which could be an additional benefit. Cationic emulsifiers in general are stable in the presence of metal salts, but they are prone to instability at high pH and when combined with negative ions.

Advantages claimed for the new class of cationic emulsifiers, which include dialkyl and amidoamine quaternaries, are substantivity to the skin, renewable raw material sources and a good cost/performance ratio. This is in addition to desirable sensorial effects – matte finish and smooth powdery after-feel. It is also stated that these two classes of cationic emulsifiers allow for high active loadings (especially with lipophilic actives), without sacrificing efficacy or skin feel [7]. They also provide high water resistance of sun-screen products. Disadvantages involve special formulation requirements regarding ionic strength and pH, and also incompatibility with anionic materials.

Soft skin feel produced by cationics is the result of their substantivity and the emulsifier chemical structure. Positive charges of cationics associate with the net negative charge on hair and skin due to the presence of carbonyl groups in the hair and skin proteins [3]. The unattached fatty tails of cationic emulsifiers give the lubricating skin feel.

There are a number of relatively recent patents which present various formulations of cationic-based emulsions for improved skin moisturization. For example, McManus et al. [8] deal with skin moisturizing compositions comprising of a liquid crystal/gel network (LCGN) emulsion system, which is made of water, cationic emulsifiers, low HLB emulsifiers and emollients. The authors claim excellent moisturization and improved esthetics, including faster absorbance and more pleasant after-feel, compared with traditional formulations. According to the patent, emulsifiers suitable for use in the above formulations include amidoamines, amine oxides, quaternaries (including dimethyldistearylammonium chloride) and alkoxylated amines.

Defined in a simple way, a moisturizer is a product designed to restore and maintain optimum hydration of the stratum corneum (SC), the outermost layer of the epidermis. Moisture is required in the SC for two basic reasons: to keep it soft, supple and flexible and to allow the action of enzymes responsible for desquamation (the regular shedding of corneocytes). There are two basic mechanisms for skin to retain its moisture:

1. natural moisturizing factor (NMF) within the protein matrix of corneocytes, consisting of a mix of low molecular weight hygroscopic compounds;
2. triple-lipid bilayers around and between corneocytes, consisting primarily of fatty acids, ceramides and cholesterol.

Skin moisturization is increasingly viewed as critical to the structural and functional integrity of the skin, as well as to fundamental skin care [9]. The results of many in-depth studies show that the SC architecture (the lipid and NMF components and the level of corneocyte maturation) is the predominant factor affecting water flux and retention in the skin. The principles of humectancy, emolliency and occlusion are still central in the moisturizer formulation [9]. Humectants promote water retention within the SC, occlusives minimize water loss and emollients have a complementary occlusive activity. In addition to this basic trio, one should consider the effects of emulsifier system on the level and duration of moisturization achieved by a given skin care emulsion.

There are very few data regarding *in vivo* skin performance of cationic emulsions against well-established emulsifying systems. In a recent study by Howe et al. [4], moisturizing effects of a cationic against a nonionic emulsion system (based on glyceryl stearate SE) were assessed. The cationic-based emulsion, containing distearyldimonium chloride (distearyldimethylammonium chloride), showed initially lower but longer lasting skin hydration effects. There are no data regarding anionic–cationic comparison.

The aim of this study was to perform a comparative assessment of skin hydration potential of two emulsion creams: a *cationic emulsion*, based on distearyldimonium chloride, and an *anionic emulsion*, based on hydrophobically modified acrylic acid polymer. In addition to emulsion vehicles ("placebo" samples), two types of herbal extracts were included in the formulations ("active" samples), with the aim of assessing any differences in the moisturizing efficacy. Propylene glycol extracts of chamomile flower and blackthorn fruit were used as moisturizing actives.

11.2
Materials and Methods

11.2.1
Materials

Distearyldimonium chloride (Figure 11.1) was chosen as a representative of a mild, long-chain dialkyl quaternary, while acrylates C_{10-30} alkyl acrylate crosspolymer is a commonly used hydrophobically modified acrylic acid polymer. Due to different chemical nature and formulation requirements for these emulsifiers, it was not possible to use exactly the same formulation (Table 11.1). However, both emulsions contained the same concentration of emollients (19%),

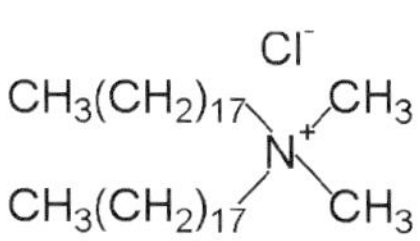

Figure 11.1 Distearyldimethylammonium chloride (distearyldimonium chloride).

Table 11.1 Composition of test vehicles. In active samples, chamomile flower extract and blackthorn fruit extract were each present at a concentration of 5% w/w.

Ingredient	Amount (g)	
	Anionic cream	Cationic cream
Distearyldimonium chloride	–	5.0
Mineral oil	19.0	
Petrolatum	–	5.0
Isopropyl myristate	–	5.0
Cetyl alcohol	–	4.0
TEA (10% solution)	q.s.	–
Citric acid (10% solution)	–	q.s.
Sodium chloride (10% solution)	–	0.5
Acrylates C_{10-30}-alkyl acrylate cross-polymer	0.7	–
Glycerine	8.0	8.0
Preservative	0.1	0.1
Purified water	Up to 100.0	Up to 100.0

humectants (8%) and preservatives (0.1%). In the case of anionic cream, the emollient consisted of mineral oil, well recognized as a highly effective moisturizing agent [10]. The cationic cream contained a mixture of emollients, including the emulsifier itself and its structure-building co-emulsifier (cetyl alcohol).

Low HLB surfactants (e.g. fatty alcohols) are commonly added as co-emulsifiers to a higher HLB main emulsifier, to add structure and increase the stability of oil/water (O/W) emulsions. The same is recommended in the case of the cationic emulsifier used in this study, whose HLB value is reported to be between 10 and 11 [7]. The low HLB cetyl alcohol contributes to the formation of the LCGN system, as suggested by McManus et al. [8]. This means that the studied emulsion system is stabilized by a combination of ionic and liquid crystalline mechanisms.

In addition to emulsion vehicles, creams containing 5% w/w herbal extracts (of chamomile flower and blackthorn fruit, respectively) in each of the emulsion vehicles were also evaluated. These two herbal actives have been chosen on the basis of our previous work [11]. They represent highly concentrated and standardized propylene glycol–water extracts, obtained by an official pharmacopoeial method of percolation (Council of Europe [12]). Both extracts were shown to possess good moisturizing properties [11, 13], which have been attributed mostly to the known presence of oligo- and polysaccharides [14].

Since the extracts are propylene glycol based, when each of them was added, the same amount (5%) of the humectant glycerol was removed from the formulation. The pH was adjusted to 4.4 in both cases, to comply with the stability requirement for cationic-based systems.

11.2.2 Methods

Emulsion vehicles were produced using standard laboratory methods. A cold/cold emulsification method was used for the anionic cream. The emulsion was obtained by dispersing a polymeric emulsifier in a mixture of water, glycerol and preservative at room temperature, by means of an overhead stirrer. After a 20-min stirring period, the oil phase was added and the stirring continued for another 30 min. The acidic dispersion obtained was then neutralized with triethanolamine solution to a pH of ~4.4.

The cationic emulsion was obtained by a hot/hot process, with both water and oil phases heated to 70 °C. The oil phase was then slowly poured into the hot water phase, with continuous and vigorous stirring for 20 min. A 10-min homogenization and additional stirring were necessary for the production of the cationic emulsion. The ionic strength and pH values were adjusted with sodium chloride and citric acid solutions, respectively.

A balanced randomized single-blind study was performed on lower legs of 10 female volunteers (mean age 34.5 years), under controlled conditions. The SC hydration level was measured by the means of a Corneometer CM 825 (Courage and Khazaka Electronics, Germany [15]; Figure 11.2). It is well established that the flow of electrical current through the skin surface (a dielectric medium) is related to its water content and thus offers a non-invasive method for assessing moisturization [16]. Three principal approaches are used, namely measurements of impedance, capacitance and conductance. Capacitance-based instruments, such as a corneometer, are well established in assessing the level of surface skin hydration [17]. Any change in the dielectric constant due to variations in skin surface hydration alters the capacitance of a precision measuring capacitor within the probe. The recorded capacitance values are automatically converted into arbitrary hydration units [relative corneometer units (rcu)], varying from 0 to 120.

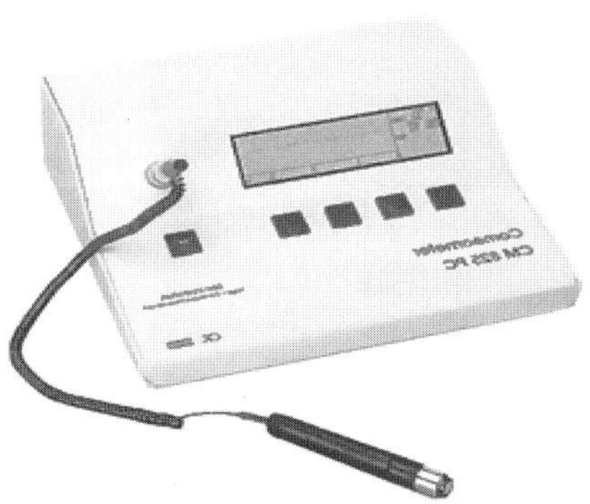

Figure 11.2 Corneometer: a capacitance-based skin hydration-measuring instrument.

This short-term study protocol was designed on the basis of the published guidelines [18]. Four square sites (3 × 3 cm, 2 cm apart) on each volunteer's shin were marked using a cardboard template, which allowed for the testing of eight sites. An experimental design including two controls (a non-treated site and 10% w/w glycerol solution) and six test samples were chosen. The room temperature was 21–22 °C and the relative humidity 53% throughout the trial.

After 30 min of acclimatization, the baseline values were measured and each designated area was treated with 2 μL cm^{-2} of a test sample (random and balanced distribution of test sites). The test samples were applied with an Eppendorf micropipette and spread homogeneously using the flattened tip of a glass rod. On the basis of our previous results and recommendations in the literature [15, 19], a period of 45 min was allowed to elapse before the first measurement was taken. Three corneometer readings of each test site were recorded, with at least 5 s between the readings. The change in skin hydration was followed for 3 h.

The values of the measured parameter are calculated as means ± SD. Following a positive testing for normality, parametric statistical tests were used ($p < 0.05$ treated as statistically significant). Data were analyzed by one-way within-subjects (repeated measurements) analysis of variance (ANOVA), followed by a Duncan multiple range test.

11.3 Results and Discussion

The study was performed as a short-term, single-application trial. Single-application tests are well established in the field of moisturizer evaluation. They assess the physical changes in skin due to the test product and can accurately predict results of long-term (e.g. 2-week) studies [9, 20]. Furthermore, the short-term measurements of skin hydration correlate well with the expert grading of dry skin [21].

The summary of the results is shown in the form of mean rcu values for all samples at all time points (Figure 11.3). Statistical analysis revealed significantly higher ($p < 0.05$) skin hydration exerted by cationic emulsion in the initial stages of the experiment, i.e. after 60 min. This applied to both the cationic vehicle and the two cationic creams with herbal extracts, showing that the type of emulsifier in this case had a prevailing effect on the measured skin parameter.

The significant difference observed between the two emulsion vehicles (cationic and anionic placebo samples) has been lost after 120 min. However, it remained borderline significant in the case of the two sets of creams with plant extracts and eventually vanished after 180 min (Figure 11.3). The hydration effect of glycerol was dominant throughout the duration of the trial, confirming that 10% glycerol was a good choice for a positive control in this study. Glycerol is generally considered as most effective humectant. Apart from the ability to bind and hold water, glycerol was shown to have an ability to prevent humidity-induced crystal phase transition in SC lipids and therefore improve the barrier function of the skin. It

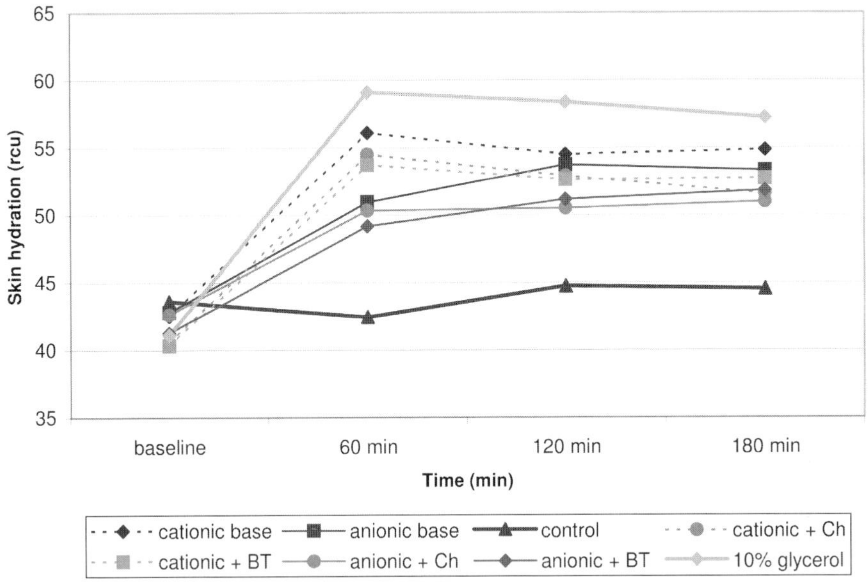

Figure 11.3 Moisturizing effects of cationic and anionic creams with and without plant extracts and the two controls, expressed in relative corneometer units (rcu) ($n = 10$).

also aids the proteolytic degradation of corneodesmosomes and thereby facilitates desquamation [9].

The application of plant extracts in cosmetics and toiletries has been a distinct trend over the last decade and, given consumers' interest in natural products, will probably continue. Both cosmetic and dermatological practices have benefited from the use of new and rediscovered plants [22]. Herbal extracts show a wide spectrum of biological activities used in skin treatments, including moisturizing, antimicrobial, anti-inflammatory, antimutagenic, antioxidant and antiacne properties [23–25]. Unlike traditional liquid extracts, which contain ethanol and where heat is used to concentrate combined batches down to the original plant weight, modern extracts are usually ambient infusions with aqueous propylene glycol at a herb/extract ratio of 1:10 [26]. The extracts used in this study were obtained by an official pharmacopoeial method and their content and physico-chemical parameters were assessed by standard analytical methods, as reported previously [11].

Hence it was of interest to assess the changes in skin hydration efficacy when part of the glycerol in the formulation was replaced with the chamomile flower and blackthorn fruit herbal extract, respectively. Despite their content of oligo- and polysaccharides and propylene glycol, the herbal extracts failed to differ significantly in efficacy from the cationic vehicle, at least within a given 3-hour study. In contrast, the sample with anionic emulsifier and blackthorn fruit ex-

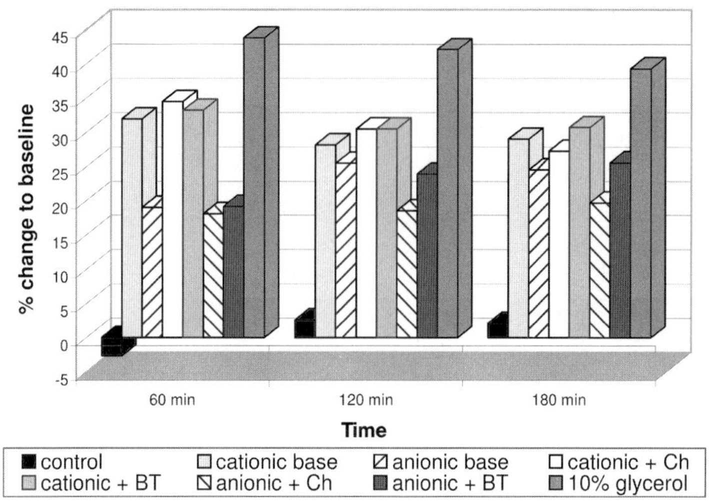

Figure 11.4 Moisturizing effects of the cationic and anionic creams with and without plant extracts and the two controls, expressed as the percentage change from baseline.

tract showed better moisturizing ability than the chamomile-containing one after 120 and 180 min. These findings can be clearly observed on the graph showing the percentage difference from baseline (Figure 11.4).

It is well documented that the type of emulsifier represents an important factor contributing to the overall product performance and especially to the skin moisturizing efficacy. According to Aikens and Friberg [27], emulsifiers play three main roles in skin care formulations: ensure their stability, make them esthetically attractive and create optimal structures to facilitate release and beneficial action of active ingredients during the structural changes after application. They argue that the last role has attracted undeservedly scant attention by researchers, who mostly focus on the structure and properties of the original dispersion. The number of stages that an emulsion goes through is bewildering and not necessary to elucidate in detail. What really matters is the final state after evaporation (normally achieved within 45 min of application), which is the one that affects the properties of the skin and the thermodynamic potential of an active. For example, it was found that the formation of lamellar liquid crystal (LLC) in a final state significantly increased the release of vitamin E acetate [27].

Dahms [28] stated that the effect of an active is closely linked to the interaction between the emulsion base, the active and the skin. This is an extremely complex interaction, which seems to be dominated by the emulsion structure. An emulsion first has to spread well and this depends on the condition of the skin, the contact angle between the skin and emulsion and the rheology of the emulsion. The more water is present in the interlamellar space, bound by LCGN structures, the longer the evaporation will take. Finally, when all water has evaporated, an occlusive film is formed, consisting of non-volatile emulsion constituents. Very often, an ideal 100% coverage by an occlusive film is not achieved and that greatly affects the release of actives. It is also very important for a hydrophilic active whether it is present in the continuous aqueous or immobilized aqueous phase. All of the above factors may have contributed to the differences seen in the two types of emulsions tested in this study.

The effect of emulsifiers on the delivery of skin active agents has also been studied by Wiechers et al. [29]. Comparing two nonionic emulsifier systems, they found a distinctly different distribution profile of an active within the upper skin layers. It is speculated that many factors may play a role in this phenomenon, including the interference of surfactants with skin lipids and the stability of the formulation.

Rawlings et al. [9] argue that the choice of emulsifier can affect the formulation's ability to moisturize. According to their data, a lotion based on a cationic emulsifier showed an immediate and sustained increase of skin moisture over 8 h. In addition, the increased moisture level was maintained over 5 days, despite repeated vigorous washing. The authors attributed this effect to the binding of positive emulsifier sites to the negative sites of the keratin proteins of the skin.

On the other hand, Warner and Boissy [30] stated that the essential influence of emulsifiers on the skin is in the interaction with the SC lipids, whose lamellar structure is important for preventing excessive drying of the skin. The intercellular lipid lamellae are not constant structures, but very dynamic in nature. They are altered by age, disease, solvent treatment and environmental conditions, but could be repaired rapidly with appropriate exogenous lipid treatment. On the basis of their microscopic study, Warner and Boissy [30] stated that moisturizers appear to enter the SC and affect the lipid structure. They presented evidence that formulated products are more effective than neat emollients in restoring normal SC lipid structure, which justifies the use of emulsions, as opposed to neat oils, as moisturizers.

Epstein and Jonasse [31] presented an explanation for an unusually high skin-hydrating efficacy of their skin care composition based on distearyldimonium chloride. They proposed that the temperature change during product application to the skin induces a phase change, which causes the humectant (glycerol) to move from the micellar interface to the external surface of the emulsion. In this way, glycerol, which was associated with the hydrophilic portion of the cationic emulsifier, migrates to the air/emulsion and skin/emulsion interfaces and becomes available for its humectant action. It is interesting that the authors suggest the use of the phase inversion temperature (PIT) method for the preparation of the proposed O/W formulation based on cationic emulsifier.

It is known that anionic creams based on acrylic polymers have a high immediate moisturizing potential, which was the reason for choosing this type of anionic system for comparison. The polymeric emulsifier used, an acrylate/C_{10-30}-alkyl acrylate cross-polymer, is claimed to be able to provide an immediate availability of the oil phase upon application. This is due to the process of deswelling of the emulsion-stabilizing microgel structures upon contact with the skin electrolytes

[32]. It was therefore surprising to see a comparable and initially better moisturizing activity exerted by the tested cationic emulsion. A number of possible explanations for the initial better performance of the cationic base could be put forward:

1. reduced amount of skin electrolytes due to the swiping of the application sites with alcohol before the start of the trial, which slowed the breakdown of the anionic emulsion structure;
2. the presence of an LCGN network structure within the cationic emulsion and the formation of favorable metaphases after evaporation of water;
3. rapid binding of positive cationic moiety of the emulsifier to the negative charge of the carbonyl groups in the skin proteins, possibly accompanied by a transient hyper-hydration effect.

It is reasonable to assume that a combination of the above processes took place. Further investigation into the mechanism of action is required in order to rationalize the difference in performance between the two emulsion systems. In any case, the results obtained confirm a high moisturizing potential of a cationic emulsifier-based vehicle, in addition to its favorable sensory properties.

11.4 Conclusion

This study has shown that the cationic emulsifier of a long-chain dialkyl quaternary type (distearyldimonium chloride) possesses considerable potential in enhancing skin hydration. A cream sample containing this type of emulsifier performed better during the first 120 min of the trial than the sample based on an anionic emulsifier of the modified acrylic acid type. This was true for both emulsion bases and samples containing 5% of either chamomile flower or blackthorn fruit extract. Further work is needed to investigate the exact role of the skin conditioning effect of cationic emulsifiers on their *in vivo* skin performance.

References

1 Goddard ED (1987). Substantivity through cationic substitution, *Cosmet. Toiletries*, **102**: 71–80.
2 Smith L Gesslein BW (1989). Multifunctional cationics for hair and skin care applications, *Cosmet. Toiletries*, **104**: 41–46.
3 Klein K (2002). Cationic emulsifiers: a most underutilized category, *Cosmet. Toiletries Mag.*, **117**: 21–23.
4 Howe A, Paez A, Meyer J, Jenni K (2006). Cationic emulsifiers: an emerging trend in skin care. In *Cosmetics and Toiletries Manufacture Worldwide*, pp. 73–77.
5 Chandler M (2004). Emulsifier selection and use, *J. Cosmet. Sci.*, **55**: 401–402.
6 Barton S (2002). Formulation of skin moisturizers. In *Skin Miosturization*, Leyden JJ, Rawlings AV (eds.), Marcel Dekker, New York, pp. 547–584.
7 Goldschmidt (2003). *Special 3: Emulsifiers for Skin Care Applications.*
8 McManus RL, Trivedi H, Murphy BP, Spengler E (2001). Cationic emulsifier-enhanced liquid crystal gel network based skin care moisturising compositions. European Patent WO 0178657.

9 Rawlings AV, Canestrari DA, Dobkowski B (2004). Moisturizer technology versus clinical performance, *Dermatol. Ther.*, **17**: 49–56.

10 Tamburic S, Abamba G, Ryan J (1999). Moisturising potential of d-α-tocopherol, *Cosmet. Toiletries Mag.*, **114**(5): 73–82.

11 Arsic I, Tamburic S, Bulatovic S, Homsek I, Vuleta G (2005). Exploring moisturising potential of naturals: the cases of St. John's wort, chamomile and blackthorn, *Euro Cosmet.*, **13**(3): 14–21.

12 Council of Europe (1997). *European Pharmacopoeia*, 3rd edn., Council of Europe, Strasbourg.

13 Tamburic S (2006) Effects of polymer entrapment of *Prunus spinosa* fruit extract on its cosmetic efficacy, *J. Appl. Cosmetol.* **24**(2): 63–76.

14 Füller E, Blashek W, Franz G (1990). Characterization of water-soluble polysaccharides from chamomile flowers, *Planta Med.*, **56**: 40.

15 Courage Khazaka Electronics (2004). *Corneometer CM 825 Operating Manual*, Courage Khazaka Electronics.

16 Grove G, Zerweck C, Pierce E (2002). Noninvasive instrumental methods for assessing moisturisers. In *Skin Moisturization*, Leyden JJ, Rawlings AV (eds.), Marcel Dekker, New York, pp. 499–528.

17 Barel AO, Clarys P (1995). Measurement of epidermal capacitance. In *Handbook of Non-Invasive Methods and the Skin*, Serup J, Jemec GBE (eds.), CRC Press, Boca Raton, FL, pp. 165–178.

18 Berardesca E (1997). EEMCO guidance for the assessment of stratum corneum hydration: electrical methods, *Skin Res. Technol.*, **3**: 126–132.

19 Marenus KD (1998). Skin conditioning benefits of moisturizing products. In *Cosmetic Claims Substantiation*, Aust LB (ed.), Marcel Dekker, New York, pp. 97–113.

20 Nole G (2002). Clinical testing of moisturisers. In *Skin Moisturization*, Leyden JJ, Rawlings AV (eds.), Marcel Dekker, New York, pp. 465–498.

21 Fang L, Conroy E, Visscher M, Wickett RR (2001). The ability of electrical measurements to predict skin moisturization. II. Correlation between one-hour measurements and long-term results, *J. Cosmet. Sci.*, **52**: 23–33.

22 Khaiat A (2000). Botanical extracts. In *Cosmeceuticals – Drugs vs. Cosmetics*, Elsner P, Maibach HI (eds.), Marcel Dekker, New York, pp. 97–106.

23 Seifert P (1997) Herbal extracts in cosmetics – opportunities and limitations. *SÖFW J.*, **123**: 28–31.

24 Buchwald-Werner S, Griesbach U (2003). High quality botanical extracts for real claim substantiation, *SÖFW J.*, **129**: 8–11.

25 Voss W, Burger C (2003). Herbal therapies from a clinical and toxicological point of view. In *Active Ingredients Conference*, Paris, *Proceedings*, pp. 151–160.

26 Whitehead (2003).

27 Aikens PA, Friberg SE (2000). Emulsifiers. In *Dry Skin and Moisturizers*, Lodén M, Maibach HI (eds.), CRC Press, Boca Raton, FL, pp. 183–201.

28 Dahms GH (1998). The impact of the emulsion structure on adsorption and release of actives on skin. In *Emulsions – Technology, Structure, Ingredients, Formulations*, H. Ziolkowsky (ed.), Verlag für Chemische Industrie, Augsburg, pp. 15–24.

29 Wiechers JW, Kelly CL, Blease TG, Dederen JC (2004). *Int. J. Cosmet. Sci.*, **26**: 173–182.

30 Warner RR, Boissy YL (2000). Effect of moisturising products on the structure of lipids in the outer stratum corneum of humans, In *Dry Skin and Moisturizers*, Lodén M, Maibach HI (eds.), CRC Press, Boca Raton, FL, pp. 349–369.

31 Epstein H, Jonasse MS (2000). Skin care composition, US Patent 6 017 548.

32 Savic S, Tamburic S, Savic M, Cekic N, Milic J, Vuleta G (2004). Vehicle-controlled effect of urea on normal and SLS-irritated skin, *Int. J. Pharm.*, **271**: 269–280.

12
The Impact of Urea on the Colloidal Structure of Alkylpolyglucoside-Based Emulsions: Physicochemical and *In Vitro/In Vivo* Characterization

Snezana Savic, Slobodanka Tamburic, Biljana Jancic, Jela Milic, and Gordana Vuleta

Abstract

Although urea is commonly used in dermatological practice, its interactions with carrier systems and the potential of these interactions to affect the efficacy of urea have not been thoroughly investigated. In particular, the effect of urea on the colloidal structure of emulsions based on novel non-irritant sugar emulsifiers is not known. If pronounced, the interactions could affect the delivery of urea into the skin and its moisturizing capacity. The aim of the study was to investigate the impact of urea on the water distribution mode within a complex lamellar gel structure of alkylpolyglucoside-based emulsions. In addition, it was of interest to evaluate the potential of this vehicle for controlled skin hydration and prolonged delivery of urea. Consequently, the study presents a comprehensive structural characterization of placebo and urea-containing (5%) samples based on an alkylpolyglucoside emulsifier with fixed emulsifier/water ratio (1:10.43) and different oil content (5, 10 and 20%). Polarization and electron microscopes, wide-angle x-ray diffraction (WAXD), rheology and thermogravimetry were employed. The release profile of urea was evaluated using enhancer diffusion cells, and its *in vivo* moisturizing capacity was assessed using a capacitance-based instrument. The results revealed that urea did not significantly affect the type of lyotropic emulsifier–water–oil interaction. However, the water distribution mode in the sample with the highest oil content was significantly altered, with an increase of interlamellarly fixed water containing dissolved urea. These findings implied a highly effective moisturizing product, which has been confirmed by a short-term *in vivo* study.

Colloids and Interface Science Series, Vol. 4
Colloids in Cosmetics and Personal Care. Edited by Tharwat F. Tadros

ISBN: 978-3-527-31464-5

12.1 Introduction

Interactions between active materials and liquid crystalline structures of topical vehicles have been investigated by a number of workers. These studies are dominated by the category of lipophilic (poorly soluble) drugs, with a limited number of them dealing with hydrophilic active materials (e.g. [1–4]). Urea is an example of a hydrophilic topical active, commonly used in cosmetic and dermatological practice [5], whose interactions with the vehicles and the potential of these interactions to affect its efficacy have not been thoroughly investigated.

In the treatment of dermatoses accompanied by dry and scaly skin, it is essential to use highly moisturizing products, most of them being oil in water (O/W) emulsions based on a mixed emulsifier system [6]. According to Eccleston [7], a lamellar crystalline gel phase is formed within those emulsion systems, due to the swelling properties of distinct entities of mixed emulsifier. The gel phase prevents the oil droplets from coalescence and controls the system consistency. This type of system is considered to be a four-phase system, with two pairs of different colloidal gel phases, the hydrophilic and the lipophilic ones [8]. Due to different swelling capacities of the hydrophilic gel phases, different ratios of two types of water, the bulk (free) and interlamellar (fixed), could appear within the system [7, 8]. It has been shown that the interlamellarly fixed or "depot" water could contribute to the moisturizing potential of dermatological and cosmetic formulations and affect the diffusion rate and penetration ability of topical actives [1–4, 11].

Recently, a novel class of nonionic sugar surfactants has been recommended for the stabilization of O/W dermocosmetic emulsions. Sugar emulsifiers are claimed to present an improvement to traditionally used polyoxyethylene derivatives and to be mild to the skin, with a large number of hydroxyl groups in their structure, capable of providing additional skin moisturization [6]. Amongst different types of sugar emulsifiers, the group of alkylpolyglucosides (APGs), also called glucolipids, has been the subject of recent research work [9, 10]. This study is focused on a particular representative of alkylpolyglucosides – cetearyl glucoside and cetearyl alcohol (INCI).

The mesomorphic behavior of the above surfactant strongly influences the colloidal structure of topical vehicles, particularly in terms of the water distribution mode within emulsion systems [11–13]. It was shown in a long-term *in vivo* efficacy study that a placebo sample based on an alkylpolyglucoside emulsifier possessed better hydration and barrier-repairing potential than a PEG emulsifier and even better than its urea-containing counterpart [5].

The aim of this study was to establish the impact of urea on the water distribution mode within a complex lamellar gel structure of cetearyl glucoside and cetearyl alcohol-based emulsions. In addition, it was of interest to evaluate the potential of this vehicle for controlled skin hydration and prolonged delivery of urea.

This chapter presents a comprehensive structural characterization of placebo and urea-containing (5%) samples based on cetearyl glucoside and cetearyl alcohol, with a fixed emulsifier/water ratio (1:10.43) and different contents of neutral oil (5, 10 and 20%). Polarization and electron microscopy, wide-angle x-ray diffraction (WAXD), rheology measurements and thermogravimetry were employed in the physicochemical characterization of the colloidal emulsion structure. In addition, the activity of urea was assessed by *in vitro* and *in vivo* studies. The *in vitro* release profile of urea was evaluated using enhancer diffusion cells, while an *in vivo* skin hydration study was employed to assess the moisturizing capacity of test samples and trans-epidermal water loss (TEWL) from human skin.

12.2 Experimental

12.2.1 Materials

The alkylpolyglucoside emulsifier cetearyl glucoside and cetearyl alcohol (Montanov 68 PHA, kindly donated by Seppic, France) at 7% (w/w) and medium-chain triglycerides (Miglyol 812, Hüls, Germany), as an oil phase of moderate polarity, were used to make model placebo and active samples. Placebo samples varied in their oil content [5, 10 and 20% (w/w)] and were denoted M7Mg5, M7Mg10 and M7Mg20, respectively. Active samples contained 5% (w/w) of urea (Merck, Germany) and were designated M7Mg5U, M7Mg10U and M7Mg20U. All samples were prepared with doubly distilled water and preserved with 0.5% (w/w) Euxyl K 300 (Schülke & Mayr, Germany).

12.2.2 Preparation of Samples

Model placebo creams were prepared according to the method presented previously [11], whereas in active samples urea was added dissolved in the part of water phase incorporated upon emulsification at 60 °C.

12.2.3 Physicochemical Characterization

12.2.3.1 Microscopy

Both groups of samples were investigated 1 week after their preparation, employing both polarizing and light microscopy (Zeiss, Germany), and also transmission electron microscopy (TEM) (Leo, Germany).

12.2.3.2 Wide-Angle X-Ray Diffraction (WAXD)

To obtain structural information on the model samples, a short-range ordering was examined using WAXD measurements. Diffraction patterns were collected using an PW-1050/25 X-ray goniometer (Philips), coupled with a xenon-filled linear counter (Fuji, Japan). X-rays were produced by a PW-1730 X-ray generator (Philips), employing a copper anode (anode current 25 mA, λ 0.154 nm, accelerating voltage 40 kV). From diffraction angle data (θ), the values of the intermolecular distances were calculated according to Bragg's law.

12.2.3.3 pH Measurements

The pH values were measured directly in the samples by a standard potentiometric method using a glass pH electrode (HI 8417 pH-meter, Hanna Instruments, USA). Prior to measurements, the pH-meter was calibrated using pH 7.0 and pH 4.0 buffers.

12.2.3.4 Conductivity Measurements

These tests were performed in order to assess the type of emulsion and to follow any changes during its storage. A CDM 230 conductivity meter (Radiometer, Denmark) was used, calibrated with 0.01 M KCl solution at room temperature (20 ± 2 °C). Measurements were made directly in the samples.

12.2.3.5 Rheological Measurements

All measurements were carried out in triplicate, under the following conditions: cone-and-plate measuring system (diameter 40 mm, angle 1°), with sample thickness of 0.030 mm, at 20 ± 0.1 °C. A frequency sweep ramp from 0.1 to 10 Hz was performed at constant shear stress (6 Pa), previously established as a linear viscoelastic region for all samples. Storage (G') and loss (G'') modulus and phase angle (δ) were used for the characterization of test samples.

12.2.3.6 Thermogravimetric Analysis (TGA)

In order to differentiate bulk from fixed (interlamellar) water, measurements were conducted using a TG 220 instrument with a 5200 H disk station (Seiko, Japan). The measurements were performed using open aluminum pans in the temperature range 20–100 °C, at a heating rate of 2 °C min^{-1} (in triplicate). Additional TGA experiments were performed with chosen formulations in the isothermal mode at 32 °C (the skin temperature) for 30 min.

This approach was aimed at differentiating the free from interlamellarly bound water [14] and to provide additional information on the colloidal structure of the samples in the presence or absence of urea. Isothermal TGA can provide an estimation of the proportion of water that evaporates during the first 30 min after application on the skin. These results could then be compared with the changes in skin hydration and trans-epidermal water loss from *in vivo* studies.

12.2.4 *In Vivo* Short-Term Study

In order to establish the hydration potential of creams based on cetearyl glucoside and cetearyl alcohol, samples with different oil contents (5, 10, 20%), with and without urea, were tested in a short-term *in vivo* study. Furthermore, an *in vitro*/*in vivo* correlation was performed on the results of TGA analysis and the dynamics of water evaporation after application to the skin.

Placebo and active samples were tested on two groups of 10 healthy human volunteers, with no signs of dermatological problems and with an average age of 22.9 and 22.5 years, respectively. The *in vivo* study was approved by the local ethics committee and took place in December 2006.

12.2.4.1 Study Design

Following a 20-min acclimatization under laboratory conditions (21 ± 1 °C and $50 \pm 5\%$ RH), a quantity of 2 mg cm^{-2} was applied to the marked square sites of the volunteers' inner forearms ($P = 9$ cm^2). Two sites on each inner forearm were used, three being test samples and the fourth an untreated control. Group I was treated with placebo whereas group II received urea-containing samples. Skin hydration (SH) (using a Corneometer CM 825, Courage and Khazaka Electronics, Germany) and transepidermal water loss (TEWL) (using a Tewameter TM 210, Courage and Khazaka Electronics) were measured prior and 30 min after sample application, following published guidelines and documents [15, 16].

12.2.5 *In Vitro* Release Study

The release profile of urea was evaluated using the rotation paddle apparatus (DT70, Erweka, Germany), modified by addition of diffusion cells ($n = 6$) (Enhancer cell, VanKel Industries, USA), with a regenerated cellulose membrane (Cuprophan, Akzo, Germany) as a permeation medium. The experimental conditions were as follows: sample weight, 2 g; phosphate buffer pH 7.4 as a receiver fluid; temperature, 32 °C; rotation speed, 100 rpm; 5-mL aliquots; sink conditions at all times (6 h). Samples were filtered using a 0.45-μm MF-Millipore membrane filter (Millipore, USA). They were assayed for urea content after enzymatic degradation using urease, according to a standard BP 1998 spectroscopic method (Specol spectrophotometer, Karl Zeiss Jena, Germany). Data obtained as the concentration of urea (μg mL^{-1}) were then expressed as a percentage of urea released during given time intervals.

12.2.6 Statistical Analysis

Whenever applicable, data are presented as mean ± SD. Rheological and TGA results were analyzed using Student's *t*-test for dependent samples. Parameters from the two *in vivo* experiments (SH, TEWL) were expressed as percentage change of the second versus first (basal) measurement, compared with untreated control, and analyzed using Student's *t*-test for independent samples.

In vitro release data were also compared by Student's *t*-test for independent samples and then fitted using appropriate mathematical models for evaluation of drug release kinetics. In all cases the statistical significance was set at $p < 0.05$.

12.3 Results and Discussion

12.3.1 Physicochemical Characterization

In order to investigate the type of interaction between urea and the colloidal structure of emulsion samples, a series of tests were performed on both placebo and urea-containing samples: light and polarizing microscopy, pH, conductivity, WAXD and a complete rheological profiling. In addition, TGA of the samples at the temperature range 20–100 °C and an isothermal test at 32 °C were performed, in order to ascertain the proportion of free and fixed water within the two types of samples.

Polarizing micrographs (Figure 12.1) of the placebo and urea-containing samples with 20% of oil (M7Mg20 and M7Mg20U, respectively) reveal a lyotropic interaction at the emulsifier/oil/water interface. Both types of samples show anisotropic droplets, i.e. deformed Maltese crosses, which points to the existence of the lamellar liquid crystalline and/or gel crystalline phases [7].

In the case of light micrographs (Figure 12.2), no major difference could be seen between placebo and test samples in the size and distribution of droplets. Layers of gel network are noticeable between and around both single and flocculated oil droplets. The micrographs are similar to those obtained in our previous studies on alkylpolyglucoside-based emulsions [11–13]. It was not clear at this stage whether urea favors or obstructs the formation of the liquid crystalline/gel crystalline structure in these systems.

However, TEM images of the same set of samples (Figure 12.3) confirm the existence of the complex lamellar gel structure, especially pronounced in the case of active sample (Figure 12.3b). This finding indicates that urea may be enhancing the gel network formation. Since TEM is mainly a qualitative method, further investigations were carried out to test this hypothesis.

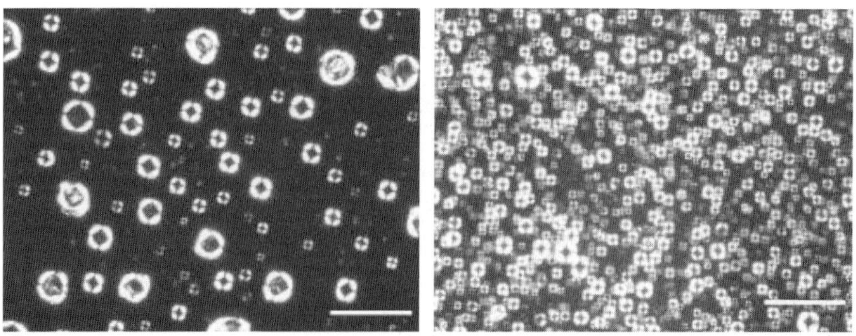

a) b)

Figure 12.1 Polarization micrographs of (a) M7Mg20 placebo and (b) active sample M7Mg20U; bar = 100 μm.

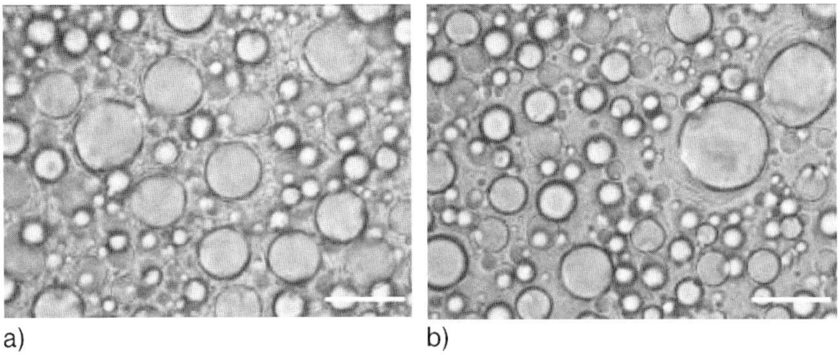

Figure 12.2 Optical micrographs of (a) M7Mg20 placebo and (b) active sample M7Mg20U; bar = 20 μm.

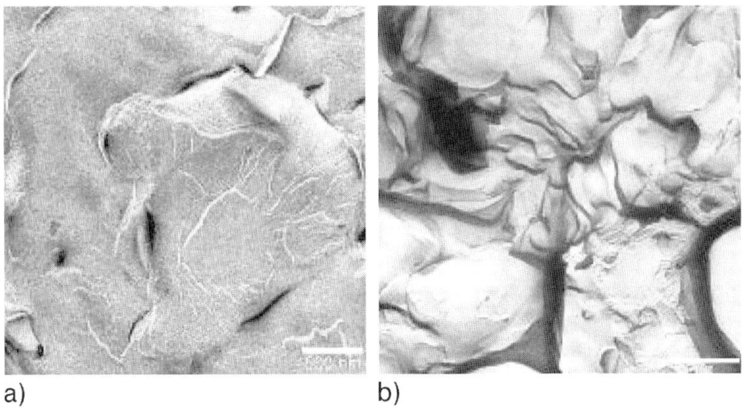

Figure 12.3 TEM images of (a) M7Mg20 placebo (bar = 500 nm) and (b) active sample M7Mg20U (bar = 100 nm).

Analysis of the WAXD patterns of the same set of samples (Figure 12.4) revealed some clear differences between them. A single sharp reflection in the range 0.41–0.42 nm is a sign of the existence of an α-crystalline gel phase within the system. In addition, a diffuse band ("halo") at 0.45 nm reveals liquid crystalline structure [18, 19]. In the placebo sample (Figure 12.4a), the WAXD pattern shows a possible overlapping of the diffuse band at 0.45 nm with a sharp interference at 0.415 nm, reflecting a complex colloidal structure, consisting of the ordered lamellar gel phase and the phase of liquid crystals. In the urea-containing sample (Figure 12.4b), the WAXD pattern reveals a more rigid structure, with fewer domains of lamellar liquid crystals, which correlates with the observations from TEM images.

The above finding also correlates with the results of specific conductivity and rheological measurements (Tables 12.1 and 12.2). From Table 12.1, it can be seen

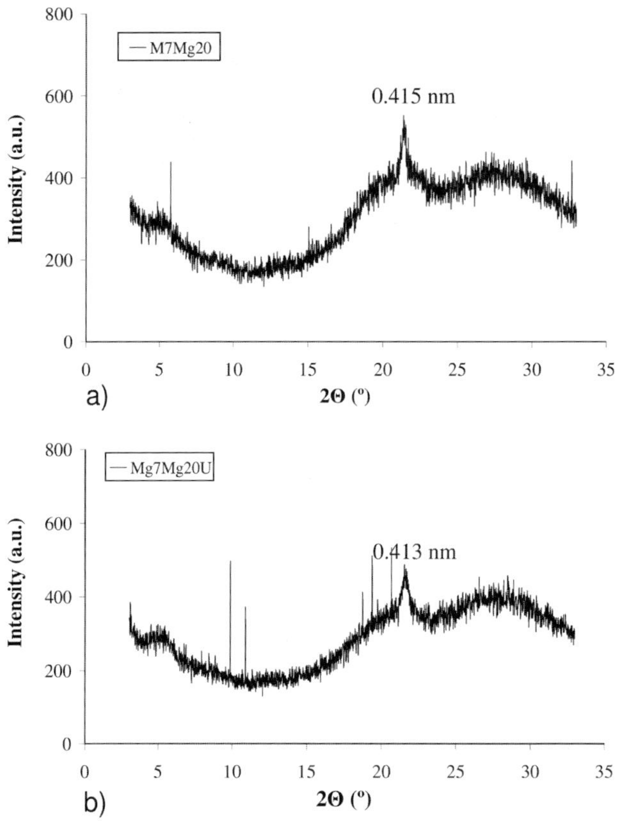

Figure 12.4 WAXD patterns of (a) urea-free sample (M7Mg20) and (b) urea-containing sample (M7Mg20U).

that the pH values of all placebo samples were between 6.07 and 6.47, whereas the pH of active samples was, as expected, higher (7.74–7.81), due to the alkaline nature of urea.

Specific conductivities of the three placebo samples were within the range 5.57–6.15 $\mu S\ cm^{-1}$. According to some authors [20, 21], conductivity values depend mainly on the fraction of free water within the system, which would mean that the three samples contain similar amounts of free water. We have shown in a previous study [11] that the amount of oil phase could strongly affect the structuring of the emulsion system and the water distribution within it. The addition of oil favors the formation of a colloidal structure, which in turn provides a higher fraction of the secondary, especially interlamellarly bound water [11]. This confirms that conductivity measurement alone is not a reliable indicator of the differences in the emulsion microstructure. As expected, the addition of urea, a water-soluble and ionizable substance, caused an increase in specific conductivity (Table 12.1), but still in the domain characteristic for multiphase/mixed systems.

Table 12.1 pH and conductivity values of samples M7Mg5–M7Mg20U.

Sample	pH	Specific conductivity (μS cm^{-1})
M7Mg5	6.07	5.57
M7Mg5U	7.74	30.75
M7Mg10	6.29	6.15
M7Mg10U	7.80	29.35
M7Mg20	6.47	6.05
M7Mg20U	7.81	14.56

Conductivity values of the active samples with 5 and 10% of oil phase were similar, but the sample with 20% of oil showed a considerably lower value (Table 12.1), despite the same content of urea. This indicates a different water distribution pattern within that system, accompanied by a different partitioning of the ionized urea. According to Kohronen et al. [20], higher conductivity values indicate a higher content of free (bulk) water, at the expense of bound water, within the system. However, because of the ionic nature of urea, we need further evidence before concluding that active samples with 20% oil contain a considerably higher fraction of bound water in comparison with the samples with 5 and 10% oil.

The results of oscillatory rheological measurements are shown in Table 12.2. The phase angle values obtained in this study were within the range reported previously [11], with the exception of the urea-containing sample with 5% oil. Relatively low phase angle values indicate a predominantly elastic, as opposed to viscous, nature of the semisolid system. In the case of M7Mg5U, the phase angle of nearly 45° points at the balance between elastic and viscous components [22] and reveals a significant difference in the structure from the corresponding placebo sample, due to addition of urea.

Table 12.2 Oscillatory parameters of placebo and urea-containing samples at a frequency of 1 Hz.

Sample	δ (°)	G'	G''
M7Mg5	10.2 ± 0.1	733.94 ± 23.23	132.68 ± 9.40
M7Mg5U	44.9 ± 0.5[a)]	172.67 ± 10.20[a)]	172.03 ± 9.11[a)]
M7Mg10	9.80 ± 0.04	860.27 ± 18.12	149.02 ± 7.10
M7Mg10U	11.80 ± 0.05	614.63 ± 11.28[a)]	128.70 ± 7.13
M7Mg20	11.30 ± 0.10	1495.30 ± 45.15	299.45 ± 10.10
M7Mg20U	11.70 ± 0.02	1186.80 ± 21.13[a)]	246.42 ± 9.21[a)]

a) Statistically significant difference from placebo, $p < 0.05$.

The remaining two active samples (M7Mg10U and M7Mg20U) both show a decrease in elastic moduli (Table 12.2), but the phase angle, as the overall measure of viscoelasticity, stayed low. This indicates that the elastic lamellar α-crystalline and/or liquid crystalline gel was still present after the addition of urea in these two systems.

It is well known that a more pronounced elastic than viscous component indicates the predominance of a lamellar liquid crystalline phase within the structure [23, 24]. A significant decrease in the elastic (G') and a lesser decrease in the of viscous modulus (G'') (not a uniform trend) recorded in urea-containing samples could implicate an interaction with urea, resulting in a less pronounced liquid crystalline phase within the system. It is possible that urea causes a certain dehydration of cetearyl alcohol semihydrates within the lipophilic gel, resulting in a transition of liquid crystalline phase to crystalline gel.

Moreover, it is possible that the balance that existed between different fractions of the fixed and bulk (free) water changed in the presence of a highly hydrophilic substance such as urea. It is expected that urea will be hydrogen bonded with water molecules. At the same time, the glucopyranoside part of the alkylpolyglucoside surfactant, incorporated mostly within the hydrophilic gel, has a greater affinity to hold water than fatty alcohols, present mostly in the lipophilic gel [11]. Consequently, it can be speculated that in all active samples, independent of the oil content, the fraction of interlamellarly fixed water (hydrophilic gel) has increased.

TGA provided additional information about the emulsion structure of the test samples. Table 12.3 shows the total and partial weight loss from all samples, and Figure 12.5 presents the total and derivative curves for the placebo and active samples with 20% oil. The results of isothermal TGA at 32 °C shown in Table 12.4 were obtained in order to estimate the percentage of water still remaining in the system after 30 min and hence available for skin hydration.

Table 12.3 Percentage water loss over specified temperature ranges for placebo and urea-containing samples.

Sample	TG (%)	20–50 °C (%)	50–70 °C (%)	70–100 °C (%)
M7Mg5	84.1 ± 3.2	27.25 ± 1.12	56.08 ± 2.34	1.61 ± 0.03
M7Mg5U	80.9 ± 2.3	21.06 ± 0.78	51.05 ± 2.12	9.40 ± 0.71[a]
M7Mg10	80.8 ± 1.9	25.45 ± 1.03	52.22 ± 0.98	3.83 ± 0.11
M7Mg10U	76.6 ± 2.6	20.71 ± 0.67	46.57 ± 0.78[a]	9.68 ± 0.24[a]
M7Mg20	65.8 ± 2.1	18.92 ± 0.69	32.81 ± 074	14.40 ± 0.81
M7Mg20U	66.2 ± 1.7	17.94 ± 0.46	25.11 ± 0.44[a]	23.54 ± 0.99[a]

a) Statistically significant difference from placebo, $p < 0.05$.

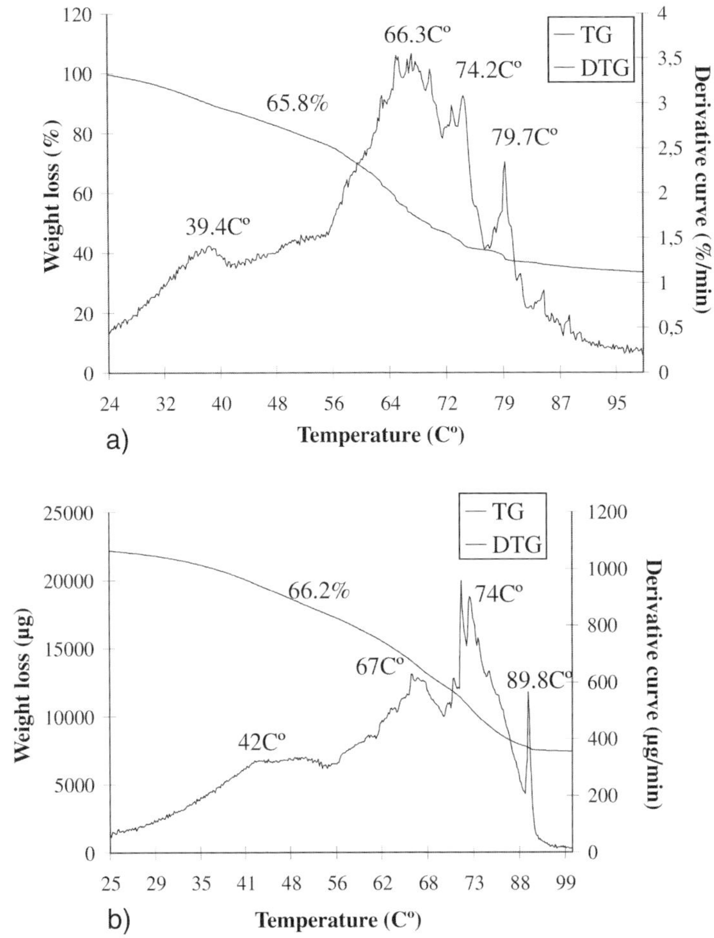

Figure 12.5 TG and DTG profiles: (a) sample M7Mg20 and (b) sample M7Mg20U.

Table 12.4 Isothermal water loss from placebo and urea-containing samples (32 °C; 30 min).

Sample	Water loss (%)	RD–FW (%)[a]	Rest of water (%)
M7Mg5	35.9 ± 0.5	+8.65	48.2
M7Mg5U	32.6 ± 0.2	+11.54	48.3
M7Mg10	30.6 ± 0.3	+5.15	50.2
M7Mg10U	39.7 ± 0.7	+19.53	36.9
M7Mg20	28.4 ± 0.5	+9.48	37.4
M7Mg20U	42.0 ± 1.3	+24.06	24.2

a) RD–FW, relative difference with respect to percentage of free water.

The TGA results reveal a clear difference in the level of structuration between placebo and active samples. The results obtained correlate well with our previous studies [11–13], which showed that the sample with the highest oil content had the highest fraction of interlamellarly bound water. It was found in this study that placebo sample M7Mg20 contained 14.4% interlamellarly bound water. The total mass (water) loss of this sample was 65.8%, of which 21.9% was bound water, 28.75% free water and the rest secondary water distributed within the lipophilic gel phase (calculated from Table 12.3). The corresponding active sample

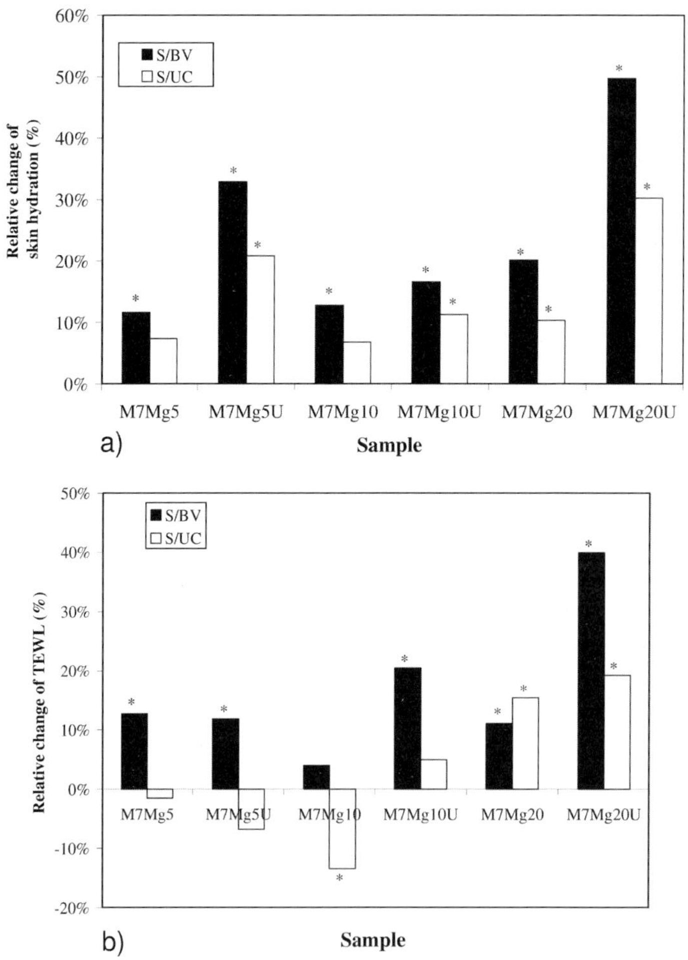

Figure 12.6 (a) Relative change in skin hydration (SH) 30 min after sample application with respect to basal value (S/BV) and to untreated control (S/UC). (b) Relative change of TEWL 30 min after sample application with respect to basal value (S/BV) and to untreated control (S/UC). * means significant difference, $p < 0.05$.

contained a significantly smaller fraction of secondary water and a higher fraction of interlamelarlly bound water (35.6% of the total water loss). These results correlate, to a certain extent, with conductivity measurements (Table 12.1).

The results of isothermal analysis (Table 12.4) revealed an unexpected trend that active creams lost proportionally more water than placebo creams. The lowest water loss during 30 min at 32 °C was detected from the placebo sample with 20% oil, whereas the highest loss was from its urea-containing counterpart. Overall, the results reveal that, in addition to bulk water, a certain amount of secondary water (from the lipophilic gel phase) is also lost and that this loss is considerably smaller from placebo samples. In the case of active sample M7Mg20U, for example, practically the whole amount of secondary water evaporates within the first 30 min after application (Table 12.4).

The results of the *in vivo* short-term application study (Figure 12.6) show a general trend of increases in both TEWL and skin hydration values 30 min after application in all samples. However, the increase in skin hydration is significantly higher after the application of active samples, especially in the case of the cream with 20% oil (Figure 12.6a). The results obtained from the *in vivo* study correlate fairly well with the isothermal TGA data (Table 12.4), although there are some exceptions. In principal, the correlation is much better for the urea-containing samples. This can be related to the hygroscopic nature of urea, whose effect on the skin is detectable 30 min after application.

Figure 12.7 and Tables 12.5 and 12.6 show the results of urea release from the active samples. The highest rate of release, and also the highest flux and highest cumulative amount released in 6 h, were achieved from the lowest viscosity formulation, with 5% oil (Table 12.5 and Figure 12.7). The second best in the release of urea was a sample with 20% oil, followed by that with 10% oil, even though the latter was less viscous. In all three cases, the release pattern follows first-order kinetics, with a correlation coefficient of $r > 0.9$ (Table 12.6). This find-

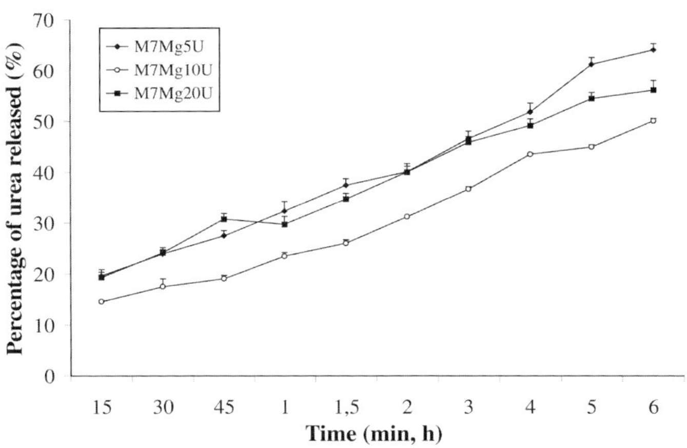

Figure 12.7 Percentage of urea released during 6 h; $n = 6$.

Table 12.5 Release parameters of urea from samples with different oil contents.

Sample	Flux ($mg\ cm^{-2}\ h^{-1}$)	Release rate ($mg\ cm^{-2}\ h^{-1/2}$)	Total amount of released urea, $t = 6$ h ($mg\ cm^{-2}$)
M7Mg5U	7.30 ± 5.30	7.57 ± 1.01	16.01 ± 3.26
M7Mg10U	5.40 ± 3.86	5.66 ± 0.64	12.52 ± 2.89
M7Mg20U	7.19 ± 5.38	7.33 ± 1.32	14.03 ± 3.21

Table 12.6 Release kinetics of urea from samples with different oil contents.

Sample	Zero-order kinetics	First-order kinetics	Hixon–Crowell model	Huguchi model
M7Mg5U	0.8956	0.9489	0.9367	0.8689
M7Mg10U	0.9799	0.9886	0.9867	0.6345
M7Mg20U	0.9613	0.9897	0.9756	0.7098

ing indicates that the processes of diffusion and release of urea were largely controlled by the colloidal structure of the system [4].

It is expected that urea, as a polar substance hydrogen bonded with water, becomes incorporated into interlamellar spaces within the lamellar liquid crystalline phase, around dispersed oil droplets and inside the hydrophilic lamellar gel phase. The results indicate that part of the urea does undergo the above process, causing an increase in the fraction of interlamellarly bound water in all active samples (Table 12.3). This means that part of urea is "fixed" within interlamellar spaces [2–4], whereas the other part is dispersed in the free (bulk) water or in the water present in the lipophilic gel phase (and partly used for the hydration of fatty alcohol). The "free" urea is readily available for release, due to its having a higher diffusion coefficient than the "fixed" urea [8].

The release profile of urea during the first 30 min (Figure 12.7) was in correlation with the results of isothermal TGA (Table 12.4) and with *in vivo* short-term hydration effects of active samples (Figure 12.6). This supports previous findings [6–8] that bulk water and part of the secondary water (from the lipophilic gel) evaporate shortly after application to the skin, drawing a portion of dissolved urea. However, depending on the oil content within the sample, between 40 and 50% of urea remained unreleased after 6 h (probably entrapped within interlamellar water) (Figure 12.7). Interlamellarly fixed water and, in the case of active samples, part of urea within it, seem to remain available for prolonged skin delivery. This is very important from the dermato-cosmetic point of view, since prolonged hydration is essential for the treatment of dermatoses accompanied by severe dryness.

12.4 Conclusion

It has been shown that a hydrophilic topical active urea interacts with the alkylpolyglucoside-type emulsifier cetearyl glucoside and cetearyl alcohol and the medium-chain triglyceride oil, affecting the colloidal structure of the emulsion system. The effect is dependent on the concentration of the oil. In principle, urea increases the fraction of interlamellar water, especially water bound to the liquid crystalline lamellae situated at the edges of oil droplets. In addition, urea affects the distribution of other water fractions within the complex colloidal system.

The overall effect of the observed structural changes has shown to be two-fold:

1. An increased amount of water was lost within the first 30 min after application, in comparison with the placebo samples. This helps achieve an efficient cooling and immediate hydrating effect, important in the treatment of very dry skin.
2. An increased amount of interlamellarly bound water was detected. This portion of water is not lost within the first 30 min and it has a potential for prolonged release of hydrophilic actives.

The study has confirmed that cetearyl glucoside and cetearyl alcohol forms a stable emulsion system, with a potential for prolonged/controlled release of hydrophilic topical actives.

References

1 Müller-Goymann CC, Frank SG Interaction of lidocaine and lidocaine-HCl with the liquid crystal structure of topical preparations. *Int. J. Pharm.* 1986; **29**: 147–159.

2 Farkas E, Zelko R, Nemeth Z, Palinkas J, Marton S, Racz I. The effect of liquid crystalline structure on chlorhexidine diacetate release. *Int. J. Pharm.* 2000; **193**: 239–245.

3 Farkas E, Zelko R, Torok G, Racz I, Marton S. Influence of chlorhexidine species on the liquid crystalline structure of vehicle. *Int. J. Pharm.* 2001; **213**: 1–5.

4 Makai M, Csanyi E, Nemeth Z, Palinkas J, Erős I. Structure and drug release of lamellar liquid crystals containing glycerol. *Int. J. Pharm.* 2003; **256**: 95–107.

5 Savic S, Tamburic S, Savic M, Cekic N, Milic J, Vuleta G. Vehicle-controlled effects of urea on normal and SLS-irritated skin. *Int. J. Pharm.* 2004; **271**: 269–280.

6 Aikens PA, Friberg SE. Emulsifiers. In *Dry Skin and Moisturizers*, Lodén M, Maibach HI (eds.), CRC Press LLC, Boca Raton, FL, 2000, pp. 183–201.

7 Eccleston GM. The importance of mesomorphic (lamellar) phases in emulsion stability. *J. Cosmet. Sci.* 2001; **52**: 142–143.

8 Junginger HE. Multiphase emulsions. In *Surfactants in Cosmetics*, Rieger MM,

Rhein LD (eds.), Marcel Dekker, New York, 1997, pp. 155–182.

9 Holmberg K. Natural surfactants. *Curr. Opin. Colloid Interface Sci.* 2001; **6**: 148–159.

10 Stubenrauch C. Sugar surfactants – aggregation, interfacial and adsorption phenomena. *Curr. Opin. Colloid Interface Sci.* 2001; **6**: 160–170.

11 Savic S, Vuleta G, Daniels R, Müller-Goymann CC. Colloidal microstructure of binary systems and model creams stabilized with an alkylpolyglucoside nonionic emulsifier. *Colloid Polym. Sci.* 2005; **283**: 439–451.

12 Savic SD, Savic MM, Vesic SA, Vuleta GM, Müller-Goymann CC. Vehicles based on a sugar surfactant: Colloidal structure and its impact on *in vitro*/*in vivo* hydrocortisone permeation. *Int. J. Pharm.* 2006; **320**: 86–95.

13 Savic S, Savic M, Tamburic S, Vuleta G, Vesic S, Müller-Goymann CC. An alkylpolyglucoside surfactant as a prospective pharmaceutical excipient for topical formulations: the influence of oil polarity on the colloidal structure and hydrocortisone *in vitro*/*in vivo* permeation. *Eur. J. Pharm. Sci.* 2007; **30**: 441–450.

14 Müller-Goymann CC, Alberg U. Modified water content containing hydrophilic ointment with suspended hydrocortisone-21-acetate – the influence of the microstructure of the cream on the *in vitro* drug release and *in vivo* percutaneous penetration. *Eur. J. Pharm. Biopharm.* 1999; **47**: 139–143.

15 Berardesca E. EEMCO guidance for the assessment of stratum corneum hydration: electrical methods. *Skin Res. Technol.* 1997; **3**: 126–132.

16 Rogiers V, EEMCO Group. EEMCO guidance for the assessment of transepidermal water loss in cosmetic sciences. *Skin Pharmacol. Appl. Skin Physiol.* 2001; **14**: 117–128.

17 Knorst MT, Neubert R, Wohlrab W. Analytical methods for measuring urea in pharmaceutical formulations. *J. Pharm. Biomed. Anal.* 1997; **15**: 1627–1632.

18 Fairhust CE, Fuller S, Gray J, Holmes MC. Lyotropic surfactant liquid crystals. In *Handbook of Liquid Crystals*, Vol. 3, Demus D, Goodby J, Gray GW, Spiess HW, Vill V (eds.), Wiley-VCH, Weinheim, 1998, pp. 341–392.

19 Krog N, Lauridsen JB. Food emulsifiers and their assocoations with water. In *Food Emulsions*, Friberg S (ed.), Marcel Dekker, New York, 1976, pp. 67–139.

20 Korhonen M, Niskanen H, Kiesvaara, J, Yliruusi J. Determination of optimal combination of surfactants in creams using rheology measurements. *Int. J. Pharm.* 2000; **197**: 143–151.

21 Korhonen M, Hellen L, Hirvonen J, Yliruusi J. Rheological properties of creams with four different surfactant combinations – effect of storage time and conditions. *Int. J. Pharm.* 2001; **221**: 187–196.

22 Adeyeye MC, Jain AC, Ghorab MKM, Reilly WJ. Viscoelastic evaluation of topical creams containing microcrystalline cellulose/sodium carboxymethyl cellulose as stabilizer. *AAPS Pharm. Sci. Technol.* 2002; **3** (2 article 8) (available at http://www.aapspharmaceutica.com).

23 Robles-Vasquez O, Corona-Galvan S, Soltero JFA, Puig JE, Tripodi SB, Valles E, Manero O. Rheology of lyotropic liquid crystals of Aerosol OT. II. High concentration regime. *J. Colloid Interface Sci.* 1993; **160**: 65–71.

24 Nemeth Z, Halasz L, Palinkas J, Bota A, Horanyi T. Rheological behavior of a lamellar liquid crystalline surfactant–water system. *Colloids Surf. A* 1998; **145**: 107–119.

13 Models for the Calculation of Sun Protection Factors and Parameters Characterizing the UVA Protection Ability of Cosmetic Sunscreens

Bernd Herzog

Abstract

For the computational simulation of sun protection factors (SPF) and parameters characterizing the UVA protection by sunscreens, above all the overall UV spectrum of the sunscreen formulation resulting from the amounts of the UV filters present and their spectroscopic properties has to be considered. The second major influence is due to the irregular profile of a sunscreen film on the skin. The optical transmission of an absorbing film with an irregular profile is always higher than that of a film with a uniform shape with the same average thickness. Because the inverse of the transmission of UV radiation through a sunscreen film is directly related to the SPF, the film profile can have a dramatic influence on sunscreen efficacy. Various models for the simulation of the irregular film structure have been suggested, such as two-step film, four-step film, quasi-continuous step film and continuous height distribution models. With some approaches, parameters characterizing the model film structure were adjusted by fitting to data from either in vitro measurements or from in vivo SPF determinations. In further refinements, the photostability of the UV-filter mixture in the sunscreen formulation has been considered, in addition to the distribution of the extinction in the oil and the water phase of the emulsion. Taking all these influences into account, the performance of most sunscreens can be simulated satisfactorily. This holds true not only for the SPF but also for parameters characterizing the UVA protection.

13.1 Introduction

It is a well-known fact that an overexposure of human skin to UV light may lead to sunburn and an increased risk of skin cancers. These effects are mainly attributed to the UVB part of the solar spectrum (290–320 nm), but the UVA portion (320–400 nm) is also an important factor. UVA contributes to the development

Colloids and Interface Science Series, Vol. 4
Colloids in Cosmetics and Personal Care. Tharwat F. Tadros

ISBN: 978-3-527-31464-5

of erythema to about 15% of the corresponding effect of the overall solar UV radiation. It penetrates deeper into the skin than UVB and therefore is considered to contribute also to premature skin aging [1–3]. Concerning the induction of skin cancers, an additional effect of UVA is also under discussion [4]. The sun protection factor (SPF) is merely a measure of sunburn prevention and is defined as the ratio of the minimal erythemal doses (MED) of simulated solar radiation directed to human skin in the presence and absence of a sunscreening agent [5]. Based on this definition, *in vivo* methods for sunscreen testing on volunteers have been established [6–8] which are applied when SPF claims on sunscreen products are made.

However, for purposes of experimental screening *in vitro* methods for the determination of the SPF have been introduced [9–12], since testing on volunteers is time consuming and expensive. *In vitro* methods are based on the assumption that the UV protection of sunscreens is merely caused by the attenuation of UV light according to the absorption characteristics and concentrations of the UV absorbers used in the sunscreen formulation. Any further effects that may be of relevance *in vivo*, such as anti-inflammatory or anti-oxidative properties, are not considered in the *in vitro* methods. In most cases *in vitro* methods model in some way the inhomogeneous surface structure of the human skin by using appropriate substrates such as Transpore tape, poly(methyl methacrylate) (PMMA) plates or quartz-plates with a rough surface [9]. This is important because the optical transmission of an absorbing film of uniform thickness is lower than that of a corresponding irregular film of the same average thickness.

Although the sun protection factor depends strongly on the amounts and efficacy of the UV absorbers used, it is not unequivocally related to the overall UV spectrum of a sunscreening formulation. For that reason, the protection against UVA has to be assessed with a further measure specific for this spectral range. The most frequently used *in vivo* method is persistent pigment darkening (PPD), where irradiation of volunteers with a pure UVA light source induces pigmentation. In analogy with the SPF, the UVA protection factor (UVA-PF) is defined as the ratio of the minimal doses of UVA leading to an effect in the presence and absence of a sunscreening agent [13]. There are several *in vitro* methods for the assessment of UVA protection. For the Australian Standard a threshold level of reduction of the UV transmission has to be achieved [14]. The UVA/UVB ratio relates the level of protection in the UVA to that in the UVB range [15]. The critical wavelength works in a similar way, while being less sensitive than the UVA/UVB ratio. The level of UVA protection may also be assessed with an *in vitro* spectroscopic measurement in relation to the SPF [16, 17]. On this basis, a new COLIPA method has been established [18]. As with *in vitro* SPF testing, the *in vitro* measurements for assessment of UVA protection are performed on substrates with an irregular surface structure.

For the computational simulation of SPF and UVA parameters, the following influences have to be taken into account:

1. the overall UV spectrum of the sunscreen formulation resulting from the amounts and spectroscopic properties of the UV filters inside;
2. the irregularity of the sunscreen film on the skin.

These two points are of major importance. In addition, two further properties have been discussed:

3. the photostability of the UV filter mixture in the sunscreen formulation;
4. the distribution of the extinction over the water and the oil phases of the emulsion, according to the presence and amounts of water- and oil-soluble (or dispersible) UV filters.

13.2
Basic Principle

The basic principle of SPF and UVA parameter calculations is the fact that the inverse of the UV transmission through an absorbing layer, $1/T$, is the factor by which the intensity of the UV light is reduced. Thus, at a certain wavelength λ, $1/T(\lambda)$ is regarded as a monochromatic protection factor (MPF). Since the spectral range relevant for the *in vivo* SPF is between 290 and 400 nm, the monochromatic protection factors have to be averaged over this range. This average must be weighted using the intensity of a standard sun, $S_s(\lambda)$, and the erythemal action spectrum, $s_{er}(\lambda)$, leading to Eq. (1), which was published first by Sayre et al. [19]

$$\mathrm{SPF} = \frac{\sum_{290}^{400} s_{er}(\lambda) S_s(\lambda)}{\sum_{290}^{400} s_{er}(\lambda) S_s(\lambda) T(\lambda)} \tag{1}$$

Data for $S_s(\lambda)$ and $s_{er}(\lambda)$ are available in the literature [9, 20], but $T(\lambda)$ has to be determined for the respective sunscreen. This can be done either via trans-

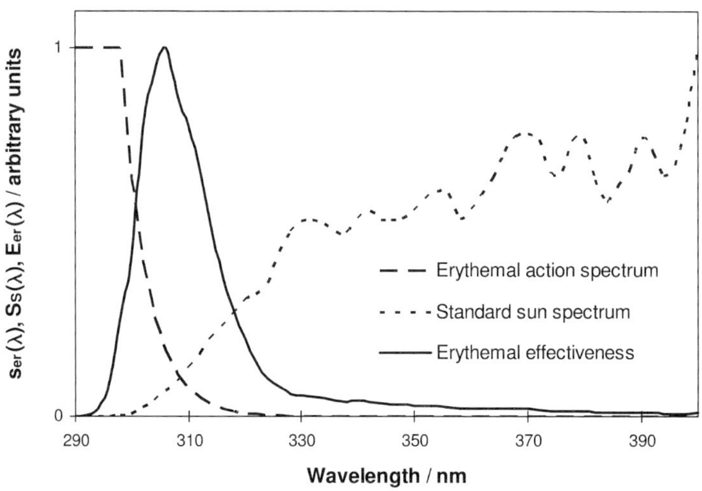

Figure 13.1 Erythemal action spectrum, standard sun spectrum and spectrum of erythemal effectiveness.

mission measurements with special UV spectrometers [21] and using substrates with a rough surface [22] or via the calculation of transmission. The product of $S_s(\lambda)$ and $s_{er}(\lambda)$ is called the erythemal effectiveness spectrum or erythemally effective irradiance, E_{er}. These three quantities are illustrated in Figure 13.1.

13.3 Calculation of the Overall UV Spectrum of a Sunscreen Agent

In order to obtain the average extinction coefficient of a UV absorber mixture, it is first necessary to calculate the average molecular weight and the average molar concentration. For this purpose, one has to know the UV spectra of the UV absorbers between 290 and 400 nm in terms of the molar decadic extinction coefficients and the concentration of the UV absorbers in weight per unit volume, β_i. The average molar weight can be obtained via

$$\bar{M} = \frac{\sum_{i=1}^{n} \beta_i}{\sum_{i=1}^{n} \frac{\beta_i}{M_i}} \tag{2}$$

and the average molar concentration, $\bar{c}$, according to

$$\bar{c} = \frac{10 \sum_{i=1}^{n} \beta_i}{\bar{M}} \tag{3}$$

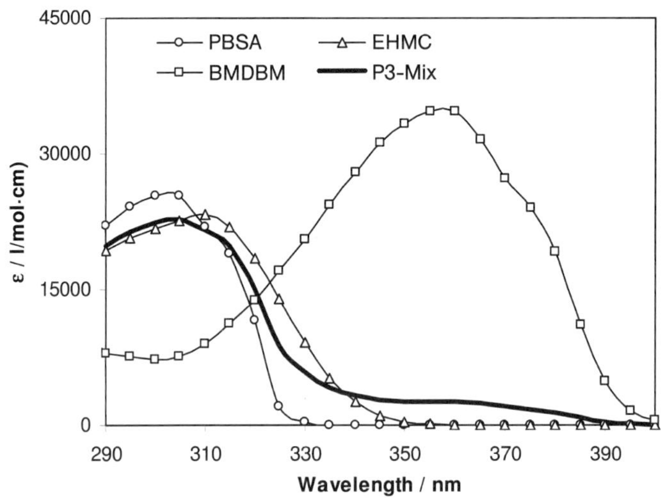

Figure 13.2 UV spectra of phenylbenzimidazolesulfonic acid (PBSA), ethylhexyl methoxycinnamate (EHMC), butylmethoxydibenzoylmethane (BMDBM) and the COLIPA P3 sunscreen standard formulation.

The average molar extinction coefficient, $\overline{\varepsilon(\lambda)}$, of this mixture then becomes [23]

$$\overline{\varepsilon(\lambda)} = \frac{\sum_{i=1}^{n} \frac{\varepsilon(\lambda)_i \beta_i}{M_i}}{\sum_{i=1}^{n} \frac{\beta_i}{M_i}} \qquad (4)$$

As an example, Figure 13.2 shows the extinction coefficients of the single filters inside the COLIPA P3 standard sunscreen formulation [24] and the average extinction coefficient of their mixture (COLIPA = European Cosmetics, Toiletries and Perfumeries Association).

13.4 Models for Film Irregularities

When measuring transmission with *in vitro* methods, in some way the irregular surface structure of the human skin has to be modeled by using appropriate substrates with a rough surface. This is important because the optical transmission of an absorbing film of uniform thickness is lower than the transmission of a corresponding irregular film of the same average thickness. This effect can be dramatic. It has also a strong impact when calculating transmissions employing the average UV spectrum of an absorber mixture according to

$$T(\lambda) = 10^{-\bar{\varepsilon}(\lambda)\bar{c}d} \qquad (5)$$

and using then the $T(\lambda)$ in Eq. (1) for calculation of the SPF. Since with an *in vivo* SPF test 2 mg cm^{-2} are distributed on the skin of the volunteer, which corresponds to a volume of approximately 2 μL cm^{-2}, the optical pathlength d should be set to 20 μm (= 0.002 cm). The resulting SPFs can be more than a factor of five higher than the corresponding *in vivo* values. Results consistent with that statement were reported by Ferrero et al. [25], who calculated the effective average film thickness for a range of 25 products to be only 14% of the thickness expected from the application amount (20 μm). This discrepancy could be resolved by introducing film irregularities by means of a mathematical operation. O'Neill was the first to publish such a model [26].

13.4.1 The Step Film Model by O'Neill

The derivation of the step film model presented by O'Neill [26] starts with a homogeneous film of absorbing material of a certain thickness *d* and a horizontal extension of 1. The derivation of the model is illustrated in Figure 13.3 as a two-dimensional sketch. A portion of the homogeneous film given by the horizontal extension g and the thickness *fd* is removed from its original position and added

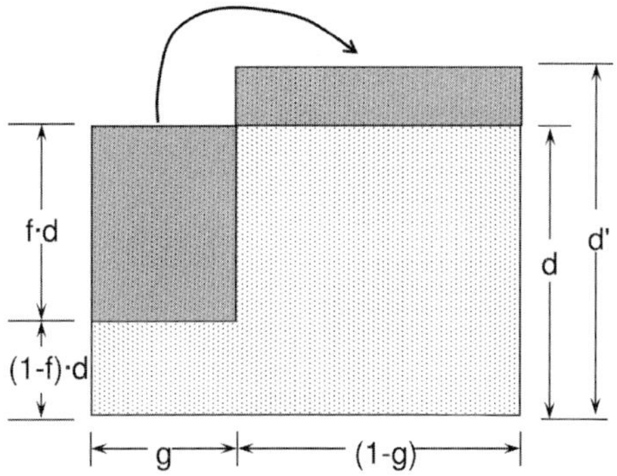

Figure 13.3 Illustration of O'Neill's step film model.

to the remaining thicker part of the film. This results in two film fractions of different thicknesses d' and $(1-f)d$ with horizontal extensions $1-g$ and g, respectively. The transformation of the film geometry is carried out under the condition that the amount of absorbing material stays constant, corresponding to a normalization condition. The expression for d' is then

$$d' = d\left(\frac{fg}{1-g} + 1\right) \tag{6}$$

The transmission of a step film $T_{SF}(\lambda)$ can be written as the sum of the transmissions through the two fractions of the film

$$T_{SF}(\lambda) = g \times 10^{-\bar{\varepsilon}(\lambda)\bar{c}\cdot d(1-f)} + (1-g) \times 10^{-\bar{\varepsilon}(\lambda)\bar{c}d[gf/(1-g)+1]} \tag{7}$$

where d is the average thickness of the step film. The expression $\bar{\varepsilon}(\lambda)\bar{c}d$ in the exponents of the terms in Eq. (7) is the extinction of the corresponding homogeneous film $E_{HF}(\lambda)$. The step film extinction is then given as

$$E_{SF}(\lambda) = \log \frac{1}{T_{SF}(\lambda)} \tag{8}$$

One can show that for $E_{HF} > 1$, Eq. (9) holds approximately (this is true for any wavelength)

$$E_{SF} = (1-f)E_{HF} - \log g \tag{9}$$

and for $E_{HF} \ll 1$, Eq. (10)

$$E_{SF} = E_{HF} \tag{10}$$

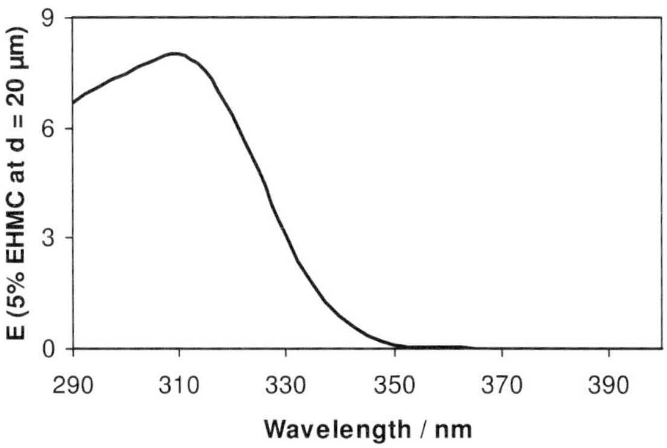

Figure 13.4 Parent film extinction spectrum of 5% EHMC.

Figure 13.4 shows the UV spectrum of 5% EHMC at a homogeneous optical pathlength of 20 µm. In Figure 13.5, a plot of E_{SF} as function of E_{HF} is shown for a range of EHMC concentrations between 0 and 5% at three wavelengths (310, 330 and 350 nm). E_{SF} was calculated with the step film parameters $g = 0.269$ and $f = 0.935$. Equations (9) and (10) are also visualized, Eq. (9) using the same values of g and f. This graph makes clear that for a given set of step film parameters, all spectral data for any type of UV absorber at any concentration or any wavelength will be located on one master curve. In other words, there is a unique function which describes the dependence of E_{SF} as a function of E_{HF} and the corresponding plot of both can be seen as the characteristic curve of the model. This very general relationship also holds true for other models of

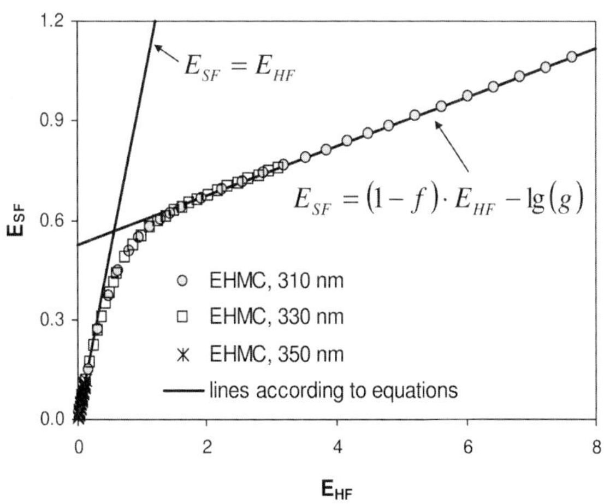

Figure 13.5 Characteristic curve of the step film model calculated with step film parameters $g = 0.269$ and $f = 0.935$.

film irregularity and was described for the first time by Ferrero and coworkers [33], who also demonstrated that experimental *in vitro* absorbance data of sunscreen products spread on roughened PMMA plates show similar behavior.

13.4.2 The Modified Version of the Step Film Model by Tunstall

Tunstall published in 2000 a modified version of the step film model. Instead of two, he introduced a model with four steps [27]. For the calculation of the transmission an expression corresponding to Eq. (5) was used. The total transmission $T_{SF}(\lambda)$ of the film is then given as

$$T_{SF}(\lambda) = \sum_n f_n T_n(\lambda) \tag{11}$$

where the $T_n(\lambda)$ is the transmittance corresponding to surface fraction f_n and thickness x_n, with $x_n \geq 0$. The following conditions must apply

$$\sum_n f_n = 1 \tag{12}$$

$$\sum_n f_n x_n = 20\ [\mu m] \tag{13}$$

the latter being a normalization condition for keeping constant the amount of absorbing material, whatever shape of film would be constructed. Hence there are eight parameters, which are needed to describe the film and two equations, thus leaving six independently adjustable parameters.

Tunstall used the approach to simulate *in vitro* SPF measurements in terms of MPF-spectra of formulations with different amounts of the UV absorber ethylhexyl methoxycinnamate (EHMC) on two different substrates, Vitro-Skin and Transpore Tape. In order to simulate the MPF spectra, film profiles were adjusted by least-squares fitting, resulting in profiles with characteristic differences for the two substrates. With Vitro-Skin a three-step model was sufficient for satisfactory curve fitting, but with Transpore Tape four steps appeared to be necessary.

Recently, Hewitt [28] published a modified four-step Tunstall model, where a thickness profile for formulations spread on skin was given. The corresponding transmission is given by the equation

$$\begin{aligned} T_{SF}(\lambda) = {} & 0.026 \times 10^{-\bar{\varepsilon}(\lambda)\bar{c}\times 0} + 0.202 \times 10^{-\bar{\varepsilon}(\lambda)\bar{c}\times 2.15\times 10^{-4}} \\ & + 0.076 \times 10^{-\bar{\varepsilon}(\lambda)\bar{c}\times 5.9\times 10^{-4}} + 0.696 \times 10^{-\bar{\varepsilon}(\lambda)\bar{c}\times 28\times 10^{-4}} \end{aligned} \tag{14}$$

A visualization of the Tunstall model making use of in Eq. (14) is given in Figure 13.6. Its characteristic curve is depicted in Figure 13.7. In order to use more general terms, the extinction of the step film is now designated $E_{\text{model film}}$ and the

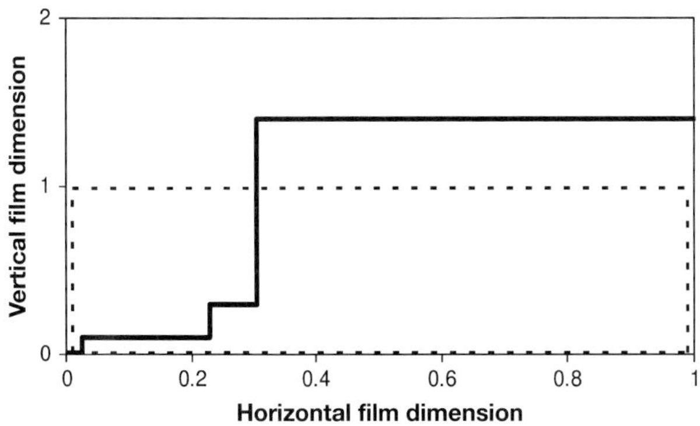

Figure 13.6 Illustration of the Tunstall model with the parameters used in Eq. (14).

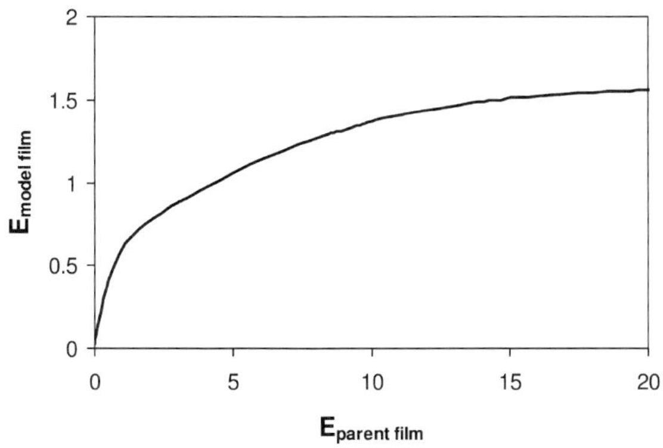

Figure 13.7 Characteristic curve of the Tunstall model with the parameters used in Eq. (14).

extinction of the corresponding homogeneous film $E_{\text{parent film}}$, a term which had already been coined by O'Neill [26]. Unlike it is the case with the two-step film model, there is no constant slope of the characteristic curve at high parent film extinctions with the Tunstall model. However, at high parent film extinctions, the extinction of the model film runs into saturation. This is due to the first term of Eq. (14), where the film height is set to zero, giving rise to a 2.6% fraction of completely uncovered film area. Hence there will remain at least 2.6% of transmission and the SPF calculated with this version of the Tunstall model can never reach higher than the inverse of 0.026, that is, 38.

13.4.3
The Calibrated Two-Step Film Model

It was suggested by Herzog in 2002 [23] to calibrate the parameters of the step film model by using *in vivo* SPF data. The basic idea is to vary the parameters g

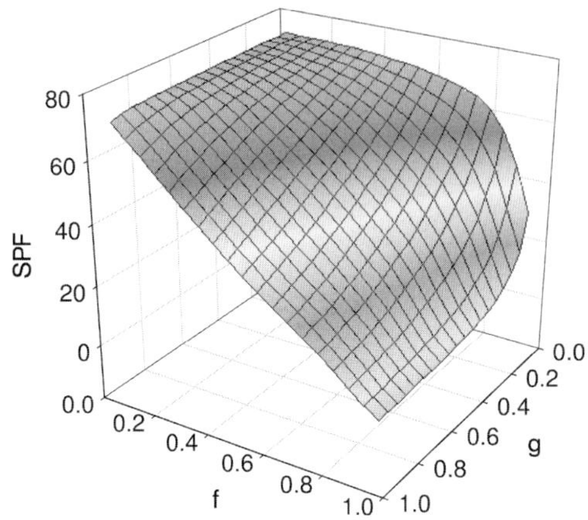

Figure 13.8 Matrix of SPF values calculated as function of step film parameters g and f for the COLIPA P3 standard.

and f of O'Neill's model and calculate the corresponding transmissions via Eq. (7). In order to perform this calculation, one needs access to quantitative UV spectroscopic data for the UV absorbers involved [23]. From the transmissions, the matrix of possible SPF values can be calculated by using Eq. (1). This is shown for the COLIPA P3 standard sunscreen formulation [24] in Figure 13.8. The term standard sunscreen here stands for a sunscreen, the SPF of which has been characterized according to the COLIPA or the International SPF method at least three times in independent investigations. For three sunscreen standards I, II and III, the matrix of calculated SPF values (SPF) is compared with the corresponding *in vivo* data, searching for the minimum of the sum of the squared deviations $\Delta_{g,f}$

$$\Delta_{g,f} = \left[\frac{SPF(\mathrm{I})_{g,f} - SPF(\mathrm{I})_{in\ vivo}}{SPF(\mathrm{I})_{in\ vivo}}\right]^2 + \left[\frac{SPF(\mathrm{II})_{g,f} - SPF(\mathrm{II})_{in\ vivo}}{SPF(\mathrm{II})_{in\ vivo}}\right]^2 + \left[\frac{SPF(\mathrm{III})_{g,f} - SPF(\mathrm{III})_{in\ vivo}}{SPF(\mathrm{III})_{in\ vivo}}\right]^2 \quad (15)$$

When using the standard sunscreen formulations described in Table 13.1, this procedure finally results in the step film parameters $g = 0.269$ and $f = 0.935$. The structure of this film is illustrated in Figure 13.9. The characteristic curve of this model has already been shown in Figure 13.5.

Reasonable agreement of calculated and *in vivo* SPF data from three different sources (in total 36 examples) was obtained when employing this model [23], showing for the first time that the simulation of *in vivo* SPF data is possible, when film irregularities are taken into account.

Table 13.1 Filter contents and SPF data for the standard sunscreens P1, P3 and CF4.[a)]

Standard sunscreen	Filter content	$SPF_{in\ vivo}$ ($\pm$CI)	$SPF_{calculated}$ with $g = 0.269, f = 0.935$
P1 (COLIPA)	2.7% EHMC	4.2 ± 0.2	5.0
P3 (COLIPA)	3% EHMC, 2.78% PBSA, 0.5% BMDBM	15.5 ± 1.5	10.9
CF4	5% EHMC, 10% MBBT	35.7 ± 3.2	38.5

a) CI = confidence interval; for other abbreviations, see Table 13.4.

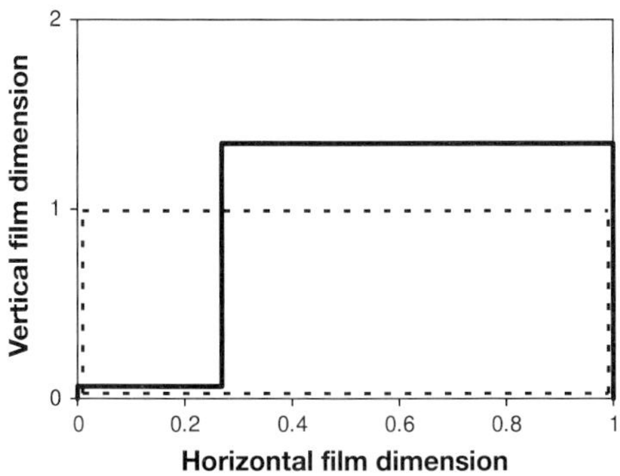

Figure 13.9 Illustration of the calibrated step film profile ($g = 0.269$ and $f = 0.935$).

13.4.4
The Calibrated Quasi-Continuous Step Film Model

Employing a quasi-continuous film model rather than a step function for describing the irregularity of the sunscreen film on human skin was realized with an exponential approach of the following kind [29, 30]

$$h(i) = A \exp\left[-B\left(\frac{i}{n}\right)^{C}\right] \tag{16}$$

with $i = 1, 2 \ldots, n$, where n is the number of steps the exponential function is cut into for numerical treatment and $h(i)$ is the height of the film at step i. B and C are parameters determining the shape of the film and A is introduced for normalization. The normalization condition is

$$\frac{A}{n}\sum_{i=1}^{n} \exp\left[-B\left(\frac{i}{n}\right)^{C}\right] = 1 \tag{17}$$

The transmission of the quasi-continuous step film can then be calculated as the sum of the transmissions through all steps of height $h(i)$ according to the equation

$$T_{\text{film}}(\lambda) = \frac{1}{n}\sum_{i=1}^{n} 10^{-\overline{\varepsilon(\lambda)}\bar{c}dh(i)} \tag{18}$$

The calibration procedure again makes use of the sunscreen standards and the shape parameters B and C are obtained by minimizing the sum of the squared deviations $\Delta_{B,C}$

$$\Delta_{B,C} = \left[\frac{SPF(\text{I})_{B,C} - SPF(\text{I})_{in\ vivo}}{SPF(\text{I})_{in\ vivo}}\right]^2 + \left[\frac{SPF(\text{II})_{B,C} - SPF(\text{II})_{in\ vivo}}{SPF(\text{II})_{in\ vivo}}\right]^2 + \left[\frac{SPF(\text{III})_{B,C} - SPF(\text{III})_{in\ vivo}}{SPF(\text{III})_{in\ vivo}}\right]^2 \tag{19}$$

An example of this model, calibrated to the P1, P3 and CF4 standard sunscreen formulations, is shown in Figure 13.10. The scale of the abscissa was reversed with respect to Eq. (17) in order to make the depiction more comparable to Figures 13.6 and 13.9. When using the mentioned standard sunscreens P1, P3 and CF4 for adjusting the model parameters, it was found that the parameter C was close to the value of 2, resulting in a Gaussian-type function [29]. Setting $C = 2$, the parameter B was determined as $B = 2.041$ with the normalization constant $A = 2.610$. The characteristic curve of this model is shown in Figure 13.11. As with the Tunstall model, there is a steady decrease in the slope with increasing parent film extinction, but no saturation.

The development of the quasi-continuous step film model is, like the approach of Tunstall, a refinement of O'Neill's two-step film model. Both are attempts to

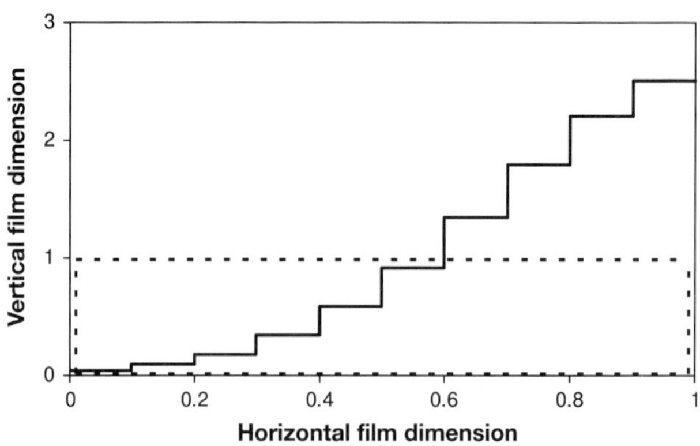

Figure 13.10 Illustration of the calibrated quasi-continuous step film profile ($C = 2$ and $B = 2.041$).

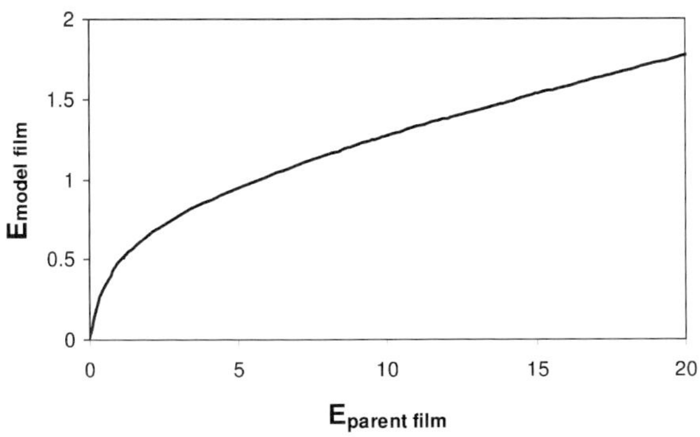

Figure 13.11 Characteristic curve of the calibrated quasi-continuous step film profile ($C = 2$ and $B = 2.041$).

create mathematical descriptions of sunscreen film irregularities on human skin that may be closer to reality. As the four-step Tunstall model has six independently adjustable parameters, it is fairly flexible, but it seems difficult to obtain an unequivocal solution. On the other hand, the quasi-continuous step film model has only two independent parameters, making an unequivocal solution easier. However, it is less flexible, since a function had been defined in advance and the question remains of whether this function is reflecting reality in a suitable way.

13.4.5 The Continuous Height Distribution Model Based on the Gamma Distribution

This model, published by Ferrero and coworkers in 2003, uses an approach common in surface metrology. In this case, the film profile becomes equivalent to the bearing area curve of Abbott and Firestone [31] and is constructed based on a cumulative distribution function F containing the film height h as a random variable [32]. Among possible probability functions, the authors chose a gamma law representing an asymmetric distribution and $f(h)$ is the associated probability density function

$$f(h) = \left(\frac{h}{b}\right)^{c-1} \mathrm{e}^{-h/b} \frac{1}{b\Gamma(c)} \tag{20}$$

where h is the random variable "relative height", c is the shape parameter to be adjusted, b is introduced for normalization and $\Gamma(c)$ is the value of the gamma function at c. The cumulative height distribution F is obtained by integration of $f(h)$ from 0 to ∞

$$F(h) = \int_0^{\infty} f(h)\,\mathrm{d}h \tag{21}$$

To realize the film thickness profile, the inverse of the gamma law is used, h being deduced from its cumulative distribution F, ranging from 0 to 1. According to the normalization to 1 of the area under the film profile, the following relationship should be checked after adjustment of the shape parameter c

$$\int_0^1 h(F)\,\mathrm{d}F = 1 \tag{22}$$

The transmission of the film is then calculated according to

$$T_{\text{film}}(\lambda) = \int_0^1 10^{-E(\lambda)h(F)}\,\mathrm{d}F \tag{23}$$

with $E(\lambda)$ representing the extinction of the regular parent film calculated according to Beer–Lambert law applied to the amount of UV filters deposited on unit area.

In Ref. [32], the model was adjusted to comply with *in vitro* transmission measurements of sunscreens spread at 1.2 mg cm^{-2} on roughened PMMA plates, which corresponds to experimental conditions previously selected to achieve similar *in vitro* and *in vivo* SPF values [33]. The best-fitting shape parameter c was determined as 1.105. The resulting film profile defined by Eq. (22) and the density function given in Eq. (20) is depicted for this case in Figure 13.12. The characteristic curve of that model with shape parameter $c = 1.105$ is depicted in Figure 13.13.

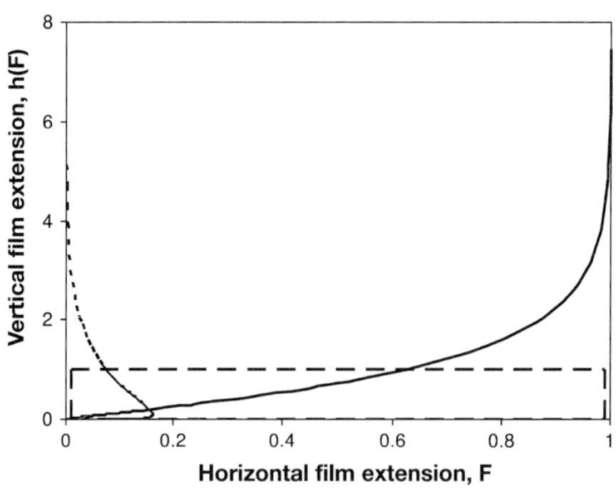

Figure 13.12 Illustration of the film profile of the gamma distribution model with $c = 1.105$. Dashed line: corresponding probability density function.

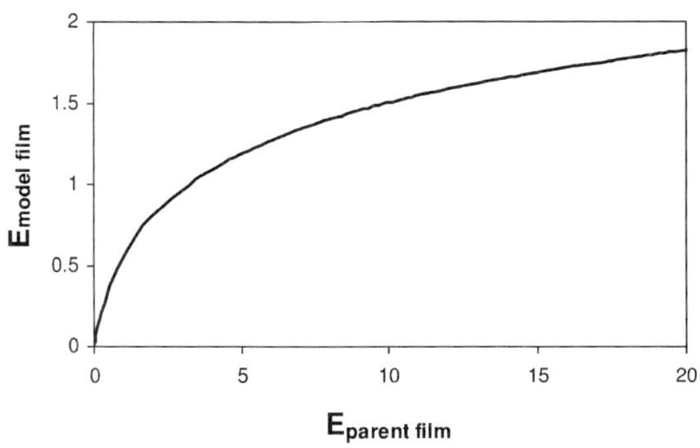

Figure 13.13 Characteristic curve of the gamma distribution model with $c = 1.105$.

13.4.6
Comparison of the Models

At this point, a first comparison of the models will be made on the basis of some sunscreen standard formulations. In addition to the compositions described in Table 13.1, the COLIPA standard P2 will be looked at and also the sunscreen formulation defined in the Japanese standard, which is used with the persistent pigment darkening (PPD) method [34]. The P2 and the JCIA standard are described in Table 13.2. The *in vivo* SPF of the JCIA standard had been determined according to the full COLIPA SPF method [6] independently in two different test centers. In one case the SPF was 8.2 and in the other 8.3.

Table 13.2 Filter contents and SPF data for the P2 and the JCIA standard sunscreens.[a]

Standard sunscreen	Filter content	$SPF_{in\ vivo}$ (±CI)
P2 (COLIPA)	7% OD-PABA, 3% B-3	12.7 ± 1.2
JCIA	3% EHMC, 5% BMDBM	8.2 ± 0.8

a) CI = confidence interval; for other abbreviations, see Table 13.4.

Table 13.3 shows the *in vivo* results for the standards P1, P2, P3, CF4 and of JCIA together with the corresponding results calculated according to the models previously described. All calculations were performed with an application rate of the sunscreen of 2 mg cm^{-2}, except with the gamma distribution, where the application rate was set to 1.2 mg cm^{-2}.

The results shown in Table 13.3 were obtained by using the models with the model parameters published by the respective authors. There have been different strategies for parameter finding, which in addition to the specific functionalities

Table 13.3 Results of SPF calculations using the models described in Sections 13.4.2–13.4.5 with examples of five sunscreen standard formulations.[a]

Standard sunscreen	$SPF_{in\ vivo}$ ($\pm$CI)	Four-step film model (Section 13.4.2)	Calibrated two-step film (Section 13.4.3)	Calibrated quasi-continuous step film (Section 13.4.4)	Gamma distribution (Section 13.4.5)
P1	4.2 ± 0.2	6.2	5.0	5.2	5.2
P2	12.7 ± 1.2	15.2	15.0	14.4	12.8
P3	15.5 ± 1.5	14.0	10.9	11.2	10.7
CF4	35.7 ± 3.2	30.8	38.5	36.4	25.2
JCIA	8.2 ± 0.8	16.6	11.3	12.9	13.4

a) CI = confidence interval.

of the characteristic curves may explain the deviations between the models. However, two features are common with all results:

1. The SPF of the JCIA standard is always overestimated. This could be explained by its strongly photounstable UV absorber composition. Photoinstabilities were so far not considered by any of the models.
2. The calculated SPF of the P2 standard sunscreen is in all cases higher than that of the P3 standard, whereas the *in vivo* observation is the opposite. This discrepancy could be attributed to the fact that the P3 standard contains filters in the oil and in the water phase, in contrast to the P2 standard, where all filters are in the oil phase. The distribution of the filters in the phases of the emulsion may give rise to different film profiles on the skin.

13.5 Taking Photoinstabilities into Consideration

Some of the frequently used UV filters undergo significant photodegradation under the conditions of use. For that reason, photoinstabilities may be taken into consideration using an approach published by Wloka et al. [35]. In this approach, the inverse of the dose-dependent SPF obtained from *in vitro* measurements, 1/SPF(dose), is plotted against the irradiation dose (in minimal erythemal doses, MED). The area under this curve is the erythemally weighted irradiation dose. Thus, when this area becomes unity, 1 MED has been transmitted through the sunscreen film and this corresponds exactly to the principle of the *in vivo* SPF measurement. The SPF can then be read from the respective UV dose given on the abscissa. The principle is demonstrated in Figure 13.14 with a photostable example sunscreen of SPF = 8. In that case 1/SPF(dose) = 0.125 at any dose

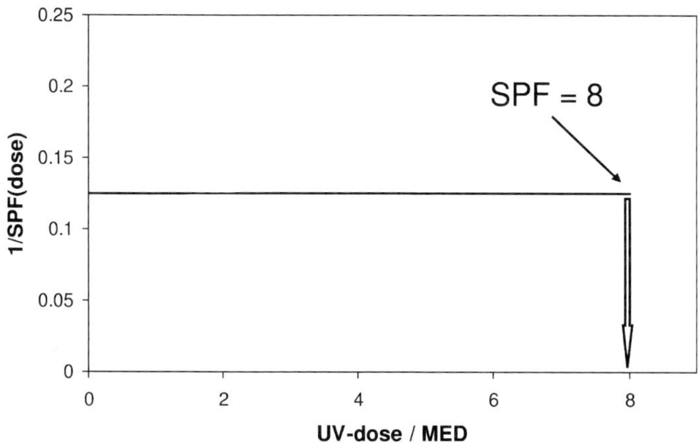

Figure 13.14 Principle of SPF evaluation with irradiation (photostable sunscreen). If the area under the curve of 1/SPF(dose) versus dose (in MED) is equal to unity, 1 MED has been transmitted; the corresponding point at the abscissa represents the externally applied dose in MED; the ratio of applied and transmitted dose is the SPF.

and with an applied UV dose of 8 MED (1 MED = 1 minimal erythemal dose) read on the abscissa, the area under the respective curve is just unity, corresponding to 1 MED transmitted through the sunscreen film.

In order to determine dose-dependent protection factors, SPF(dose), with a simulation program, first a protection factor without irradiation is calculated, SPF(0) from the given filter concentrations. This protection factor corresponds to the maximum dose to which the sunscreen would be exposed only in the case of complete photostability and defines the dimension of the abscissa scale. In the next step, the maximum dose is divided into certain dose increments representing different UV doses of irradiation. At each dose increment the filter concentrations are readjusted according to their recovery and the SPF(dose) is calculated using the respective filter concentrations [30]. Next, 1/SPF(dose) is formed and plotted against the applied dose. From this graph, the SPF can be evaluated in the previously described way, which is shown for an unstable sunscreen in Figure 13.15.

In order to calculate the filter concentrations after the various UV doses, one needs to know the respective decay rates. The decay rates of the individual UV filters are known from irradiation studies under realistic conditions with HPLC analysis of the parent compounds [36, 37]. Photodegradation of the filter molecules as a function of dose can be approximated with a simple exponential decay function

$$\text{Recovery} = \frac{c_{\text{dose}}}{c_0} = \exp(-k \times \text{dose}) \qquad (24)$$

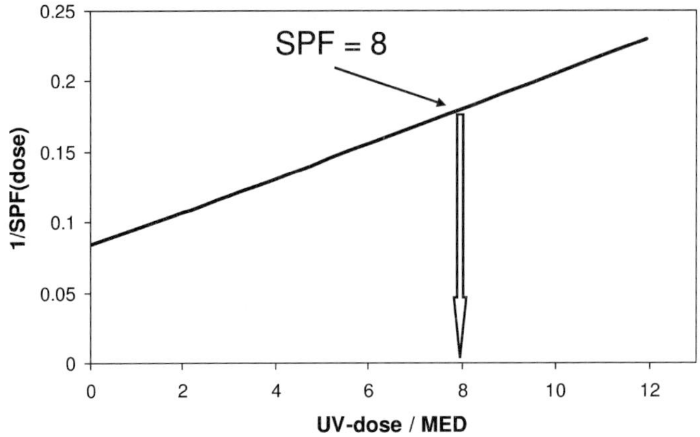

Figure 13.15 Principle of SPF evaluation with irradiation (photounstable sunscreen). If the area under the curve of 1/SPF(dose) versus dose (in MED) is equal to unity, 1 MED has been transmitted; the corresponding point at the abscissa represents the externally applied dose in MED; the ratio of applied and transmitted dose is the SPF.

The recovery of the parent molecule is given as the ratio of the concentrations at a certain dose, c_{dose}, and the concentration without irradiation, c_0. The dose in the argument of the exponential term is proportional to the irradiation time and the constant k is a "rate" constant with units MED^{-1} (1 MED = 1 minimal erythemal dose). The rate constants were obtained by least-squares fits to experimental recovery data. With the corresponding value of k once determined, the recovery of each individual UV absorber can be calculated for the respective formulation at a certain dose. Values for $k = k_0$ (k_0 refers to individual filters without interaction with other ingredients) are given in Table 13.4.

The rate constants may change when at least two filters show stabilizing or destabilizing interactions. A destabilization will be reflected by an increase in the rate constant. This can be simulated with the following approach, shown for the example of EHMC interacting with BMDBM

$$k_{\mathrm{EHMC}} = k_{0,\,\mathrm{EHMC}}(1 + r\beta_{\mathrm{BMDBM}}) \tag{25}$$

The constant r is a measure of the strength of the interaction and has to be determined experimentally and β is the concentration (% w/v) of the respective compound. For the particular case of Eq. (25), $r = 0.244$ [30]. In case of stabilization, a Stern–Vollmer-like approach may be followed. This is shown with the example of BMDBM, stabilized by octocrylene (OCR)

$$k_{\mathrm{BMDBM}} = [1/k_{0,\,\mathrm{BMDBM}} + q\beta_{\mathrm{OCR}}/(\beta_{\mathrm{BMDBM}} + \beta_{\mathrm{EHMC}})]^{-1} \tag{26}$$

The constant q, which characterizes the efficacy of the stabilization, must be determined experimentally. In the case of Eq. (26), $q = 12.4$ [30].

Table 13.4 UV dose constants for photdegradation of individual UV filters.

UV filter	$k = k_0/\text{MED}^{-1}$
Ethylhexyl methoxycinnamate (EHMC)	0.0417
4-Methylbenzylidenecamphor (4-MBC)	0.0379
Ethylhexyl salicylate (EHS)	0.0010
Octocrylene (OCR)	0.0010
Ethylhexyltriazone (EHT)	0.0057
Dioctylbutamidotriazone (DBT)	0.0057
Ethylhexyldimethyl-PABA (OD-PABA)	0.0139
Polysilicone-15 (BMP)	0.0026
Phenylbenzimidazolesulfonic acid (PBSA)	0.0047
Homomenthyl salicylate (HMS)	0.0379
Butylmethoxydibenzoylmethane (BMDBM)	0.1280
Diethylaminohydroxybenzoyl hexylbenzoate (DHHB)	0.0021
Terephthalidenedicamphorsulfonic acid (TDSA)	0.0057
Disodium phenyldibenzimidazoletetrasulfonate (DPDT)	0.0057
Benzophenone-3 (B-3)	0.0010
Benzophenone-4 (B-4)	0.0010
Drometrizoletrisiloxane (DTS)	0.0010
Bisethylhexyloxyphenolmethoxyphenyltriazine (BEMT)	0.0010
Methylenebisbenzotriazolyltetramethylbutylphenol (MBBT)	0.0002
Titanium dioxide (TiO_2)	–
Zinc oxide (ZnO)	–

Table 13.5 Comparison of a photostable and a photounstable sunscreen composition and corresponding simulation results.[a)]

Composition	**Photostable sunscreen with SPF = 20, 7% UV filters: 4% DBT, 3% BEMT**	**Photounstable sunscreen with SPF = 20, 25% UV filters: 10% EHMC, 5% BMDBM, 10% OCR**
SPF *in vivo*	20	21
Calculation with quasi-continuous step film without photostability treatment	21	58
Calculation with quasi-continuous step film with photostability treatment	21	23

a) CI = confidence interval; for other abbreviations see Table 13.4.

For implementation of the photodegradation behavior of UV filters, the photoinstabilities have to be included in the process of parameter calibration, since the standard sunscreens used there are mostly not completely photostable.

In order to demonstrate the impact of photoinstabilities on the calculation result, Table 13.5 shows the comparison of a photostable and a photounstable sunscreen composition of the same *in vivo* SPF, together with the calculation results when using the quasi-continuous step film model with and without photoinstabilities taken into consideration.

13.6
Consideration of the Distribution of the UV Extinction in the Water and the Oil Phases of the Formulation

When comparing the COLIPA standard formulations P2 and P3, it is striking, that the *in vivo* SPF of P2 (12.7) is smaller than that of P3 (15.5), although P2 contains 10% of UV filters (7% OD-PABA and 3% B-3), whereas P3 contains only 6.3% (3% EHMC, 2.8% PBSA, 0.5% BMDBM), giving rise to a smaller overall extinction of P3 compared with P2 (Figure 13.16). The smaller overall extinction of P3 compared with P2 is also reflected in the model results (see Table 13.3). A reason for this discrepancy is seen in the distribution of the UV filters in the phases of the emulsions: with P2, all filters are located in the oil phase of the emulsion, whereas in the case of P3, the UVB filters are approximately equally distributed in water and oil. The more even phase distribution in case of P3 may result in a different film structure, which obviously favors higher efficiency of the protection.

This effect was studied and confirmed by systematic variation of the phase distribution of filters in similar formulations such as P2 and P3, followed by *in vitro*

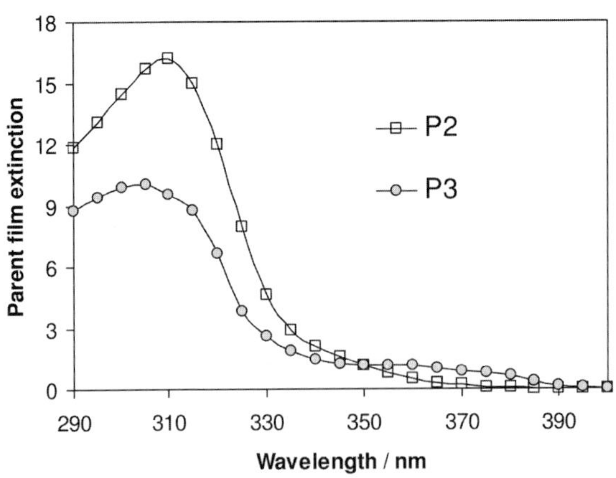

Figure 13.16 Parent film extinction spectra of COLIPA P2 and P3 standard sunscreens.

Table 13.6 Adjustment of quasi-continuous step film model parameters with standard formulations.[a]

Calibration formulation	Filter mixture	$SPF_{in\ vivo}$	Model parameters	Calculated SPF after adjustment
	Filters only in oil (case A):			
P1	2.7% EHMC	4.2	$B = 3.435$,	4.5
P2	7% OD-PABA, 3% B-3	12.7	$C = 1.243$	13.2
CF7	4% BEMT	9.0		8.3
	Filters in oil and water (case B):			
P3	3% EHMC, 2.8% PBSA, 0.5% BMDBM	15.5	$B = 1.174$,	13.3
CF4	5% EHMC, 10% MBBT	35.7	$C = 8.546$	35.1
CF6	5% EHMC, 4% MBBT	21.9		19.2

a) For other abbreviations see Table 13.4.

SPF measurements [38]. It was included in the quasi-continuous step film model by adjustment of two sets of parameters B and C in Eqs. (16) to (19), one set for the case where all filters are either in the water or the oil phase and another set for the case where they are approximately evenly distributed in the oil and water phases. For that purpose, two further calibrating formulations were introduced, CF6 and CF7, the compositions of which are given in Table 13.6, which also summarizes the results of the adjustments. The results are shown in Figure 13.17 in terms of the model film structure. It is important to note that in these adjust-

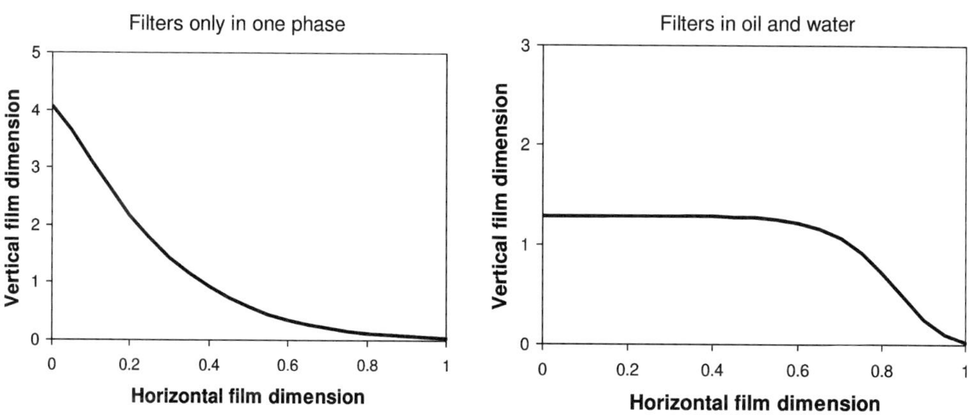

Figure 13.17 Illustration of film profiles of the quasi-continuous step film model adjusted using standard sunscreens with filters only in the oil phase of the emulsion and filters in the oil and the water phases.

ments the photodegradations of the filters inside the calibration formulations were already taken into account as described in the previous section (P2 is less prone to photodegradation than P3).

The two sets of model parameters given in Table 13.6 represent extreme cases. In practical examples, formulations will probably be somewhere in between. A relevant parameter for characterizing a formulation in this respect is the relative erythema active extinction in the oil phase, $REAE(\mathrm{o})$. For a single filter i with concentration β_i the erythema active extinction EAE_i is calculated via the equation

$$EAE_i = \beta_i \sum_{\lambda=290}^{400} E_{11}(i,\lambda) s_{er}(\lambda) S_s(\lambda) \tag{27}$$

where $E_{11}(i,\lambda)$ is the specific extinction of filter i at wavelength λ and the other quantities have the same meaning as in Eq. (1). The relative erythema active extinction due to all filters in the oil phase is given via the equation

$$REAE(\mathrm{o}) = \frac{\sum_i EAE_i(\mathrm{o})}{\sum_i EAE_i(\mathrm{o}) + \sum_j EAE_j(\mathrm{w})} \tag{28}$$

In the calculations it is assumed that $REAE(\mathrm{o}) = 0.5$ represents optimum conditions. The average $REAE(\mathrm{o})$ of the standards used in case B is 0.57 and therefore deviates only slightly from this assumption. For any given filter mixture two SPF values according to cases A and B (see Table 13.7) are calculated and an interpolation is performed according to the equation

$$SPF(x) = SPF_A + \frac{xSPF_A(1-x)SPF_B}{xSPF_A + (1-x)SPF_B} \tag{29}$$

where $x = REAE(\mathrm{o})$. The interpolated data are normalized, such that SPF_A is the minimum and SPF_B the maximum value for the function $SPF(x)$ [30]. Table 13.7

Table 13.7 Comparison of calculations with the quasi-continuous step film model with and without the distribution effect.[a)]

UV filters in formulation	$SPF_{in\ vivo}$	$SPF_{simulated}$ (no treatment of phase distribution)	$SPF_{simulated}$ (with treatment of phase distribution)
P2 standard: 7% OD-PABA + 3% B-3	12.7	14.4	13.2
P3 standard: 3% EHMC + 2.8% PBSA + 0.5% BMDBM	15.5	11.2	13.3

a) For other abbreviations see Table 13.4.

shows a comparison of the results for the P2 and the P3 standard using the quasi-continuous step film model with and without the distribution effect. The result is improved in terms of simulation of the *in vivo* data.

13.7 Calculation of UVA Parameters

With the tools developed for the model calculations of the SPF it is also possible to simulate the common parameters used for assessment of UVA protection by sunscreens.

13.7.1 Australian Standard

The Australian Standard [39] describes an *in vitro* method for broad-spectrum assessment. For the simulation the purpose is to calculate the transmission spectrum of a parent film with 8 μm optical thickness for a given UV filter composition. If the transmission is below 10% at any wavelength between 320 and 360 nm, the criterion is fulfilled. Here only the tool for calculation of the overall UV spectrum is used.

13.7.2 UVA/UVB Ratio and Critical Wavelength

These quantities may be obtained by *in vitro* UV spectroscopic measurements with the sunscreen spread on a substrate with a rough surface. As they are static parameters, no irradiation is applied in the experimental procedure and, thus, photoinstabilities must not be considered in the calculations. The UVA/UVB ratio [15] relates the average extinction in the UVA range to that in the UVB range. The critical wavelength [15] is defined as the wavelength at which the area under the extinction curve just becomes 90% of the total area under that curve between 290 and 400 nm. Both measures can be calculated with any irregular film model using the respective film extinction as given in Eq. (8). However, the result depends on the shape of the specific characteristic curve of the respective model. The effect that the model film extinction is always smaller than the parent film extinction leads to certain deformation of the sunscreen spectra at higher extinctions, which can influence the value of the UVA/UVB ratio and of the critical wavelength [40]. This is demonstrated in Figure 13.18, where the spectra of 1% and 8% EHMC, calculated with the continuous height distribution model based on the gamma law ($c = 1.105$), are shown. With 1% EHMC a UVA/UVB ratio of 0.17 is obtained and with 8% EHMC a value of 0.23. Hence there is a broadening of the UV spectra due to the film irregularity, which is also observed with *in vitro* transmission measurements, when roughened substrates are used. It has been demonstrated recently by Ferrero et al. that the UVA/UVB ratio increases with increasing substrate roughness [22].

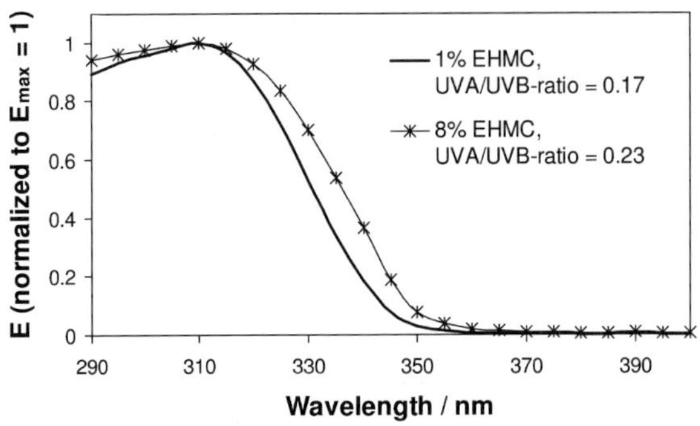

Figure 13.18 Broadening of extinction spectra of the model film (gamma distribution model) at higher UV filter concentration. Both spectra are normalized with respect to their maximum extinction.

In Ref. [37], a very good correlation of calculated and experimental results was obtained for the UVA/UVB ratio. The experimental data were obtained from measurements on roughened quartz plates as substrates and the calibrated step film model was employed for the calculations. However, the quality of such correlations depends on how the film irregularity of the model fits to the roughness of the surface of the *in vitro* substrate.

Concerning the critical wavelength, it was reported by several authors that this parameter is not suitable for the characterization of the UVA performance of sunscreens, since its dynamic range is extremely small [41, 42]. This was also confirmed with calculations using a calibrated step film model [37].

13.7.3
UVA Protection Factor (UVAPF)

The UVAPF can be simulated like the SPF with an analogues approach to that given in Eq. (1) using the calculated transmission spectrum. The UVAPF is based on the persistent pigment darkening effect (PPD), the action spectrum of which is known (s_{PPD}), and also the spectrum of the UVA source (S_{UVA}) [17, 43]. Its value ($UVAPF$) can be calculated according to the equation

$$\text{UVAPF} = \frac{\sum_{320}^{400} s_{PPD}(\lambda) S_{UVA}(\lambda)}{\sum_{320}^{400} s_{PPD}(\lambda) S_{UVA}(\lambda) T(\lambda)} \tag{30}$$

However, the treatment of photoinstabilities is even more important for the UVAPF calculation than for the SPF calculation, since one of the commonly used UVA filters is rather photounstable [37] and, in addition, the UVA dose corresponding to 1 MPD (minimal pigmenting dose) is about five times higher than the corresponding UVA dose of 1 MED [45] and the factor of five has also to be considered in the simulation.

This is demonstrated with the example of the Japanese Standard, the JCIA sunscreen (5% BMDBM, 3% EHMC): if the photoinstability of the filters is not taken into consideration, the calculated UVAPF result is 19. When the photoinstabilities and the UVA dose are considered, the result of the UVAPF comes to 4.1, which is very close to *in vivo* data, where an average of 4.2 is reported [45].

13.7.4 The COLIPA Method for Assessment of UVA Protection

This method is basically an *in vitro* determination of the UVAPF taking photoinstabilities into account. It makes use of the *in vivo* SPF in order to adjust the *in vitro* measurement. The adjustment is carried out by fitting a parameter C as an exponent to the transmission data such that the adjusted *in vitro* SPF ($SPF_{in\ vitro,\ \mathrm{adj}}$) matches the *in vivo* SPF (or the SPF which is labeled on a product). The other quantities in Eq. (31) have the same meaning as in Eq. (1).

$$SPF_{in\text{-}vivo} = SPF_{in\text{-}vitro,\,\mathrm{adj}} = \frac{\sum_{290}^{400} s_{\mathrm{er}}(\lambda)S_{\mathrm{S}}(\lambda)}{\sum_{290}^{400} s_{\mathrm{er}}(\lambda)S_{\mathrm{S}}(\lambda)T(\lambda)^{C}} \tag{31}$$

The resulting transmission spectrum is then used to calculate an *in vitro* UVA protection factor using the same approach as in Eq. (30). This is designated UVAPF_0, with the zero indicating that no irradiation step has been applied so far

$$UVAPF_0 = \frac{\sum_{320}^{400} s_{\mathrm{PPD}}(\lambda)S_{\mathrm{UVA}}(\lambda)}{\sum_{320}^{400} s_{\mathrm{PPD}}(\lambda)S_{\mathrm{UVA}}(\lambda)T(\lambda)^{C}} \tag{32}$$

Up to this point, the procedure is identical with the methodology for the determination of the UVA-balance, where the static $UVAPF_0$ is related to the labeled SPF [17, 46]. With the COLIPA method [18, 47, 48], an irradiation step is added. The amount of the irradiation dose D to which the sample will be exposed is obtained from $UVAPF_0$

$$D = UVAPF_0 \times D_0 \tag{33}$$

where D_0 is a unit UVA dose, which was determined in a COLIPA round-robin test and is fixed at 1.2 J cm^{-2}. After irradiation, the UVAPF value ($UVAPF$) is obtained with the equation

$$UVAPF = \frac{\sum_{320}^{400} s_{\mathrm{PPD}}(\lambda)S_{\mathrm{UVA}}(\lambda)}{\sum_{320}^{400} s_{\mathrm{PPD}}(\lambda)S_{\mathrm{UVA}}(\lambda)T(\lambda)^{C}} \tag{34}$$

For numerical simulation of this procedure, it is important, that the COLIPA method requires *in vitro* transmission measurements on special PMMA substrates, on which an amount of 0.75 mg cm^{-2} of the sunscreen is applied. The irregular film extinction of such substrates in terms of the characteristic curve can be simulated via Eq. (35), the parameters of which were obtained by parameter fitting to experimental data [30]

$$E_{\mathrm{PMMA}} = 0.247 E_{\mathrm{parent}}^{0.552} \quad \text{and} \quad T_{\mathrm{PMMA}} = 10^{-E_{\mathrm{PMMA}}} \tag{35}$$

with E_{parent} representing the extinction of the regular parent film calculated according to the Beer–Lambert law applied to the amount of UV filters deposited on unit area. With T_{PMMA} given in Eq. (35) the *in vitro* SPF can be simulated. The constant C is now determined by adjusting this simulated *in vitro* SPF to the previously simulated *in vivo* SPF with application of Eq. (31). $UVAPF_0$ is then obtained by employing Eq. (32). For simulation of the irradiation step, D_0 can be translated into 0.2 MED corresponding to 1.2 J cm^{-2}.

The kinetic data for photodegradation of UV filters reported here, which were used for simulations of SPF and UVAPF, were obtained from measurements on roughened quartz substrates with an application rate of 2 mg cm^{-2}. It has been shown that these degradation results are comparable to the degradation kinetics of sunscreens on human skin [49]. However, the photodegradation kinetics depends on the roughness of the substrate [50] and also on the amount of sunscreen applied [51]. For that reason, it may be necessary to introduce a further correction factor for the dose applied when simulating the kinetics on PMMA plates. In order to achieve a UVAPF of 4.1 for the JCIA standard formulation, D_0 had to be set to 0.66 MED instead of 0.2 MED. This indicates that photodegradation on the PMMA plates specified in the COLIPA method is faster by a factor of about three compared with the kinetic data in Table 13.4.

It was recommended by the European Commission in 2006 that the $\mathrm{SPF_{label}}$/UVAPF ratio should not exceed a value of 3 [52].

13.8 Correlations

13.8.1 Correlation of *In Vivo* SPF Data with SPF Calculations Using the Quasi-Continuous Step Film Model

Calculations for this correlation were performed using the quasi-continuous step film model, with and without considering the photostabilities and phase distribution of the UV filters inside the sunscreen formulations. The results are given in Table 13.8 and depicted in Figures 13.19 and 13.20. With the improved version of the calculation, which takes into account photoinstablities and the phase distribution of the UV filters, a higher correlation coefficient r and a slope closer to 1 are obtained. The other version of the model tends to overestimate the SPF of photounstable sunscreens, therefore leading to a slope higher than 1 in the correlation.

Table 13.8 Correlation of *in vivo* SPF data and SPF calculations using the quasi-continuous step film model, with (SPF_{calc}) and without (SPF_{calc}*) considering the photostabilities and phase distribution of the UV filters.

UV filter composition	$SPF_{in\ vivo}$	SPF_{calc}*	SPF_{calc}
1% BEMT	2.9	3.4	3.1
2% BEMT	5.3	5.1	4.7
3% BEMT	6.2	6.7	6.4
4% BEMT	7.3	8.5	8.3
3% EHMC, 5% BMDBM	8.2	12.9	7.7
2.78% PBSA, 2% OCR, 1.1% DBT, 0.5% BMDBM	13	11.1	15.3
0.9% DBT, 2.9% OCR, 5% BMDBM	11	12.9	10.2
3% BEMT, 4% B-3	15	11.1	10.8
3% BEMT, 4% EHT	30	20.3	19.6
3% BEMT, 4% Padimate O	13	15.5	14.5
3% BEMT, 4% MBC	16	14.9	12.8
3% BEMT, 4% EHMC	18	14.4	12.7
3% BEMT, 4% DBT	20	21.3	20.6
3% BEMT, 1% BMDBM	9.4	7.8	7.3
3% BEMT, 2% BMDBM	9.4	8.8	8.1
3% BEMT, 5% MBBT, 1% BMDBM	16	16.4	21.3
3% BEMT, 5% MBBT, 2% BMDBM	14	17.9	23.1
2% BEMT, 2.5% MBBT	10	8.6	10.9
2% BEMT, 5% MBBT	14	12.5	15.4
2% BEMT, 7.5% MBBT	16	17.3	20.1
5% BEMT, 2.5% MBBT	15	15.2	19.4
5% BEMT, 5% MBBT	30	20.7	27.8
5% BEMT, 7.5% MBBT	31	27.4	35.5
5% BEMT	8.8	10.4	10.4
2% MBBT	3.0	4.3	3.2
4% MBBT	5.0	6.9	5
8% MBBT	11	13.1	9.1
1% MBBT, 5% EHMC	12	11.4	10.4
2% MBBT, 5% EHMC	12	13.7	13.4
4% MBBT, 5% EHMC	19	18.3	19.2
8% MBBT, 5% EHMC	30	29.4	30.9
5% EHMC	7.6	7.6	6.5
2% BEMT, 4% MBBT, 5% EHMC	29	25	26.7
2% BEMT, 4.1% BMDBM, 1.8% EHT, 9.6% EHMC, 2.7% TiO_2	30	73.1	29.1
2% BEMT, 4.5% BMDBM, 3.5% OCR, 2.8% TiO_2	20	21.1	22.4
0.9% BEMT, 0.9% BMDBM, 4% OCR, 1.8% TiO_2	12	10.8	12.3
3% BEMT, 4.9% BMDBM, 1.9% PBSA, 1.6% EHT, 2% DBT, 4.6% TiO_2	50	72.5	47.9
0.7% BMDBM, 1% PBSA, 0.5% EHT, 0.5% MBC	10	6.5	8.2
7.5% BEMT, 5% EHS, 10% EHMC, 5% HMS	40	56.1	40.7

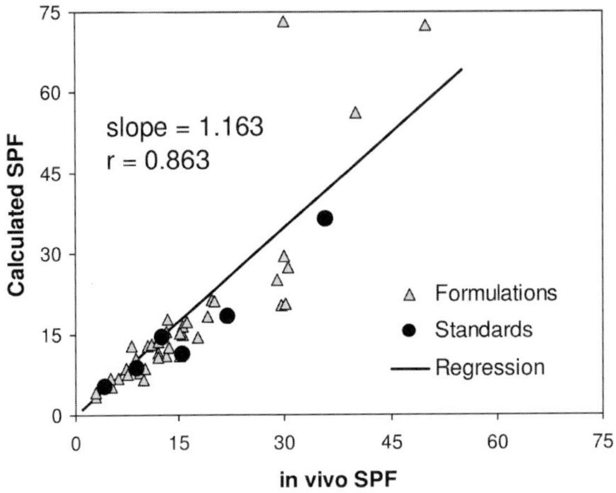

Figure 13.19 Correlation of calculated SPF (quasi-continuous step film model) without consideration of photoinstabilities, and *in vivo* SPF.

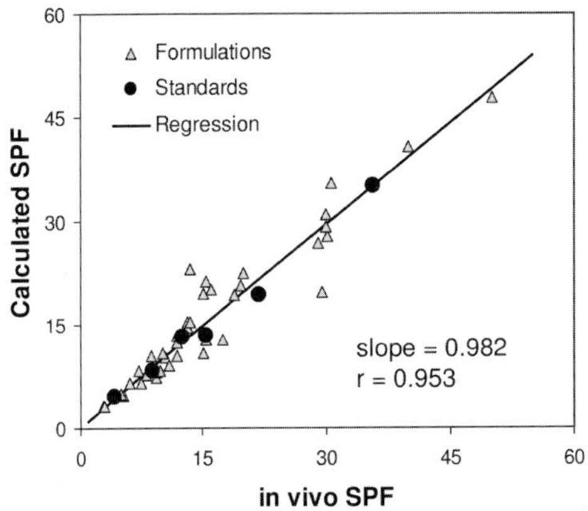

Figure 13.20 Correlation of calculated SPF (quasi-continuous step film model) with consideration of photoinstabilities, and *in vivo* SPF.

13.8.2
Correlation of *In Vivo* UVAPF Data with UVAPF Calculations

The UVAPF calculations were performed in two ways: one using exactly the same parameters of the quasi-continuous step film model with photoinstabilities and phase distribution of UV filters taken into account as described in Section 13.7.3, and the other involving the simulation according to the COLIPA method as described in Section 13.7.4. The simulated results are listed in Table 13.9 together with the *in vivo* data and plotted against the *in vivo* data in Figures 13.21 and 13.22.

Table 13.9 Correlation of *in vivo* UVAPF data and UVAPF calculations simulating PPD and COLIPA method conditions using the quasi-continuous step film model with consideration of photoinstabilities and phase distribution of the UV filters.

UV filter composition	$UVAPF_{in\ vivo}$	$UVAPF_{calc}$ (PPD)	$UVAPF_{calc}$ (COLIPA)
2% BMDBM	2.2	3	3.8
4% BMDBM	3.9	3.8	4.1
5% BMDBM	5.0	4.1	3.9
2% BMDBM, 5% OCR	5.7	5.3	5.7
4% BMDBM, 5% OCR	5.5	6.8	8
5% BMDBM, 5% OCR	7.6	7.3	8.2
5% OCR, 5% EHMC	3.4	1.8	1.6
2% BMDBM, 5% OCR, 5% EHMC	4.0	4.5	4.5
4% BMDBM, 5% OCR, 5% EHMC	5.0	5.4	6
2% BMDBM, 5% EHS	4.4	3.1	3.7
4% BMDBM, 5% EHS	4.1	3.9	4.2
5% BMDBM, 5% EHS	4.2	4.1	4
5% OCR + 5% EHS	2.7	1.7	1.5
2% BMDBM, 5% OCR, 5% EHS	5.7	5.3	5.5
4% BMDBM, 5% OCR, 5% EHS	6.9	6.9	7.9
5% BMDBM, 5% OCR, 5% EHS	7.3	7.3	8.2
2% MBBT, 5% EHMC	2.7	4.3	3.6
4% MBBT, 5% EHMC	5.1	6.4	6.5
8% MBBT, 5% EHMC	11.5	11.5	13.5
2% MBBT, 5% EHS	3.3	3.5	4.4
4% MBBT, 5% EHS	6.1	5.6	7.5
8% MBBT, 5% EHS	8.0	10.6	14.1
2% ZnO	3.0	2.2	1.9
4% ZnO	2.8	3.1	2.6
8% ZnO	3.8	4.7	3.8
16% ZnO	4.2	8.3	6.2
2% ZnO, 5% EHMC	2.8	2.4	1.8
4% ZnO, 5% EHMC	3.2	3.3	2.3
8% ZnO, 5% EHMC	5.0	4.9	3.4
16% ZnO, 5% EHMC	6.1	8.5	5.6
1% BEMT	2.9	2.7	2.7
2% BEMT	4.0	4	3.8
3% BEMT	5.1	5.2	4.9
4% BEMT	7.6	6.5	6.1
5% TiO_2	5.3	2.8	2.3
3% BEMT + 5% TiO_2	11.0	8.8	8.3
3% EHMC + 5% BMDBM	4.2	4.1	4.1

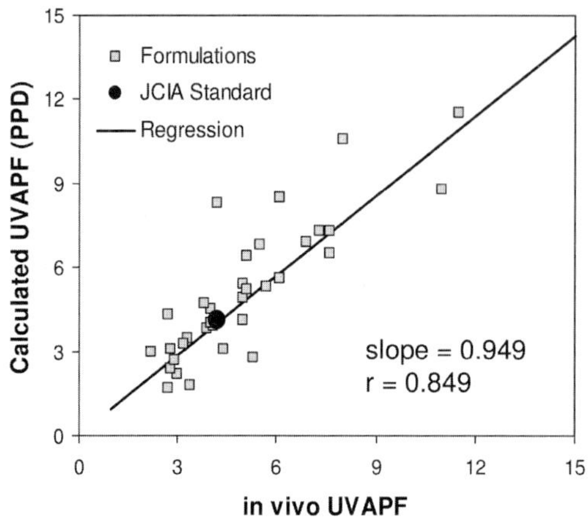

Figure 13.21 Correlation of calculated UVAPF (quasi-continuous step film model) simulating the PPD (*in vivo*) method, and *in vivo* SPF.

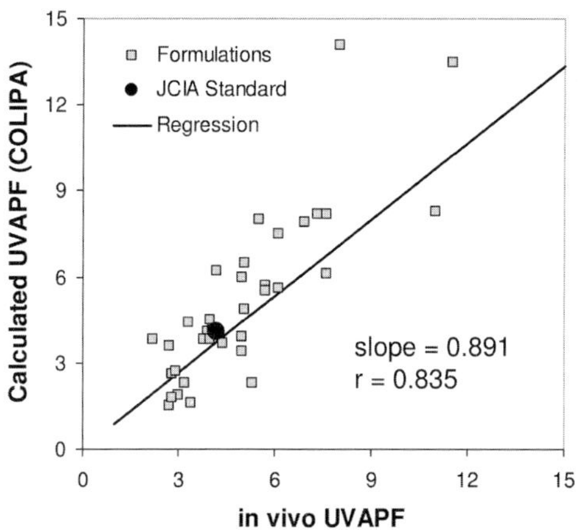

Figure 13.22 Correlation of calculated UVAPF (quasi-continuous step film model) simulating the COLIPA (*in vitro*) UVA method, and *in vivo* SPF.

The quality of the correlations shown in Figures 13.21 and 13.22 in terms of the correlation coefficient r are comparable. The slope of the correlation with calculated UVAPFs when simulated according to the PPD method is slightly closer to 1 than for the other case. The fact that both correlations are of similar quality confirms the good correlation of *in vivo* UVAPFs (measured by PPD) and the *in vitro* UVAPFs (measured with the COLIPA method) [47].

13.9 Conclusion

When aiming to describe the performance of sunscreens with mathematical models, one faces two kinds of difficulties: First, the reality to be described is extremely complex, and second, the quality of the experimental data to be simulated such as *in vivo* SPF and *in vivo* UVAPF is rather poor.

O'Neill's step film idea was a breakthrough in understanding an important part of the physical basis of how sunscreens perform. A lot of effort has been put into the improvement of sun protection factor calculations since his early work. The concept of an irregular film structure has been confirmed meanwhile with direct observations [53] and has proven to be relevant and extremely valuable. The progress made in the mathematical description of the sunscreen film has allowed more realistic predictions of the SPF.

As the irregular film concept applies very generally, other refinements, such as the treatment of photoinstabilities or the phase distribution of the UV filters in the emulsion, seem important to be considered only in specific cases, but nevertheless are necessary in order to describe as many sunscreens as possible in a satisfactory way.

Today, the SPF and UVA parameter simulation models are extremely helpful tools, not only in the development of sunscreen formulations, but also for understanding how sunscreens work.

Acknowledgments

Special thanks are due to Louis Ferrero (Coty-Lancaster) for his valuable contributions and comments. Stimulating discussions with Stefan Mueller (Ciba Specialty Chemicals), Uli Osterwalder (Ciba Specialty Chemicals) and Marc Pissavini (Coty-Lancaster) are gratefully acknowledged.

References

1 Motoyoshi K, Ota Y, Takuma Y, Takenouchi M, Wrinkles from UVA exposure. *Cosmet. Toiletries* 1998; **113**: 51–58.

2 Scharfetter K, Wlaschek M, Hogg A, Bolsen K, Schothorst A, Goerz G, Plewig G, UVA irradiation induces collagenase in human fibroblasts *in vitro* and *in vivo*. *Arch. Dermatol. Res.* 1991; **283**: 506–511.

3 Lavker R, Kaidbey K, The spectral dependence for UVA induced cumulative damage in human skin. *J. Invest. Dermatol.* 1997; **108**: 17–21.

4 Seité S, Moyal D, Verdie MP, Hourseau C, Fourtanier A, Accumulated p53 protein and UVA protection level of sunscreens. *Photodermatol. Photoimmunol. Photomed.* 2000; **16**: 3–9.

5 Schulze R, Einige Versuche und Bemerkungen zum Problem der handelsüblichen Lichtschutzmittel. *Parfüm. Kosmet.* 1956; **37**: 6, 7, 310–315, 365–372.

6 Ferguson J, Brown M, Alert D, Bielfeldt S, Brown J, Chardon A, Hourseau C, Mazilier C, Cuthbert J, Arcy-Burt D, Jolley J, Murdoch M, Finkel P, Masson P, Merot F, Maclennon A, Poret J, Siladji S, Collaborative development of a sun protection factor test method, a proposed European Standard. *Int. J. Cosmet. Sci.* 1996; **18**: 203–218; COLIPA, COLIPA Sun Protection Factor Test Method, Ref. 94/289. *European Cosmetic Toiletry and Perfumery Association – COLIPA*, Brussels, 1994.

7 Department of Health, Education and Welfare, US FDA, Sunscreen drug products for over-the-counter human drugs; proposed safety, effective and labeling conditions. *Fed. Regist.* 1978; **43/166**: 38206–38269.

8 Japan Cosmetic Industry Association (JCIA), *Standard SPF Test Method.* JCIA, Tokyo, 1991.

9 Diffey BL, Robson J, A new substrate to measure sunscreen protection factors throughout the ultraviolet spectrum. *J. Soc. Cosmet. Chem.* 1989; **40**: 127–133.

10 Reece BT, Deeds D, Rozen M, An *in vitro* method for screening sunscreen formulations for sun protection factor using a full-thickness skin model. *J. Soc. Cosmet. Chem.* 1992; **43**: 307–312.

11 Tronnier H, Kockott D, Meick B, Hani N, Heinrich U, Zur *In Vitro*-Bestimmung des SPF. *Parfüm. Kosmet.* 1996; **77**: 326–329.

12 Springsteen A, Yurek R, Frazier M, Carr KF, *In vitro* measurement of sun protection factor of sunscreens by diffuse transmittance. *Anal. Chim. Acta* 1999; **380**: 155–164.

13 Chardon A, Moyal D, Hourseau C, Persistent pigment-darkening response as a method for evaluation of ultraviolet A protection assays. In *Sunscreens, Development, Evaluation and Regulatory Aspects*, 2nd edn., Lowe, NJ, Shaath, NA, Pathak, MA (eds.), Marcel Dekker, New York, 1997, pp. 559–582.

14 AS/NZS, Australian/New Zealand Standard, AS/NZS 2604, 1998.

15 Diffey BL, A method for broad spectrum classification of sunscreens. *Int. J. Cosmet. Sci.* 1994; **16**: 47–52.

16 Wendel V, Klette E, Wittern KP, Gers-Barlag H, Measurement of UVA protection. *SÖFW J.* 2002; **128**: 34–38.

17 DIN. *Characterization of UVA Protection of Dermal Suncare Products by Measuring the Transmittance with Regard to the Sun Protection Factor.* DIN 67502. Normenausschuss Lichttechnik (FNL) at DIN Deutsches Institut für Normung, Berlin, 2004.

18 COLIPA, COLIPA Method for the *In Vitro* Determination of UVA Protection Provided by Sunscreen Products. European Cosmetic Toiletry and Perfumery Association – COLIPA, Brussels, 2007.

19 Sayre RM, Agin PP, LeVee GJ, Marlowe E, A comparison of *in vivo* and *in vitro* testing of sunscreening formulas. *Photochem. Photobiol.* 1979; **29**: 559–566.

20 McKinlay AF, Diffey BL, A reference action spectrum for ultraviolet-induced erythema in human skin. *CIE J.* 1987; **6**: 17–22.

21 Sellers RL, Carpenter FG, An instrument for *in vitro* determinations of SPF. *Cosmet. Toiletries* 1992; **107**: 119–122.

22 Ferrero L, Pissavini M, Dehais A, Marguerie S, Zastrow L, Importance of substrate roughness for *in vitro* sun protection assessment. *IFSCC Mag.* 2006; **9**: 97–108.

23 Herzog B, Prediction of sun protection factors by calculation of transmissions with a calibrated step film model. *J. Cosmet. Sci.* 2002; **53**: 11–26.

24 Industrieverband Körperpflege und Waschmittel, *Die Methode zur Bestimmung des Lichtschutzfaktors.* Industrieverband Körperpflege und Waschmittel, Frankfurt/Main, 1995.

25 Ferrero L, Orcet AM, Zastrow L, *The effective film thickness of sun products – how is it calculated?* In *Proceedings of the 19th IFSCC Congress*, Sydney, 1996.

26 O'Neill JJ, Effect of film irregularities on sunscreen efficacy. *J. Pharm. Sci.* 1984; **73**: 888–891.

27 Tunstall DF, A mathematical approach for the analysis of *in vitro* sun protection factor measurements. *J. Cosmet. Sci.* 2000; **51**: 303–315.

28 Hewitt JP, *Integrating theory and practice in suncare formulations. Layer thickness models for inorganic susnscreens.* In

Proceedings of the 24th IFSCC Congress, Osaka, 2006.

29 Herzog B, Prediction of sun protection factors and UV-A Parameters by calculation of UV transmissions through sunscreen films of inhomogeneous surface structure. In *Sunscreens – Regulation and Commercial Development*, 3rd edn., Shaath N (ed.), Taylor and Francis, Boca Raton, FL, 2005.

30 Herzog B, Müller S, Neuenschwander A, Deshayes C, Acker S, Osterwalder U, Improved simulation of sun protection factors and UVA-parameters – a useful tool for the development of sunscreen formulations. In *Proceedings of the 24th IFSCC Congress*, Osaka, 2006.

31 (a) Thomas RT, *Rough Surfaces*, 2nd edn. Imperial College Press, London, 1998, pp. 91–104. (b) Abbott EP, Firestone FA, Specifying surface quality. *Mech. Engng.* 1933; **55**: 569–572.

32 Ferrero L, Pissavini M, Marguerie S, Zastrow L, Efficiency of a continuous height distribution model of sunscreen film geometry to predict a realistic sun protection factor. *J. Cosmet. Sci.* 2003; **54**: 463–481.

33 Pissavini M, Ferrero L, Alard V, Heinrich U, Tronnier H, Kockott D, Lutz D, Tournier V, Zambonin M, Meloni M, Determination of the *in vitro* SPF. *Cosmet. Toiletries* 2003; **118**: 63–71.

34 Japan Cosmetic Industry Association, *JCIA Measurement Standard for UVA Protection Efficacy*. JCIA, Tokyo, 1995, p. 105.

35 Wloka M, Lange RFM, Flösser-Müller H, An *in vitro* SPF screening approach considering the photostability of the UV Filters. In *Proceedings of International Sun Protection Conference*, London, 2005.

36 Herzog B, Sommer K, Investigations on photostability of UV-absorbers for cosmetic sunscreens. In *Proceedings of the 21st IFSCC Congress*, Berlin, 2000.

37 Herzog B, Mongiat S, Quass K, Deshayes C, Prediction of sun protection factors and UVA parameters by using a calibrated step film model. *J. Pharm. Sci.* 2004; **93**: 1780–1795.

38 Neuenschwander A, For the prediction of sun protection factors – comparison of *in vitro* measurements and model calculations. *Diploma Thesis*, Institute of Pharmaceutical Technology, University of Basel, 2005.

39 AS/NZS, *Australian/New Zealand Standard*, AS/NZS 2604, 1998.

40 Ferrero L, Orcet AM, Zastrow L, Spectroscopy of sunscreen products. How to explain the special shape of UV curves obtained from *in vitro* SPF tests. In *Proceedings of the 20th IFSCC Congress*, Cannes, 1998.

41 Herzog B, Mongiat S, Deshayes C, Neuhaus M, Sommer K, Mantler A, *In vivo* and *in vitro* assessment of UVA protection by sunscreen formulations containinmg either butyl methoxy dibenzoyl methane, methylene bis-benzotriazolyl tetramethylbutylphenol or microfine ZnO. *Int. J. Cosmet. Sci.* 2002; **24**: 170–185.

42 DGK Task Force "Sun Protection", The reproducibility of an *in vitro* determination of the UVA index describing the relative UVA protection of sun care products. *IFSCC Mag.* 2002; **5**: 161–166.

43 Chardon A, Moyal D, Hourseau C, Persistent pigment-darkening response as a method for evaluation of ultraviolet A protection assays. In *Sunscreens, Development, Evaluation and Regulatory Aspects*, 2nd edn., Lowe NJ, Shaath NA, Pathak MA (eds.), Marcel Dekker, New York, 1997, pp. 559–582.

44 Cesarini JP, personal communication.

45 Moyal D, Wichrowski K, Tricaud C, *In vivo* persistent pigment darkening method, a demonstration of the reproducibility of the UV-A protection factors results at several testing laboratories. *Photoderm. Photoimmunol. Photomed.* 2006; **22**: 124–128.

46 Wendel V, Klette E, Gers-Barlag H, A new *in vitro* test method to assess the UVA protection performance of sun care products. *SÖFW J.* 2001; **127**: 12–15.

47 Gers-Barlag H, Harmonized *in vitro* determination of UVAUV-A protection. In *Proceedings of International Sun Protection Conference*, London, 2005.

48 Ferrero L, Protection UVAUV-A, methodes *in-vitro*. In *Actualités et Futur de la Photoprotection*, Conference SFC, Paris, 2005.

49 Giesinger J, Eine neue, *in vivo* kalibrierbare Methode zur Bestimmung des Photoabbaus kosmetischer UV-Filter. *Techniker-Arbeit*. Carl Engler Schule, Karlsruhe, 2006.

50 Ferrero L, Pissavini M, Dehais A, Marguerie S, Zastrow L, Importance of substrate microrelief for *in vitro* sun protection assessment of non photostable sunscreen products. In *Proceedings of the 24th IFSCC Congress*, Osaka, 2006.

51 Herzog B, Müller S, Sohn M, Osterwalder U, New insight and prediction of photostability of sunscreens. In *Proceedings of Cosmetics Science Conference*, Paris, 2007.

52 European Commission, Recommendation on the efficacy of sunscreen products and the claims made relating thereto. *Off. J.* 2006; L265: 2006/7647/EC; **49**: 39–43.

53 Lademann J, Rudolph A, Jacbi U, Weigmann HJ, Schaefer H, Sterry W, Influence of nonhomogeneous distribution of topically applied UV filters on sun protection factors. *J. Biomed. Opt.* 2004; **9**: 1358–1362.

Subject Index

Colloids and Interface Science Series, Vol. 4
Colloids in Cosmetics and Personal Care. Edited by Tharwat F. Tadros

ISBN: 978-3-527-31464-5